# Sample Preparation for Hyphenated Analytical Techniques

***Dedication:*** *To Shmilke and Roise Rosenfeld. Their idealism and belief that scholarship and books are the building blocks of a better world were the inspirations of a lifetime.*

# Sample Preparation for Hyphenated Analytical Techniques

*Edited by*

**J. M. Rosenfeld**
Faculty of Health Sciences
McMaster University
Hamilton, Canada

**Blackwell**
Publishing

**CRC Press**

Editorial offices:
Blackwell Publishing Ltd, 9600 Garsington Road, Oxford OX4 2DQ, UK
Tel: +44 (0)1865 776868
Blackwell Publishing Asia Pty Ltd, 550 Swanston
Street, Carlton, Victoria 3053, Australia
Tel: +61 (0)3 8359 1011

ISBN 1-4051-1106-2

Published in the USA and Canada (only) by
CRC Press LLC, 2000 Corporate Blvd., N.W.,
Boca Raton, FL 33431, USA
Orders from the USA and Canada (only) to CRC Press LLC

USA and Canada only:
ISBN 0-8493-2374-6



First published 2004

Library of Congress Cataloging-in-Publication Data:
A catalog record for this title is available from the Library of Congress

British Library Cataloguing-in-Publication Data:
A catalogue record for this title is available from the British Library

Set in 10.5/12 pt Times
by TechBooks
Printed and bound in Great Britain
by MPG Ltd, Bodmin, Cornwall

The publisher's policy is to use permanent paper from mills that operate a sustainable forestry policy, and which has been manufactured from pulp processed using acid-free and elementary chlorine-free practices. Furthermore, the publisher ensures that the text paper and cover board used have met acceptable environmental accreditation standards.

For further information on Blackwell Publishing, visit our website:
www.blackwellpublishing.com

# Contributors

**Hossein Ahmadzadeh** Department of Chemistry, University of Minnesota, 207 Pleasant Street SE, Minneapolis, MN 55455, USA

**Edgar A. Arriaga** Department of Chemistry, University of Minnesota, 207 Pleasant Street SE, Minneapolis, MN 55455, USA

**Damià Barceló** Department of Environmental Chemistry, IIQAB-CSIC, c/Jordi Girona 18–26, E-08034 Barcelona, Spain

**Malonne Davies** Department of Pharmaceutical Chemistry, University of Kansas, 2095 Constant Avenue, Lawrence, KS 66047, USA

**Miral Dizdaroglu** Chemical Science and Technology Laboratory, National Institute of Standards and Technology, Bldg 227/A243, Gaithersburg, MD 20899-8311, USA

**Kathleen Heppert** Department of Pharmaceutical Chemistry, University of Kansas, 2095 Constant Avenue, Lawrence, KS 66047, USA

**Bryan Huynh** Department of Pharmaceutical Chemistry, University of Kansas, 2095 Constant Avenue, Lawrence, KS 66047, USA

**Jonathan A. Karty** Department of Chemistry, Indiana University, 800 E. Kirkwood Avenue, Bloomington, IN 47405-7102, USA

**Tae-Young Kim** Department of Chemistry, Indiana University, 800 E. Kirkwood Avenue, Bloomington, IN 47405-7102, USA

**Jenny Kingston** Physical Methods Section, MSD Neuroscience Research Centre, Eastwick Road, Harlow, Essex CM20 2QR, UK

**Qi Liang** Department of Cellular Pathology, Armed Forces Institute of Pathology, 1413 Research Blvd, Bldg 101, Rockville, MD 20850, USA

**Jack H. Lichy** Department of Cellular Pathology, Armed Forces Institute of Pathology, 1413 Research Blvd, Bldg 101, Rockville, MD 20850, USA

**Sara Logan** Department of Pharmaceutical Chemistry, University of Kansas, 2095 Constant Avenue, Lawrence, KS 66047, USA

**Susan Lunte** Department of Pharmaceutical Chemistry, University of Kansas, 2095 Constant Avenue, Lawrence, KS 66047, USA

**Desmond O'Connor** Drug Metabolism Section, MSD Neuroscience Research Centre, Eastwick Road, Harlow, Essex CM20 2QR, UK

**Hélène Perreault** Department of Chemistry, Room 517 Parker Building, University of Manitoba, Winnipeg, Manitoba, Canada R3T 2N2

**Mira Petrovic** Department of Environmental Chemistry, IIQAB-CSIC, c/Jordi Girona 18–26, E-08034 Barcelona, Spain

**James P. Reilly** Department of Chemistry, Indiana University, 800 E. Kirkwood Avenue, Bloomington, IN 47405-7102, USA

**Jack M. Rosenfeld** Faculty of Health Science, McMaster University, 1200 Main Street W, Hamilton, Ontario, Canada L8N 3Z5

**Tim Sparey** Automated MedChem Group, MSD Neuroscience Research Centre, Eastwick Road, Harlow, Essex CM20 2QR, UK

**Steven Thomas** Physical Methods Section, MSD Neuroscience Research Centre, Eastwick Road, Harlow, Essex CM20 2QR, UK

# Contents

# 1 Introduction: current techniques and issues in sample preparation

JACK M. ROSENFELD

The purpose of sample preparation is to move an analyte from its original matrix into one that is more amenable to analysis and/or to produce within the new matrix a derivative that is detected at higher sensitivity or selectivity. It can be a time consuming, labor intensive process, and successful implementation requires integration of physical, organic and inorganic chemistry, the chemistry of analysis, as well as the sciences of instrumentation and measurements. Sample preparation is a non-trivial process. It follows that one focus in development of many analytical techniques and methods is that no, or minimal, sample preparation be required to produce the results. In the necessary interests of speed and economics, reduction in technical complexity and requirements of manipulation within a method are advantageous. Recent literature, however, suggests that such laudable simplification is neither always desirable nor efficient. Several authors [1,2] identified that even a combination such as high performance liquid chromatography (HPLC)–tandem mass spectrometry (MS/MS), while inherently specific, still requires sample preparation. Matrix effects produce ion suppression in HPLC-MS/MS [1] and the authors suggest that high quality sample preparation, particularly for multiple analyte methods, would effectively reduce such interferences. It is not surprising that problems are particularly acute with more complex matrices. Simple dilution of urine preparatory to HPLC–atmospheric pressure chemical ionization (APCI) MS/MS produced acceptable results. In contrast, analysis of oral fluids required protein precipitation prior to analysis, whereas determination of organics from plasma required solid phase extraction (SPE) [2].

Utility of sample preparation is evident from the importance attached to these techniques by investigators on new and developing instrumentation [3–5]. Matsoskva and co-workers argue that useful application of fast gas chromatography (GC) is in no small measure dependent on rapid effective sample preparation [3,4]. Investigators in miniaturization incorporate sample preparation on the micro scale [5] in order to achieve the quality of data necessary for studies in the discipline of biology.

Developments in proteomics further highlight the practicality of expending some effort in isolation and modification of analyte prior to analysis. Arguably, proteomics is one of the most complex techniques in the separation sciences; arguably, analytical derivatization is one of the most complicated techniques in sample preparation [6,7] and investigators often try to avoid it. Nevertheless, groups under Reilly [8] and Keough [9] reported substantial increases in information content of proteomic analyses resulting from incorporating analytical derivatizations into their techniques. Zhou *et al.* [10] and Sechi [11] adopted different derivatization methods in their approaches to functional genomics. This discipline seeks to identify the production of new proteins as a result of physiological or pathological modification of a cell, using isotope dilution techniques to differentially label normal and modified cells. A combination of isotope dilution, affinity tagging and a photolabile group provided the photoactive isotope coding affinity tag (ICAT) that enabled both isotope labeling and affinity tagging for sample purification. Sechi exploited the problem that when preparing polyacrylamides gels, there are residual acrylamide monomers which react with the thiols of proteins and peptides [11], and managed this matter by pre-derivatization of the proteins with acrylamide before polyacrylamide gel electrophoresis. Sechi tagged the normal cell line with normal acrylamide and the modified cell proteins with deuterated acrylamide, thus enabling a comparison of the cell lines. Elegant as these approaches may be, Kjellstrom and Jensen have demonstrated that proteomic analysis benefits from something as 'simple' as liquid–liquid extraction [12].

Finally, consideration must be given to the wide range of domains in which chemical analysis has import. It is to be expected that there will be numerous samples in which concentrations are exceedingly low [13,14] and/or matrices are very complex [15]. In such instances, sample preparation to fortify, and preferably selectively fortify [15], analyte concentrations [13–16] in the final isolate will be a *pro forma* requirement.

There exists a strong theoretical basis for development of efficient sample preparation techniques. Pawliszyn's recent and extensive overviews [16,17] provide considerable insight into the underlying principles and mechanisms which comprise the concepts of equlibria, kinetics of phase transfer and diffusion processes. Sophisticated knowledge of the underlying processes [18] has led to improvements in classical liquid–liquid extraction (LLE) [19,20] and encouraged its use in proteomics [12]; it has also led to the discovery and widespread use of numerous solid phase processes. Recent reports describe such techniques as immuno SPE [15,21], solid phase micro-extraction (SPME) [16,17,20], stir bar sorbtive extraction (SBSE) [22,23], matrix dispersion solid phase extraction (MDSPE) [24], capillary extraction (CE)–capillary GC (CE-CGC) [25] and solid phase analytical derivatization (SPAD) [26]. A combination of need and innovation provided the technique of membrane extraction with sorbtive interface (MESI) for measuring volatiles [27] in air and breath.

A considerable proportion of effort in sample preparation is focused on quantitative analysis. Nevertheless, problems in qualitative analysis concern the identification of complex molecular structures of compounds found in complex mixtures [28,29]; these also require high quality sample preparation before HPLC-MS. An elegant example of the importance of sample preparation is the unravelling of the structure of misfolded rhodopsin in retinosa pigmentosa reported by Khorana's group [30–32].

These needs for such a multiplicity of sample preparation techniques arise from the diverse matrices analysed and the different needs of the client sciences (e.g. biomedical, pharmaceutical and environmental sciences) and institutions (pharmaceutical industry, regulatory agencies, the legal system). The nature of matrices that the separation scientists need to address can vary from the interior of a cell, to plasma in a test tube, to manure and environmental waters.

Expectations of modern analytical methods that deal with these matrices are stringent. End-users need and expect high sensitivity and selectivity in quantitative analysis. In addition, both quantitative *and* qualitative analyses are expected to have high throughput capability and economic efficiency. Increasingly, meeting these expectations requires the combination of at least two techniques which often operate under radically different principles or conditions.

In classical methodology, samples would be subjected to a technique that usually separated or isolated an analyte which was then transferred for analysis by a second technique that provided qualitative or quantitative information on the components. Techniques become hyphenated when such transfer is eliminated, resulting in higher sensitivity, selectivity and throughput, but above all, higher quality data.

In analytical chemistry the theoretical and the applied are inextricably and closely linked, all the more so because new methods are very rapidly incorporated in a wide variety of fields. A common analytical problem comprises a need from another discipline that defines the requirements of analysis but also constrains the approaches that can be taken. For instance, both environmental and pharmacokinetic analyses require at least micromolar sensitivity. Whilst it is reasonable to use tens to hundreds of milliliters of urine or a natural water sample, in a pharmacokinetic analysis not only are large volumes not available at any given time-point, but the total volume from multiple samples cannot exceed a certain physiological, and tightly regulated, limit. This book addresses the interaction between the needs of the 'client' sciences with the capabilities of sample preparation, in transferring analytes from an original matrix to a matrix and/or derivative suitable for hyphenated analytical instrumentation.

The client's needs addressed in this book are those of investigators who seek to advance our understanding of biology, pharmacology and toxicology, because these affect the human population and the regulators who seek to protect it. The organization of the book is based on the biological principle. Accordingly, the first section deals with biological macromolecules because these control the development of an organism. The discussion begins with a chapter on the analysis of genetic material, and describes genomic measurements in the context of the emerging scientific and clinical discipline of molecular pathology, the applications of which are the study and detection of genetic disease. The next chapter addresses the determination of damage to genetic material brought about by oxidative processes, giving rise to lethal mutations or development of neoplastic disease. The function of genes is to make proteins, so the chapter on proteomics follows those related to the genome. Many cellular functions, such as recognition, require glycosylation of proteins; the need to understand this biochemistry has led to the rapidly expanding field of glycomics. The fifth chapter addresses the analysis of oligosaccharides involved in glycosylation and completes the section on biological macromolecules.

The middle section of the book addresses the quantitative analysis of small molecules. One of the main applications of the sciences dealing with biological macromolecules is drug development. By and large, and at least at present, most drugs are small molecules. The emphases on speed, high throughput, economics and regulatory requirements generate some of the daunting problems in analytical chemistry, instrumentation and, particularly, sample preparation. Chapter 6 describes the techniques of analysis from early discovery to pharmacokinetic investigations. Organic contaminants are also small molecules ingested involuntarily by humans and their feedstock animals, possibly with toxic or mutagenic consequences. Determination of environmental contaminants, as described in Chapter 7, is another client field that makes very high demands of the analytical and instrumental sciences and one that deals with the most highly variable matrices.

The last section of the book recognizes that the ultimate purpose of the investigations discussed previously is to understand biology as it applies to living organisms and, ultimately, humans. The increasing level of complexity ranges from subcellular components, to intact cells, to tissues, to organs, and finally to the living mammalian organism, with small mammals serving as surrogates for humans. Life is not compatible with analytical instrumentation, nor is it readily amenable to removal of the compounds that mediate its functions. Determination of analytes directly from an organism requires the use of coupled techniques and sampling should be appropriate to the needs of biological study. Because living systems are dynamic the most important information will come from studying them over time and will require multiple sampling. This leads to the second consideration, which is that sampling removes material required for life; samples should therefore be small to minimize perturbation. Selective sampling is highly desirable because it would limit the removal of compounds not required within the analysis but needed for functioning of the organism.

Ideally, sampling should be from a live organism (this includes the cell and its components). Consideration of the literature on single cell analyses in Chapter 8 shows that the field is moving in this direction, with particular support from investigations and successes in miniaturization. There are several challenges in sample preparation for single cell analysis. The first is the formidable problem of removing materials from a matrix as small as a single cell when the sample is the complete cell, its secretions or internal fluids; appropriate nano sampling and often derivatization are required. Second is the lack of compensatory or mediating responses available to multicellular systems. A single cell losing a mediator or a metabolite can release a signal that it requires restoration of some mediators, hormones or nutrients, but it cannot receive them from other cells. Because the mediator or metabolite is also the analyte, replacing it without interfering with the measurement can be a difficult. It may very well be that sampling from a living organism, even as small as a cell, may require replenishment compounds as much as isolating analyte.

Sampling from live and often mobile animals, including humans, is an accomplished fact [33] and is amply described in Chapter 9. Use of very sophisticated instrumentation is combined with knowledge of anatomy and biology, as well as of analytical chemistry. The animal itself is the sample, and preparation requires a number of techniques. There is a premium on skilful surgery; proper anaesthesia requires that it have minimal effect on the physiology of the system under study and in all cases the appropriate replacement of physiological fluid is essential. Furthermore, mobile animals require restraint to within reach of the sampling device. Students of this discipline must, in addition, be aware of the ethical considerations and regulatory restrictions involved in experimentation on live animals. Nevertheless, techniques have advanced to the stage where they can be used in the clinic. Here, investigators must again be prepared to deal not only with the difficult problems of analysis, but also with ethical issues, such as informed consent, and the regulatory matters arising when undertaking studies in humans.

The power and sophistication of modern hyphenated analytical techniques can provide the sensitivity and specificity required for determination of numerous and diverse analytes from numerous and diverse matrices. High throughput methods are also well within the capabilities of modern instruments. Appropriate sample preparation techniques enhance the quality of data that an instrument can produce. On the macro scale analyses have been well studied and automated, thus making entire processes more rapid and economic, so alleviating concerns of cost. The ability to sample on the nano and pico scales has expanded the concept of hyphenation to include analysis from single cells and/or from living organisms. As detection limits drop to the single molecule, the client sciences and disciplines will extend the number of questions that they ask. This will require new sampling techniques and creation of the means to deliver such small samples to the instrument in a reproducible, reliable and interpretable manner.

## References


1. Matuszewski, B.K., Constanzer, M.L. & Chavez-Eng, C.M. (2003) Strategies for the assessment of matrix effect in quantitative bioanalytical methods based on HPLC-MS/MS. *Anal. Chem.*, **75**(13), 3019–3030.
2. Dams, R., Huestis, M.A., Lambert, W.E. & Murphy, C.M. (2003) Matrix effect in bio-analysis of illicit drugs with LC-MS/MS: influence of lonization type, sample preparation, and biofluid. *J. Am. Soc. Mass. Spectrom.*, **14**(11), 1290–1294.
3. Matisova, E. & Domotorova, M. (2003) Fast gas chromatography and its use in trace analysis. *J. Chromatogr. A*, **1000**(1-2), 199–221.
4. Mastovska, K. & Lehotay, S.J. (2003) Practical approaches to fast gas chromatography–mass spectrometry. *J. Chromatogr. A*, **1000**(1-2), 153–180.
5. Saito, Y. & Jinno, K. (2003) Miniaturized sample preparation combined with liquid phase separations. *J. Chromatogr. A*, **1000**(1-2), 53–67.
6. Rosenfeld, J. (2002) Recent developments in the chemistry and application of analytical derivatizations. In: *Sample Preparation in Laboratory and Field* (ed. J. Pawliszyn), pp. 609–668. Elsevier, Amsterdam.
7. Rosenfeld, J. (2003) Derivatisation in current practise of analytical chemistry. *Trends Anal. Chem.*, **22**, 785–798.
8. Beardsley, R.L. & Reilly, J.P. (2002) *Anal. Chem.*, **74**, 1884–1890.
9. Keough, T., Younquist, R.S. & Lacey, M.P. (2003) Sulfonic acid derivatives for peptide sequencing by MALDI MS, *Anal. Chem.*, **75**, 156A–165A.
10. Zhou, H., Ranish, J.A., Watts, J.D. & Aebersold, R. (2002) Photoactivated ICAT. *Nat. Biotechnol.*, **20**, 512–515.
11. Sechi, S. (2002) A method to identify and simultaneously determine the relative quantities of proteins isolated by gel electrophoresis. *Rapid Commun. Mass Spectrom.*, **16**, 1416–1424.
12. Kjellstrom, S. & Jensen, O.N. (2003) In-situ liquid–liquid extraction as a sample preparation method for matrix-assisted laser desorption/ionization MS analysis of polypeptide mixtures. *Anal. Chem.*, **75**(10), 2362–2369.
13. Lopez de Alda, M.J., Diaz-Cruz, S., Petrovic, M. & Barcelo, D. (2003) Liquid chromatography–(tandem) mass spectrometry of selected emerging pollutants (steroid sex hormones, drugs and alkylphenolic surfactants) in the aquatic environment. *J. Chromatogr. A*, **1000**(1-2), 503–526.
14. Huybrechts, T., Dewulf, J., Dewulf, J. & Van Langenhove, H. (2003) State-of-the-art of gas chromatography-based methods for analysis of anthropogenic volatile organic compounds in estuarine waters, illustrated with the river Scheldt as an example. *J. Chromatogr. A*, **1000**(1-2), 283–297.
15. Grant, G.A., Frison, S.L. & Sporns, P.J. (2003) A sensitive method for detection of sulfamethazine and *N*4-acetylsulfamethazine residues in environmental samples using solid phase immunoextraction coupled with MALDI-TOF MS. *Agric. Food. Chem.*, **51**(18), 5367–5375.
16. Pawliszyn, J. (2002) *Sampling and Sample Preparation for Field and Laboratory*. Elsevier, Amsterdam.
17. Pawliszyn, J. (2003) Sample preparation–quo vadis? *Anal. Chem.*, **75**, 2543–2558.
18. Cantwell, F. & Losier, M. (2002) Liquid–liquid extraction. In: *Sampling and Sample Preparation for Field and Laboratory* (ed. J. Pawliszyn). Elsevier, Amsterdam.
19. Wen, X., Tu, C. & Lee, H.K. (2004) Two-step liquid–liquid–liquid microextraction of nonsteroidal antiinflammatory drugs in wastewater. *Anal. Chem.*, **76**(1), 228–232.
20. Ugland, H.G., Krogh M. & Reubsaet, L. (2003) Three-phase liquid-phase microextraction of weakly basic drugs from whole blood. *J. Chromatogr. B*, **798**(1), 127–135.
21. Hennion, M.C. & Pichon, V. (2003) Immuno-based sample preparation for trace analysis. *J. Chromatogr. A*, **1000**(1-2), 29–52.

22. Kawaguchi, M., Inoue, K., Sakui, N., Ito, R., Izumi, S., Makino, T., Okanouchi, N. & Nakazawa, H. (2004) Stir bar sorptive extraction and thermal desorption–gas chromatography–mass spectrometry for the measurement of 4-nonylphenol and 4-*tert*-octylphenol in human biological samples. *J. Chromatogr. B*, **799**(1), 119–125.
23. Verhe, R., Sandra, P. & De Kimpe, N. (2003) Analysis of volatiles of malt whisky by solid-phase microextraction and stir bar sorptive extraction. *J. Chromatogr. A*, **985**(1-2), 221–232.
24. Blasco, C., Font, G. & Pico, Y. (2002) Comparison of microextraction procedures to determine pesticides in oranges by liquid chromotography–mass spectrometry. *J. Chromatogr. A*, **970**(1-2), 201–212.
25. Nardi, L. (2003) Guidelines for capillary extraction–capillary gas chromatography: preparation of extractors and analysis of aromatic compounds in water. *J. Chromatogr. A*, **1017**(1-2), 1–15.
26. Kuklenyik, Z., Ekong, J., Cutchins, C.D., Needham, L.L. & Calafat, A.M. (2003) Simultaneous measurement of urinary bisphenol A and alkylphenols by automated solid-phase extractive derivatization gas chromatography/mass spectrometry. *Anal. Chem.*, **75**, 6820–6825.
27. Lord, H., Yu, Y., Segal, A. & Pawliszyn, J. (2002) Breath analysis and monitoring by membrane extraction with sorbent interface. *Anal. Chem.*, **74**, 5650–5657.
28. Lin, Z., Crockett, D.K., Lim, M.S. & Elenitoba-Johnson, K.S. (2003) High-throughput analysis of protein/peptide complexes by immunoprecipitation and automated LC-MS/MS. *J. Biomol. Tech.* **14**(2), 149–155.
29. Schild, L.J., Divi, R.L., Beland, F.A., Churchwell, M.I., Doerge, D.R., Gamboa da Costa, G., Marques, M.M. & Poirier, M.C. (2003) Formation of tamoxifen–DNA adducts in multiple organs of adult female cynomolgus monkeys dosed with tamoxifen for 30 days. *Cancer Res.*, **63**(18), 5999–6003.
30. Cai, K., Itoh, Y. & Khorana, H.G (2001) Mapping of contact sites in complex formation between transducin and light-activated rhodopsin by covalent crosslinking: use of a photoactivatable reagent. *Proc. Natl Acad. Sci. USA*, **98**, 4877–4882.
31. Hwa, J., Klein-Seetharaman, J. & Khorana, H.G. (2001) Structure and function in rhodopsin: mass spectrometric identification of the abnormal intradiscal disulfide bond in misfolded retinitis pigmentosa mutants. *Proc. Natl Acad. Sci. USA*, **98**, 4872–4876.
32. Itoh, Y., Cai, K. & Khorana, H.G. (2001) Mapping of contact sites in complex formation between light-activated rhodopsin and transducin by covalent crosslinking: use of a chemically preactivated reagent. *Proc. Natl Acad. Sci. USA*, **98**, 4883–4887.
33. Djurhuus, C.B., Gravholt, C.H., Nielsen, S., Pedersen, S.B., Moller, N. & Schmitz, O. (2003) Additive effects of cortisol and growth hormone on regional and systemic lipolysis in humans. *Am. J. Physiol. Endocrinol. Metab.*, **286**, E488–E494, 2004. First published Nov 4, 2003.

# 2 Molecular pathology: applications of genomic analyses to diagnosis of genetic diseases

JACK H. LICHY AND QI LIANG

## 2.1 Overview

This chapter is broadly divided into two parts, one devoted to an introduction to the field of molecular genetic pathology as a rapidly emerging field of clinical medicine, the other to a discussion of methods and instrumentation used in the genomic studies. We first present a discussion of the properties of nucleic acids that molecular diagnostics assays aim to detect, and then give examples to illustrate how these assays have become important components of patient care. Subsequent sections focus on the technological aspects of molecular diagnostics, beginning with brief overviews of the traditional methods for carrying out each step in the analysis and then proceeding to a description of the more automated – and in the above context more hyphenated – methods that have now begun to integrate this technology into the routine work of the clinical laboratory.

Since the mid-1980s nucleic acid analysis has become an integral component of patient management in virtually all areas of medical practice. Nucleic acid based testing methods have been implemented both in the context of traditional clinical laboratories – typically including microbiology, chemistry, hematology and the blood bank – and in specialized molecular diagnostics laboratories devoted to those tests that involve the analysis of nucleic acids purified from a tissue specimen, blood, or body fluid. Nucleic acid analysis for diagnosis and management has grown into a subspecialty of laboratory medicine now known as molecular genetic pathology. In acknowledgment of the emergence of this new field, the American Board of Pathology and American College of Medical Genetics jointly developed an examination, offered for the first time in the fall of 2001, for board certification in molecular genetic pathology for physicians trained in pathology or genetics. The Association for Molecular Pathology, established in 1994, represents the field internationally. The reader is referred to the web site of this association, www.molecularpathology.org, for more detailed information about molecular pathology than will be presented in this chapter.

As the co-sponsorship of the certification exam indicates, the field of nucleic acid testing evolved from the distinct fields of genetics and pathology. Traditionally, a department of medical genetics would run a group of laboratories using a variety of methodologies, but all focusing on the diagnosis of inherited disease. Individual laboratories in such a department would typically be devoted to particular methodologies: a cytogenetics laboratory would analyze chromosomes for abnormalities in copy number and for structural abnormalities such as deletions, duplications, and translocations. A biochemical genetics laboratory would offer tests involved in the detection and quantitation of analytes in body fluid specimens, usually blood or urine, associated with specific inherited diseases. Finally, a molecular genetics laboratory would be involved in the diagnosis of inherited disease by analyzing purified preparations of nucleic acids

**Table 2.1** Examples of genetic diseases amenable to diagnosis by DNA testing.

| Disease or syndrome | Inheritance | Gene(s) | Comments |
|---|---|---|---|
| Achondroplasia | Autosomal dominant | *FGFR3* | Single point mutation, G1138A in 98% of cases |
| Hereditary breast and ovarian cancer | Autosomal dominant | *BRCA1*, *BRCA2* | Familial cancer syndrome |
| Duchenne muscular dystrophy | X-linked | | |
| Familial adenomatous polyposis | Autosomal dominant | *APC* | Familial cancer syndrome |
| Fragile X syndrome | X-linked | *FMR1* | Trinucleotide repeat expansion |
| Hemophilia | X-linked | *F8* and *F9* | Chromosome inversions common |
| Huntington disease | Autosomal dominant | *HD* | Trinucleotide repeat expansion |
| Li–Fraumeni syndrome | Autosomal dominant | *TP53* | Familial cancer syndrome |
| Polycystic kidney disease | Autosomal dominant | *PKD1*, *PKD2* | |
| Prader–Willi syndrome | Uniparental disomy | | Imprinting defect: loss of expression of paternal gene |
| Familial retinoblastoma | Autosomal dominant | *RB1* | Familial cancer syndrome |
| Sickle cell anemia | Autosomal recessive | *Beta globin* | Glu6Val point mutation |
| Tay–Sachs disease | Autosomal recessive | *HEXA* | |
| von Hippel–Lindau syndrome | Autosomal dominant | *VHL* | Point mutations and larger deletions |

for a specific abnormality responsible for the disease. As the specific molecular defects have been discovered, diagnostic testing based on direct analysis of DNA has become possible for an ever increasing number of genetic syndromes. Table 2.1 lists some of the genetic diseases and syndromes it is now possible to diagnose by such methods.

At the same time as research was revealing the specific genetic abnormalities responsible for inherited disease, developments in microbiology and oncology led to the development of nucleic acid based tests with clinical utility in the diagnosis and management of infectious disease and cancer. As the genomes of bacteria and viruses were analyzed at the level of DNA (or RNA) sequencing, it became possible to develop highly sensitive diagnostic tests for specific infectious agents based on detection of their nucleic acid genomes. In oncology, specific genetic abnormalities were associated with specific tumors, in some cases leading to improved subclassification of tumors based more on genetic composition than the traditional method of morphological examination of stained sections by light microscopy. In some cases, the distinction between genetics and oncology became blurred, as it was discovered that genetic abnormalities that predispose to the familial cancer syndromes (see Table 2.1) also occur in sporadic cancers. Similar DNA based diagnostic tests therefore applied to both conditions.

Molecular genetic pathology thus represents a fusion of those aspects of genetics and pathology that share common methodologies for analyzing nucleic acids with the aim of diagnosing or managing disease. Table 2.2 presents a list of tests commonly offered in molecular diagnostics laboratories, spanning the three areas of inherited disease, infectious disease, and cancer. The technological sophistication of the field has undergone a process of evolution from assays performed with the most basic methods of molecular biology, often in a research setting, to tests performed on dedicated instruments designed specifically for high throughput applications in the clinical

**Table 2.2** Tests commonly performed by molecular diagnostic laboratories.

| Test | Gene name(s) | Indications/comments |
|---|---|---|
| I. Tests for inherited disease | | |
| Factor V Leiden | *Factor V* | Hypercoagulability, venous thrombosis |
| Prothrombin G20210A | *Prothrombin (Factor II)* | Hypercoagulability, venous thrombosis |
| Cystic fibrosis | *CFTR* | Screen for carrier status; confirmation of diagnosis |
| Hemochromatosis | *HFE* | Iron overload; family history |
| II. Tests for cancer diagnosis and management | | |
| Gene rearrangement assays for clonality assessment: | | Lymphoma diagnosis |
| Ig heavy chain, T cell receptor beta, T cell receptor gamma | | |
| Chromosome translocations associated with leukemias and lymphomas: | | |
| t(14;18) | *IgH/Bcl2* | Follicular lymphoma |
| t(9;22) | *BCR/Abl* | Chronic myelogenous leukemia diagnosis and detection of minimal residual disease |
| t(11;14) | *IgH/Cyclin D1* | Mantle cell lymphoma |
| t(15;17) | *PML/RARA* | Promyelocytic leukemia |
| t(2;5) | *ALK/NPM* | Anaplastic large cell lymphoma |
| Numerous other entity specific translocations | | Diagnosis, minimal residual disease detection, monitoring response to therapy |
| Chromosome translocations associated with sarcomas: | | |
| t(X;18) | *SYT/SSX* | Synovial sarcoma |
| t(1;13) and t(2;13) | *PAX/FKHR* | Alveolar rhabdomyosarcoma |
| t(11;22) | EWS/FLI-1 | Ewing's, PNET |
| III. Tests for agents of infectious disease | | |
| Viral: | | |
| Human papillomavirus(HPV) | | Atypical pap smear results |
| Human Immunodeficiency virus (HIV) | | Both qualitative and quantitative assays available<br>Drug resistance testing by DNA sequencing |
| Hepatitis C virus | | Qualitative and quantitative assays available; subtyping to predict response to treatment |
| Epstein–Barr virus | | |
| Cytomegalovirus | | |
| Herpes simplex virus | | |
| Bacterial: | | |
| *Neisseria gonorrhoeae* (NG) | | NG and CT are high volume tests used for the diagnosis of common STDs |
| *Chlamydia trachomatis* (CT) | | |
| Group A *Streptococci* | | |
| Group B *Streptococci* | | |
| *Mycobacterium tuberculosis* | | |

laboratory. 'Hyphenation' in the context of molecular diagnostics will refer in this chapter to instrumentation that combines multiple steps in the analytical process, producing the advantages of higher throughput, faster turnaround times, greater accuracy, and decreased cost.

## 2.2 Molecular diagnostics: types of assay offered and clinical examples

### 2.2.1 *Properties of nucleic acids relevant to clinical diagnosis*

In general, a molecular diagnostics assay addresses one of only a small number of questions about a specific nucleic acid sequence, referred to here as the 'target sequence'. These basic questions include the following: (1) Is the target sequence present or absent? (2) How much of the target sequence is present in the specimen? (3) Has the target sequence undergone large scale structural alterations? (4) Are there smaller scale alterations, such as single nucleotide changes, in the target sequence?

The field of infectious disease provides good examples of the first type of assay. The situation will typically be that the clinical presentation of the patient suggests a diagnosis of a specific entity. For example, symptoms and signs of hepatitis raise the possibility of hepatitis C virus (HCV) infection. A positive result on testing for the presence of the genomic RNA of HCV on a blood specimen would support this diagnosis [1].

Assays for quantitation of the target sequence find clinical application in infectious disease and oncology. In patients infected with HCV or the human immunodeficiency virus (HIV; the virus that causes AIDS), the concentration of viral particles in the blood is a critical measure of the patient's therapeutic response, and plays a major role in determining the appropriate course of therapy. These two tests, both directed at quantitating a viral genomic RNA in a plasma or serum specimen, are probably those most frequently ordered in molecular diagnostics. In the area of oncology, quantitation can be relevant both at the level of gene copy number and gene expression. In breast tumors, the copy number of the gene for the growth factor receptor HER2 is an important predictor of prognosis and the patient's chance of responding to therapy with herceptin, an antibody directed against the HER2 molecule [2–6]. In some cases, overexpression of a specific gene is so characteristic of a specific type of tumor that quantitative expression assays can solve otherwise intractable diagnostic problems. One example of this situation is overexpression of the cell cycle regulator cyclin D1 in mantle cell lymphoma [7,8]. Detection of cyclin D1 overexpression distinguishes this aggressive malignancy from other subclassifications of lymphoma that appear similar by classical methods of pathological analysis, such as light microscopy and immunohistochemistry [9,10]. Current research on gene expression in cancer makes use of high density arrays capable of analyzing the expression of thousands of genes simultaneously. It is hoped that such genome-wide analyses of gene expression will provide a new system for molecular classification of tumors that will lead to improved diagnosis and treatment.

Assays for large scale structural alterations of the genome include tests for specific chromosomal translocations characteristic of many leukemias, lymphomas, and sarcomas. As with cyclin D1 overexpression, the utility of these assays often lies in their ability to distinguish pathological entities that cannot be resolved by other methods. The most commonly performed of these tests detect the chromosome translocations t(14;18) associated with follicular lymphoma [11,12] and t(9;22), the 'Philadelphia chromosome' translocation of chronic myelogenous leukemia [13,14]. The latter assay has recently assumed additional importance because of the development of an effective drug, STI571 or Gleevec, that targets the fusion protein BCR/Abl resulting from the translocation of chromosomes 9 and 22 [15]. Such entity-specific translocations have been identified for many types of leukemias and for a group of sarcomas most commonly seen in the pediatric age range. Examples of the latter group include the t(11;22) translocation of the Ewing's sarcoma/PNET family of tumors [16,17], the t(1;13) and t(2;13) translocations of alveolar rhabdomysarcomas [18–20], and the t(X;18) translocation of synovial sarcoma [21,22]. Assays for detecting such translocations have proven extremely useful in arriving at a correct diagnosis in individual cases, as well as in improving tumor classification in general.

Large scale insertions, deletions, and rearrangements of the human genome have also been causatively associated with a variety of genetic diseases. The most common cause of hemophilia A, for example, is an inversion of the X chromosome that results in the inactivation of the coagulation factor VIII gene [23,24].

The fourth category of molecular diagnostics tests to be considered is that of those detecting smaller scale alterations in the genome. This category includes single base changes and short deletions and insertions, which may range in size from a single nucleotide to a few tens of nucleotides. Although many genetic diseases result from alterations of only one or a few nucleotides, many, in fact thousands, of such sequence variations occur naturally in populations and have no pathological significance. Nevertheless, these genetic variations, or polymorphisms, have tremendous utility because they provide a genetic fingerprint unique to each individual. Formally, a polymorphism is defined as the presence in the population of two or more alternative variants (alleles) of a DNA sequence, with the most common allele having a frequency of 99% or less. While generally not associated with disease, assays that determine the genotype of a DNA specimen at a group of polymorphic loci have become an essential tool used in criminal investigations for associating tissues and body fluids found at a crime scene with a specific individual [25,26]. The methodology has also found extensive use in paternity testing, where the property of Mendelian inheritance of polymorphic loci can be used to support or refute paternity based on genotyping of the alleged father and the child [27,28].

Two types of polymorphism have played an especially important role in genetics: single nucleotide polymorphism (SNP) and the short tandem repeat (STR). A SNP is a system of alleles distinguished by the specific nucleotide present at a single position. SNPs constitute the most common type of genetic variation in humans, accounting for about 90% of all sequence differences. Sequencing data obtained from the Human Genome Project have resulted in the identification of more than 5 million SNPs in the human genome as of October 2003. A SNP database is available at the website of the National Center for Biotechnology Information: http://www.ncbi.nlm.nih.gov/SNP [29]. An STR, also referred to as a 'microsatellite,' consists of repeating units of 2–5 nucleotides. Alleles of an STR differ in the number of repeats present. The dinucleotide G-T is one of the most commonly occurring building blocks of STR polymorphisms.

In addition to these very common polymorphisms, genomic alterations affecting a small segment of DNA sequence, and often only a single base (point mutations), constitute the molecular basis for numerous genetic diseases. Although often indistinguishable from benign polymorphisms at the analytical level, sequence alterations with pathogenic potential, i.e. believed to cause a disease, are generally referred to as 'mutations'. Common diseases usually resulting from point mutations include hereditary hemochromatosis [30], hypercoagulability due to factor V Leiden or the prothrombin mutation G20210A [31,32], cystic fibrosis [33], and achondroplasia [34]. Mutations in STR polymorphisms, particularly those having a trinucleotide repeat, are responsible for a group of genetic diseases, including several neuromuscular degeneration syndromes and the fragile X syndrome, one of the more common causes of mental retardation [35]. In this class of genetic diseases, the mutation consists of an expansion of the trinucleotide repeat. For example, the trinucleotide repeat of normal alleles of the X chromosome *FMR* gene responsible for fragile X contains less than 60 repetitions of the sequence CAG. Mutant alleles associated with the syndrome usually contain more than 200 copies.

### 2.2.2 *Two clinical examples*

For a test to be useful in the diagnosis of genetic diseases, it must efficiently detect the full spectrum of mutations associated with the disease. Thus, it is important to characterize the nature of the mutations that may occur in any given disease entity. For the physician and genetic counselor, data on the specific prognostic implications of specific mutations are useful in diseases in which multiple mutations can lead to the disease. Two specific diseases in which molecular diagnostic testing plays a role, hereditary hemochromatosis and cystic fibrosis, will serve as prototypical examples of medical and analytical questions. The discussion of these two conditions will provide a sense of the nature of the technical issues that need to be addressed in developing a diagnostic test for a genetic disease, and some of the clinical considerations that go into deciding when it is appropriate to order such a test.

2.2.2.1 *Hereditary hemochromatosis (HH)* HH is an autosomal recessive disease resulting from the inability to properly regulate body iron stores, leading to accumulation of iron in many organs of the body [36,37]. Iron accumulation eventually leads to loss of function in the organs in which it occurs, which can include the liver, pancreas, thyroid, adrenal, testes, heart, skin, spleen, and bone marrow. However, organ damage occurs late in the disease process; early recognition of the syndrome allows for proper treatment which can prevent later organ damage. Furthermore, the treatment is relatively simple: periodic phlebotomy to deplete iron stores. Thus the importance of a genetic test for the syndrome is that it provides a means to render a diagnosis before symptoms develop, thereby permitting a more favorable outcome for the patient [30,38]. The gene responsible for HH, designated *HFE*, was identified in 1996 [39]. Two mutations were found to be commonly associated with the disease (Fig. 2.1). A majority comprising 60–90% of HH patients were found to be homozygous for a mutation that replaces a cysteine residue at amino acid (aa) number 282 of the protein with a tyrosine. Using the single letter code for amino acids, this mutation is usually designated C282Y.

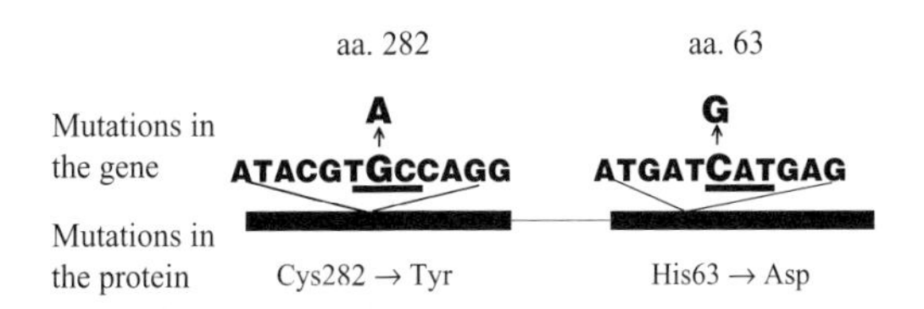


**Fig. 2.1** Hereditary hemochromatosis mutations. The gene sequence around the sites coding for amino acids 282 and 63 is shown, with the normal and mutant nucleotides indicated at each site.

An alternative nomenclature based on the numbering of the nucleotides in the mRNA coding for the *HFE* protein labels the same mutation 845G→A, indicating that the G residue present in the normal allele is replaced by an A in the mutant form. A small fraction of patients were found to be compound heterozygotes, carrying the C282Y mutation on one copy of chromosome 6 and a second mutation, H63D or 187C→G, on the other. These findings raised the possibility that the development of a diagnostic assay for HH would be a simple matter of developing assays for determining the genotype of an individual at two nucleotides, numbers 187 and 845 in the *HFE* mRNA. Several methodological approaches to making this determination are readily available, and will become the focus of the later sections of this chapter. Given the frequency of HH in the population, estimated to be 1/200 to 1/400, serious consideration has been given to performing HH testing as a population screening test with the idea that affected individuals could be identified and offered appropriate treatment. If this approach were to be adopted, it would create a need for a very high throughput technology capable of performing thousands of tests in a timely and cost efficient manner. In the case of HH, research studies aimed at testing the efficacy of population screening have suggested that the penetrance of the mutations, meaning in particular the percentage of C282Y homozygotes who have clinical evidence of HH, is lower than originally thought, and therefore that the predictive value of a positive result would be too small to be useful [40]. Population screening therefore is not recommended at the present time. The diagnostic test for the C282Y and H63D mutations remains useful in distinguishing HH from other causes of iron overload.

2.2.2.2 *Cystic fibrosis (CF)* In contrast to the case of HH, current clinical recommendations do create a requirement for high throughput testing for mutations in the gene responsible for CF [41]. This disease, inherited in an autosomal recessive manner, has a prevalence of about one in 2000 to 2500 individuals in the Caucasian population, making it one of the most common genetic diseases. Somewhat lower prevalences are found in other racial and ethnic groups. Among Caucasians the frequency of mutation carriers, most of whom have no manifestations of the disease, is one in 25 to 30 individuals. The disease affects multiple organs, most notably the lung and pancreas, both of which suffer a progressive loss of function due to abnormally viscous secretions characterized by an elevated chloride level. The disease is caused by a mutation in the gene *CFTR*, an abbreviation for 'cystic fibrosis transmembrane conductance regulator,' which encodes a chloride channel. Although the disease nearly always leads to a shortened life span, improvements in therapy have extended the life expectancy of CF patients. Early diagnosis allows for optimal treatment and prolonged survival. Although the carrier state – meaning one normal and one mutant allele – generally is not associated with disease, detection of carriers is important for informed reproductive decision making and for the earliest possible detection of the disease in an affected child. The clinical utility of the test has led to recommendations published by the American College of Medical Genetics (ACMG) and the American College of Obstetrics and Gynecology (ACOG) that CF testing be offered to (1) individuals with a family history of CF, (2) reproductive partners of individuals with CF, and (3) couples planning a pregnancy or seeking prenatal care [33,41]. These recommendations have created a tremendous demand for CF testing, and manufacturers have responded with numerous creative approaches to high throughput screening.

The development of a diagnostic assay for mutations in the *CFTR* gene presents much greater challenges than assays for HH [42]. Whereas HH assays need to detect only two point mutations, CF assays must detect a wide spectrum of possible disease associated mutations. To date, over 900 distinct mutations in the CFTR gene have been identified. However, the frequency of these mutations varies widely. One mutation, a three nucleotide deletion at codon 508, resulting in the loss of a phenylalanine residue (DeltaF508) accounts for approximately 70% of all CF mutations. Another 24, including a mixture of point mutations and small deletions, each account for 0.1% or more of all CF mutations. Based on mutation frequency, the ACOG/ACMG CF testing panel made a recommendation, now widely accepted as standard of care, that assays for CF carrier screening should be able to detect the 25 most common mutations. This approach results in an expected detection rate of about 80% in Caucasians of western European ancestry, 97% in Ashkenazi Jews, 69% in African Americans, and 57% in Hispanics. The detection rate in the Asian American population is unknown.

## 2.3 Methods of molecular pathology

This and the following sections of the chapter will shift the focus from the problems of molecular pathology to the methods available for detecting the properties of nucleic acids important for diagnostic testing and research. All molecular diagnostic assays share the common need for several steps in the analytical process: sample acquisition, nucleic acid extraction, amplification, and signal detection. We consider hyphenation in this field to refer to technologies that combine two or more of these basic processes on a single instrument. The development of the field has resulted in the introduction of instruments capable of hyphenating ever larger numbers of steps in the assay process.

The following sections will each focus on one of these steps in the analytical process, starting from the classical methods of molecular biology and proceeding to a discussion of some of the instrumentation currently on the market for automating these steps. Examples of the technology as applied to testing for the two clinical applications of HH and CF mutation detection are presented. What we refer to as classical molecular biology generally refers to the methods popularized in the research community by the book *Molecular Cloning: A Laboratory Manual* written by Tom Maniatis *et al.* [43], first published in 1982, that defined the standard operating procedures of molecular biology laboratories for many years. This basic body of technique includes a variety of methods for purification of DNA and RNA from viruses, bacteria, fungi, and eukaryotic cells, for sequencing DNA, for cutting DNA with restriction enzymes, for cloning DNA into plasmid vectors and propagating it in bacteria, and for separating nucleic acids by gel electrophoresis and detecting specific sequences by hybridization to probes – the Southern and Northern blotting procedures, referring to DNA and RNA, respectively. The development of the polymerase chain reaction (PCR) in the mid to late 1980s provided geneticists and molecular biologists with an additional method of fundamental importance, and in addition markedly facilitated the implementation of DNA testing in clinical laboratories for diagnostic use [44–48].

### 2.3.1 *Methods for isolation of nucleic acids from body fluids and solid tissues*

2.3.1.1 *Specimen types* Molecular diagnostics assays may be requested on any tissue or body fluid. The specimen may be fresh, frozen, fixed in formaldehyde, or fixed and then embedded in a block of paraffin wax. The last type of specimen results from the standard procedure used for preparation of tissue for histological analysis, and is the common way to store tissue specimens in departments of pathology. Nucleic acids recovered from paraffin embedded tissue are usually severely degraded into fragments no greater than a few hundred nucleotides in length. Nevertheless, procedures have been developed to isolate both DNA and RNA from such specimens in a form suitable for certain diagnostic assays that do not require the presence of larger fragments [49–51]. Instrumentation to automate isolation procedures from all types of tissue and fluid specimens is now available.

For molecular inherited disease detection, the mutation will be present in every nucleated cell in the body. Therefore, the specimen of choice is the one most convenient to obtain, which is usually a tube of blood or a buccal swab. Recommendations for blood specimens for molecular diagnostic testing usually call for blood drawn into an EDTA or acid–citrate–dextrose tube, as these preservatives prevent clotting and do not contain inhibitors of the enzymes used in these assays [52,53]. Heparin tubes are not recommended, and are not accepted by some laboratories, because heparin inhibits the activity of Taq polymerase [54]. However, our experience has been that current methods of DNA purification remove the heparin sufficiently for the resulting DNA preparations to amplify well enough for use in most molecular diagnostics assays.

2.3.1.2 *Research laboratory methods* Manual methods for nucleic acid isolation from tissue specimens generally involve physical separation of the tissue into small fragments, either with a scalpel or a tissue homogenizer, followed by digestion with a protease to disaggregate the cells from the connective tissue matrix. Lysis of the cells with a detergent, usually sodium dodecyl sulfate, releases the nucleic acids into solution. Further purification steps include removal of insoluble material by centrifugation, extraction of proteins and lipids with a 1:1 mixture of phenol/chloroform, and alcohol precipitation with ethanol or isopropanol. The precipitated nucleic acids are resuspended in water or a standard buffer such as TE (10 mM Tris, pH 8.0, 1 mM EDTA). The samples may be incubated at a temperature of 37–55°C for a period of several hours to overnight to allow the DNA to dissolve. This step is especially important for the high molecular weight DNA purified from fresh or frozen tissue specimens, as opposed to the fragmented molecules averaging a few hundred nucleotides in length obtained from fixed and embedded tissue. If desired, specific removal of DNA or RNA from the preparation may be accomplished enzymatically, by DNAase or RNAase treatment. A commonly used alternative procedure for RNA purification involves lysis of the cells in a concentrated guanidinium chloride solution followed by extraction with phenol/chloroform at an acidic pH [55]. Under these conditions, the DNA partitions predominantly into the organic phase, leaving the RNA in the aqueous phase. While most of the DNA is removed from the solution, sufficient remains in the final preparation to be detectable by sensitive techniques such as polymerase chain reaction (PCR). For certain applications, particularly those that quantitate gene expression, this residual DNA could lead to erroneous results, and therefore must be removed enzymatically before the assay is performed.

2.3.1.3 *Manual methods designed for clinical laboratories* As clinical applications of DNA testing have become more common, a variety of commercial kits have been made available to streamline the process of nucleic acid purification. For the diagnosis of genetic disease, the specimen of choice is often blood, and several of the manual kits are designed specifically for blood specimens. Two such methods in use in our laboratory will be described here: the Gentra generation capture column and the Qiagen QIAamp DNA Blood Mini Kit. Both of these systems involve the use of small columns that fit into 1.5 ml microcentrifuge tubes, so that the loading, washing, and elution steps can all be accelerated by centrifugation.

In the Gentra system, 200 μl of blood is applied to the column directly and allowed to absorb into the column matrix. The column is washed twice with 400 μl of a proprietary DNA purification solution, which is forced through the column matrix quickly by brief (~10 s) centrifugation. Following a third wash, a DNA elution solution is added, the column is heated to 99°C for 10–15 min, and the eluate containing the genomic DNA recovered by centrifugation. In the Qiagen system, the blood specimen is subjected to protease digestion and mixed with ethanol before loading onto a spin column. The loading step is performed by centrifugation of the lysed blood sample through the column. After two washes, the specimen is eluted in water or a low salt buffer.

Both the Gentra and the Qiagen systems offer advantages over the older methods. The organic extraction step, which involves toxic chemicals and requires special ventilation, has been eliminated. There are no time consuming incubations, and the final preparation does not require alcohol precipitations or the prolonged resuspension step associated with it. Both are also available as high throughput kits designed to process samples in 96 well plate format, providing for a further increase in efficiency when large numbers of specimens need to be processed simultaneously.

2.3.1.4 *Instrumentation* The next step in automation of nucleic acid purification procedures has been the introduction of a variety of instruments that automate the entire sequence of steps between the tissue or blood specimen and the purified DNA or RNA preparation.

*Roche MagNA Pure LC* One of the earliest instruments to be introduced for this purpose, the Roche MagNA Pure, employs magnetic bead technology in a robotic system that can be used for isolation of DNA or RNA from both fresh and fixed tissue specimens as well as blood and other body fluids [56,57]. This instrument can process up to 32 specimens per run of 1–2 h. Unlike most dedicated DNA purification devices currently on the market, the MagNA Pure LC can be set to load the final preparations into specialized reaction vessels – PCR tubes, capillaries used in PCR on the Roche LightCycler, or Cobas A-rings used for PCR in the Roche Cobas Amplicor automated PCR device [58] – resulting in a hyphenation of the isolation and amplification steps in the assay.

The purification technology employed by the instrument makes use of magnetic particles that bind nucleic acids. The instrument adds a binding/lysis buffer containing proteinase K to each sample, initially transferred manually to a well of a special sample cartridge. Following incubation for a prespecified time, the instrument adds a slurry of magnetic glass particles which bind the DNA released from the specimen. The automated pipetting system aspirates the lysate containing the magnetic particles into a pipet tip which then serves as the reaction vessel for subsequent steps in the procedure. The instrument applies a magnet to the side of the pipet tip, immobilizing the particles and bound DNA molecules. The particles, bound to the side of the pipet tip, are then washed several times by repeated cycles of aspiration and ejection of a wash buffer. Finally, the DNA is eluted from the particles into an elution buffer and expelled into a tube. The glass particles remain in the tip and are discarded.

Varying the lysis procedure and properties of the magnetic particles permits a wide variety of specimen types to be processed, and both DNA and RNA to be isolated either together as total nucleic acids, or separately. The instrument can also purify polyadenylated mRNA by replacing the particles that bind nucleic acids nonspecifically with ones coated with streptavidin and adding biotinylated oligo-dT to the lysate.

*Applied Biosystems PRISM 6100 and 6700* Applied Biosystems (ABI) has introduced two automated nucleic acid purification devices into their PRISM line of instruments that support molecular diagnostics and DNA sequencing [59,60]. Both instruments, the PRISM 6100 and 6700, make use of a column-based purification scheme. The passage of liquid through the columns is accelerated by the use of a vacuum manifold. Of the two instruments, the 6100 is the more basic model, permitting high sample throughput due to a 96 well format, but requiring significant user input during the purification process. The 6700 provides fully automated nucleic acid purification, requiring no user input after loading of samples and reagents. This instrument couples with the company's real-time PCR instruments by setting up PCR reactions in a 96 well plate format after completing the nucleic acid isolation protocol. The assay set-up features include the ability to add master mix and DNA, and then to seal the plate with an optically clear cover so that it only needs to be placed into the PCR instrument with no additional manual manipulations. In addition, the sample identification information can be transferred electronically from the 6700 to the amplification and detection instrument.

*Gentra Autopure and Versapure* Gentra manufactures two robotic devices, the Autopure and the Versapure, that automate DNA isolation from liquid specimens including blood. These instruments automate a technology marketed by the same company as the 'Puregene' system [61]. This method involves red blood cell lysis, pelleting of white cells by centrifugation, lysis of the pelleted cells, a salting out step to precipitate protein from the lysate, leaving the DNA in the supernatant, and finally an alcohol precipitation of the DNA. The Autopure processes large samples, ranging from 1–10 ml, in batches of up to 8 or 16 specimens per run of 1–1.5 h. Up to 96 samples can be processed per 8 h shift. The Versapure can process up to 400 samples of 50–300 μl per day, and can purify both DNA and RNA.

*Qiagen BioRobot M48, M96, and 9604* Several instruments manufactured by Qiagen have become available in the past few years. Two of these, the BioRobot M48 and M96, work on a magnetic particle based technology similar to that used by the MagNA Pure. A third device, the BioRobot 9604, automates a column based technology popular among basic science researchers for purification of plasmid DNA from bacteria, and allows simultaneous purification of DNA from 96 200 μl blood samples.

### 2.3.2 *Methods for amplification*

The basic problem that must be overcome in developing a molecular diagnostic assay is that the focus of the analysis – one single gene, or even one particular nucleotide – exists in a background of the entire human genome of 3 billion nucleotides. Therefore some method must be used to separate the specific signal from a potentially overwhelming background. Any methodology for detecting nucleic acid targets with the requisite specificity therefore requires some mechanism to amplify the signal to the point where it is readily detected over background noise. The methods fall into two groups: those that involve amplification of the signal from the specific target gene, and those that involve amplification of the target sequence itself.

At the root of all such methods is the ability of a DNA sequence to bind with great specificity to its complement. One of the classic methods of signal amplification, which has now been in use for almost 30 years, is the Southern blot [62,63]. This technique provides a method for detecting a single target sequence in a background of genomic DNA through specific hybridization to a labeled probe. Typically, a sample of genomic DNA is digested with a restriction enzyme and the resulting fragments separated by gel electrophoresis. The separated nucleic acids are transferred to a nitrocellulose or nylon membrane which is then incubated with the labeled probe, usually for at least several hours and more typically overnight. After the blot is washed free of unbound probe, the bound probe is detected by a method appropriate for the label used. Originally, Southern blot probes were labeled with $^{32}P$, and the signal detected by direct exposure of the blot to photographic film. In recent years, there has been a trend to avoid use of radioactivity wherever possible, leading to the development of signal amplification and detection systems based on chemiluminescence or color generation.

As a method for application in a clinical laboratory, the genomic Southern blot suffers from several disadvantages. First, it is very labor intensive, requiring numerous steps from gel preparation through gel loading to transfer, hybridization and washing the blot. Second, it is not very amenable to the high throughput required for a clinical assay since the number of samples that can be loaded per gel is quite limited. Third, the method requires relatively large amounts of genomic DNA – with the desired amount being in the 5–10 μg per sample range – to produce a detectable signal.

2.3.2.1 *The polymerase chain reaction (PCR)* Many of the obstacles to clinical implementation of nucleic acid based tests were overcome with the invention of the PCR, the first and most important of the target amplification methods to be described [48]. The method amplifies target sequences in the input DNA ranging from fewer than 100 up to several thousand nucleotides (nt) in length. Two oligonucleotides, oriented in opposite directions and complementary to the termini of the target sequence, serve as the specific probes in the reaction. These oligonucleotides function as the primers required for DNA synthesis catalyzed by a thermostable DNA polymerase. The reaction containing the sample DNA, the two primers, the polymerase, and the four deoxynucleoside triphosphates is run in a thermocycler, an instrument that alters the reaction temperature such that repetitive cycles of denaturation, annealing of primer to separated DNA strands, and extension of the primer occur, resulting in a doubling of the number of molecules of the target sequence with each cycle. The method is simple to set up, works with great reliability, and is so sensitive that it allows the detection of a specific target sequence in as little as a single molecule of template, although routine applications usually employ 10–100 ng of genomic DNA per reaction, corresponding to the genomic DNA from 1500–15000 cells, based on 6.6 pg DNA per diploid human cell. A further advantage of PCR is that the amplified target can be used, often without further purification, for additional analytical procedures, such as DNA sequencing, restriction enzyme digestion, or, as a probe, for hybridization to immobilized target DNA (such as a Southern blot).

In addition to PCR, several other target amplification methods have been developed and are marketed for clinical use in molecular diagnostics assays. The transcription mediated amplification (TMA) assay employs the enzymes reverse transcriptase and RNA polymerase to amplify a viral RNA target [64–67]. The reaction occurs at a constant temperature, without thermocycling, and the sensitivity of the method rivals that of PCR. The linked linear amplification (LLA) assay amplifies a target sequence in a reaction that requires thermocycling but employs a series of nested pairs of oligonucleotide primers in contrast to the single pair used in PCR [68].

2.3.2.2 *Hyphenation of PCR amplification with signal detection* The detection of the products generated in a PCR reaction traditionally required gel electrophoresis, sometimes even followed by a Southern blot. For application in a clinical laboratory, the amount of labor and technical expertise required in this approach prevents its adoption outside highly specialized laboratories.

A major technical advance that addressed and largely overcame these problems was the development of a variety of techniques for detecting PCR products in the reaction tube without any post-PCR manipulations. Since these techniques allow for product detection during thermocycling, they are often referred to as 'real-time PCR' [69]. The instruments designed for real-time PCR directly detect fluorescence from the PCR reaction tubes. The chemical strategies that result in the generation of a fluorescent signal only in the presence of the PCR product fall into two categories. One strategy makes use of a dye such as SYBR Green, which fluoresces when bound to double stranded DNA. As the PCR product forms, the dye binds and is detected as a fluorescent signal which is quantitated by the instrument. The disadvantage of this method is that the production of a signal does not prove specificity, since any double stranded DNA molecule would fluoresce.

The second category of techniques makes use of the property of fluorescence resonance energy transfer (FRET) in which two dyes, a reporter and quencher, are coupled to a probe. When the two dyes are in close proximity, the quencher absorbs the light emitted from the reporter dye on laser stimulation. One popular application of FRET probes is in the 5′ nuclease, or 'TaqMan,' assay [69,70]. The probe in this case is an oligonucleotide complementary to a region of the PCR product lying somewhere between the PCR primer sequences. Two dyes covalently linked to the probe, the reporter at the 5′ end and the quencher at the 3′ end, provide the basis for signal detection. In the presence of the specific PCR product, the probe binds to its complementary sequence. During the next polymerization step, a 5′ nuclease activity, specific for double stranded DNA and intrinsic to the Taq DNA polymerase molecule, digests the probe, thereby separating the reporter from the quencher, resulting in increased fluorescence detectable in real time. Other real time detection methods also employ FRET to generate a fluorescent signal only in the presence of a specific product. One such method employs two oligonucleotide probes, one labeled with the reporter, the other with the quencher, designed to bind to adjacent sites on the PCR product such that binding leads to a reduction in fluorescence detectable by the instrument [71]. Another variant makes use of a probe called a 'molecular beacon,' consisting of an oligonucleotide with complementary 5′ and 3′ ends linked to reporter and quencher dyes, and a central section complementary to the PCR product [72–74]. The complementary termini of the probe bind to each other, causing the probe to fold into a hairpin structure, leading to fluorescence quenching when free in solution. When the molecular beacon binds to the PCR product, the probe opens into a linear structure, thereby separating the reporter from the quencher, resulting in increased fluorescence.

Real-time PCR, essentially a hyphenation of the amplification and detection steps of a diagnostic assay, improves on the basic PCR technology on several levels. As a hyphenated method in the sense defined here, the time consuming and labor intensive steps of gel electrophoresis followed by staining or Southern blot have been eliminated altogether; when the PCR is done, the assay results are read immediately. Since the PCR tube does not need to be opened after the reaction, a major source of contamination in PCR is eliminated. Finally, by detecting product in real time, the technology greatly simplifies quantitation of the amount of target in a test sample. This capability results from the ability to determine the PCR cycle number at which the fluorescent signal reaches a threshold level, often abbreviated as the $C_T$ value, and the observation that the greater the number of molecules of target in the sample, the lower the cycle number required to reach the threshold. Thus, a standard curve can be generated relating target quantity to $C_T$. Figure 2.2 illustrates real-time fluorescence data in which a specimen and positive control both yielded positive results, with $C_T$ values of about 26 and 25, respectively, while the signal from the negative control reaction never rises above the threshold line.

Several manufacturers now market real-time PCR instruments. Probably the two most commonly used in molecular diagnostic laboratories are the ABI PRISM 7700 Sequence Detection System [75] and the Roche LightCycler [71]. The ABI instrument consists of a standard thermocycler modified for real-time fluorescence detection. With

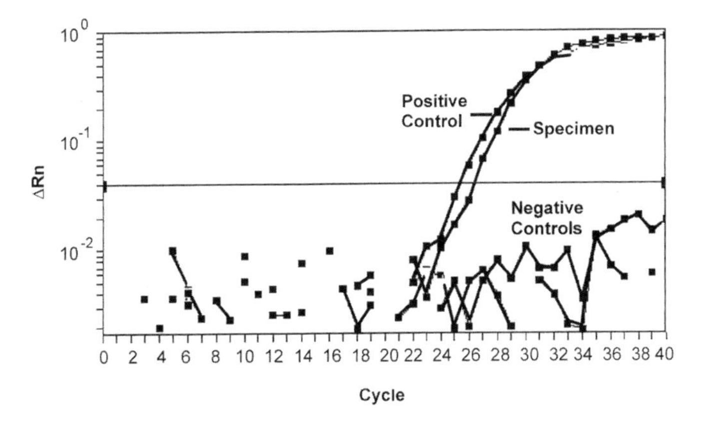


**Fig. 2.2** Real-time PCR. Plots of fluorescence versus cycle number are shown for a positive control, a test specimen, and a negative control. The horizontal line across the middle of the graph is the threshold line calculated by the instrument control software as a multiple of the baseline fluorescence averaged over the first 10 cycles of PCR. The points at which the fluorescence tracings cross this line define the $C_T$ values, which can be related to the quantity of target in the starting material by using a standard curve.

the LightCycler, reactions occur in glass capillaries rather than plastic microcentrifuge tubes. The high surface/volume ratio of the capillary allows for very rapid temperature changes in the reaction mixtures, allowing the thermocycling steps to be performed more rapidly than with most instruments, resulting in reduction of the cycling time from 2–3 h to 15–20 min.

2.3.2.3 *Signal amplification methods* Whereas real-time PCR represents a hyphenation of the amplification and detection steps in target amplification assays, hyphenated methods have also come into play in the area of signal amplification. Two such methods will be described here because of their frequent use in clinical laboratories: the branched chain and the Invader assays. In the branched chain assay, a method originally developed by Chiron and now marketed by Bayer [76,77], oligonucleotide probes (capture probes) immobilized on the surface of a well in a multiwell plate bind to target molecules present in the sample. After allowing time for the target molecules to bind to the capture probes, residual sample is removed, the wells washed, and a second probe added. If the target is present, the second probe binds to the target molecule now bound to the surface via its interaction with the capture probes. Coupled to the second probe is a complex branched DNA molecule – the 'branched chain' – covalently linked to numerous enzyme molecules. A chromogenic substrate is added after washing off the unbound probe, and color generation indicates a positive result. Color intensity varies linearly with the concentration of target in the specimen, making this method useful for quantitation. The branched chain assay is commonly used for detection and quantitation of HIV and hepatitis C virus (HCV) in blood [78,79].

The Invader assay, manufactured by Third Wave Technologies, which applies the substrate recognition requirements of the enzyme cleavase to signal amplification assays, is particularly effective in detecting SNPs and point mutations in human genomic DNA [80–83]. Figure 2.3 diagrams the fairly complex interactions of probes and specimen

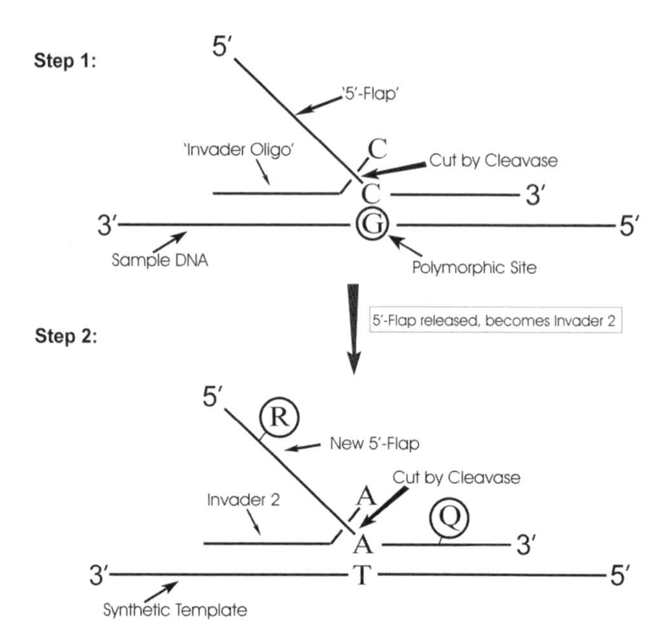


**Fig. 2.3** Invader assay. Diagram of the structures formed by probes and template. 'R' and 'Q' in step 2 refer to reporter and quencher dyes, respectively.

that occur in this assay. The target specific probes consist of two oligonucleotides. In step 1 of the assay, the Invader probe binds to the target such that the second nucleotide from the 3′ terminus binds to the nucleotide immediately adjacent to the polymorphic site of interest. The 3′ terminal nucleotide will contribute to the tertiary structure recognized by cleavase. The second probe oligonucleotide consists of a 5′ flap sequence not related to the target sequence followed by either the normal or mutant nucleotide at the polymorphic site, followed in turn by sequence complementary to the template. Cleavase will cut the signaling probe and release the 5′ flap when, and only when, the nucleotide at the polymorphic site in the template binds to the corresponding nucleotide in the signaling probe. The release of the 5′ flap allows it to function as an invader probe (step 2 in Fig. 2.3) for a synthetic template bound to a signaling probe labeled with reporter and quencher dyes. In the presence of the free 5′ flap, cleavase cuts the signaling probe, separating the dyes, and resulting in an increase in fluorescence. Steps 1 and 2 occur in a single homogeneous reaction.

Although complex to visualize, the method proves to be relatively simple to set up. As with real-time PCR, once set up no additional manipulation of the reaction is required. Invader assays require much larger amounts of template than PCR, with the usual recommendation being 150–300 ng per Invader assay compared to 10–100 ng for PCR. This must be taken into account when deciding on which sample preparation method to use; tiny tissue fragments or buccal swabs may not yield sufficient material for a successful assay. Invader assays are commonly used for the frequently ordered tests for the factor V Leiden and the prothrombin G20210A mutations associated with venous thrombosis, a common clinical problem. A multiplexed Invader assay reagent for detecting the 25 mutation panel recommended for CF carrier screening is also available.

### 2.3.3 *Methods of separation and detection*

2.3.3.1 *Gel electrophoresis* The classic method for separating mixtures of nucleic acids is gel electrophoresis. This method separates DNA fragments based on their length, with larger molecules moving more slowly than small ones, as they run through a porous medium, which in molecular biology is almost always a gel consisting of agarose or polyacrylamide [43]. The separated DNA species can be visualized very simply by staining the gel in a dilute solution of the intercalating dye ethidium bromide for about 5 min, and then transilluminating the gel from below with a UV light. The fluorescence from ethidium bound to the DNA molecules is visualized as a band on the gel which can be photographed to create a permanent record. Agarose is used for separation of larger DNA molecules in the size range 100–20 000 nt, whereas polyacrylamide is used for high resolution separations of molecules 1–500 nt in length. The information obtained is an estimate of size and quantity for each DNA species present. The gel is also the starting point for the Southern blotting procedure, described above.

2.3.3.2 *Capillary electrophoresis* The shortcomings of Southern blotting as a signal amplification method for clinical use apply equally to assays requiring separation of nucleic acids based on size. Some of the problems inherent to gel electrophoresis have been overcome by the introduction of the first of the automated technologies to be considered for separation, namely capillary electrophoresis [84–87]. The principle of separation is similar to that for gel electrophoresis; however, the separation occurs within a capillary typically 50 $\mu$m in internal diameter and 36–50 cm in length, filled with a polymeric sieving medium.

Capillary electrophoresis instruments automate the processes of sample loading, electrophoretic separation, and detection. These instruments sequentially and automatically load each sample, electrophorese it through the capillary and detect the fluorescent signal arising from the separated DNA molecules at the distal end of the capillary. Thus, rather than requiring an additional procedure such as gel staining after the run, the user reads results directly from a computer file generated by the instrument. The sample preparation method for capillary electrophoresis depends on the specific requirements of the instrument used, but in many cases involves incorporating a fluorescent dye into the DNA molecules to be analyzed. DNA size standards, labeled with a different fluorescent dye, must be mixed with each sample if the assay requires an accurate determination of product size. These instruments can separate the signals from multiple dyes, an important property for DNA sequencing applications, where four different dyes are used for labeling the four different nucleotides, allowing all the sequencing data to be obtained from a single sample run through the capillary. Samples are organized in a special holder plate, most commonly organized in an $8 \times 12$ grid of 96, although instruments capable of handling plates of 384 samples or more are available for higher throughput applications.

The throughput of a capillary electrophoresis system depends on the instrument used and the specific application. One of the common models used in molecular diagnostics laboratories, the ABI 310, contains a single capillary and takes 30–40 min to process each specimen [88]. A run of 96 samples on this instrument would therefore take up to 64 h. To shorten the processing time, multiple capillary instruments have been developed, one of the more popular of which, the ABI 3100, contains 16 capillaries, allowing a corresponding reduction in processing time. A 96 capillary instrument, the ABI 3700, was developed and used extensively for sequencing the human genome.

In addition to reducing the manual labor involved in electrophoretic separation of DNA molecules, capillary electrophoresis instruments allow for greater accuracy and precision in size determination and quantitation than are possible with slab gels. For applications requiring an accurate measurement of size, molecular weight standards are included with each sample. Quantitative recording of fluorescence intensity also permits an accurate determination of relative quantities of the various molecular species present in a mixture.

Capillary electrophoresis can be applied to the analysis of point mutations by several strategies. If a point mutation creates or eliminates a restriction enzyme recognition sequence, then the size of the restriction fragments arising from the normal and mutant alleles, readily determined by capillary electrophoresis, will differ. Figure 2.4 shows

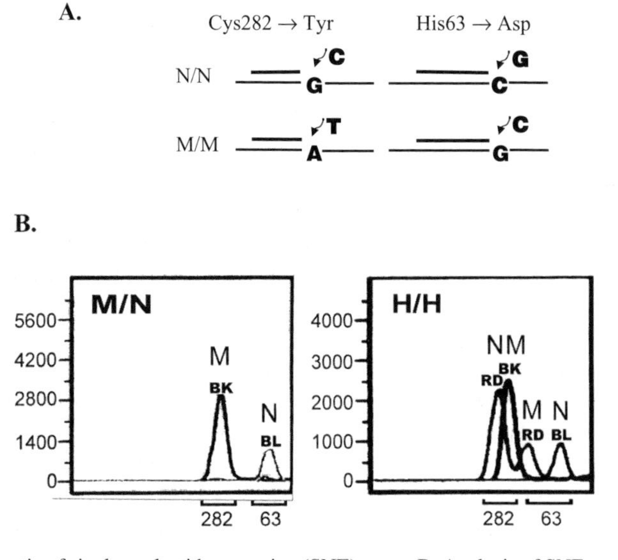


**Fig. 2.4** A. Schematic of single nucleotide extension (SNE) assay. B. Analysis of SNE products by capillary electrophoresis. PCR products containing the amino acid 63 and 282 coding sequences, of the *HFE* genes were generated in a single reaction. Oligonucleotides were annealed to the PCR products and incubated with four dideoxynucleoside triphosphates (ddNTPs), each labeled with a different fluorescent dye. The samples were analyzed on an ABI 310 capillary electrophoresis instrument. The peaks on the electropheregrams are labeled to indicate their original color, corresponding to the ddNTP added to the end of the primer in the SNE reaction: black (BK), T; blue (BL), G; red (RD), C. N, normal allele; H, heterozygous for the mutation; M, homozygous for the mutation. Reprinted with permission from [89].

results obtained with an ABI 310 instrument in an assay designed to detect the two common HH mutations [89] by an alternative strategy, single nucleotide extension (SNE) of oligonucleotides bound to PCR products containing the two potentially mutated sites.

SNE involves extension of a primer immediately adjacent to a possible mutation site with dideoxynucleoside triphosphates (ddNTPs), the standard chain terminators used in DNA sequencing. Under the conditions of the SNE reaction, the primer is extended by only one nucleotide, and only by the nucleotide complementary to the base present in the test sample. In the assay shown in Fig. 2.4B, each ddNTP was labeled with a different fluorescent dye. This technology, developed for DNA sequencing, allows the identification of the specific nucleotide added by analysis of its fluorescence spectrum, a process performed automatically by the ABI 310. In the case illustrated, the addition of a C residue indicates a normal allele at aa 282 and a mutant allele at aa 63. Similarly, addition of a T indicates the presence of a mutant allele at aa 282, and addition of a G at aa 63 indicates the presence of a normal allele at that position.

An alternative approach that uses capillary electrophoresis to detect SNPs and other sequence variants is the oligonucleotide ligation assay (OLA) [90–92]. In this method, a short segment of DNA containing the polymorphic site is amplified by PCR. Next, oligonucleotides are added together with a DNA ligase. The oligonucleotides are designed to bind to the DNA at the polymorphic site, such that the 3′ end of one is complementary to either the normal or mutant nucleotide, and the 5′ end of the second binds immediately adjacent to the 3′ end of the first. Ligation of the two oligonucleotides will occur only if the 3′ end of the first is complementary to the template. By using oligonucleotides of different sizes for the normal and mutant variants, therefore, the size of the ligated products, assessed by capillary electrophoresis, is determined by the alleles present in the template. A commercial product using the OLA to detect mutations has been developed for the ACOG/ACMG panel of mutations in the CF assay [93]. The assay involves a highly multiplexed PCR reaction followed by an OLA reaction that simultaneously analyzes all mutations in the panel. Thus, one PCR reaction, one oligonucleotide ligation reaction, and one run of capillary electrophoresis, reveals the genotype at over 30 loci within the *CFTR* gene. The ability of the capillary electrophoresis instruments to process samples arrayed in 96 well plates allows high throughput to be achieved with this system.

2.3.3.3 *Denaturing HPLC (dHPLC)* As an alternative to gel electrophoresis, HPLC may be used to separate DNA molecules by size. One particularly useful application of this technique has been incorporated into the denaturing HPLC (dHPLC) WAVE system manufactured by Transgenomic [94–96]. The methodology takes advantage of the ability of the HPLC column to separate perfectly complementary double stranded DNA molecules from ones that contain an internal mismatch. The usefulness of the method lies in its ability to detect mutations in a DNA molecule when the specific mutation is not known in advance. This method, termed ‘mutation scanning,’ becomes useful when attempting to determine the genetic abnormality responsible for a disease in which all mutations affect the same gene but do not occur in any consistent or predictable location.

Sample preparation for a mutation scanning procedure involves preparation of PCR products from a normal specimen and from the test specimen. Figure 2.5A is a schematic

A.

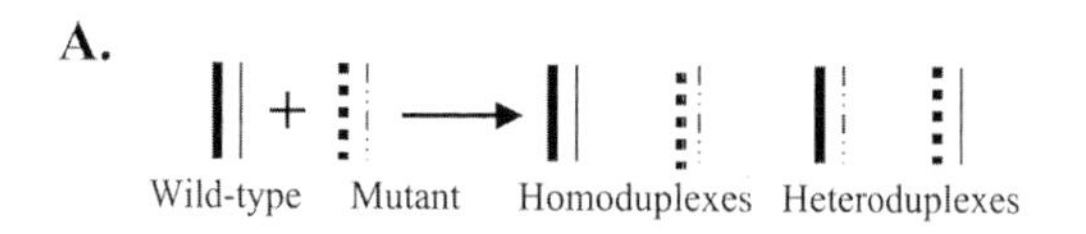


B.

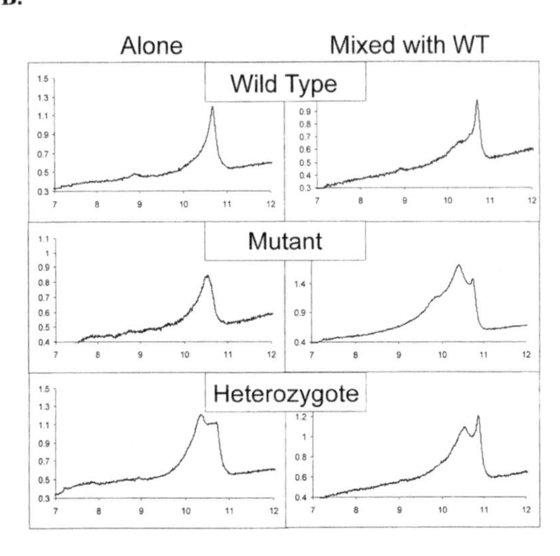


**Fig. 2.5** Heteroduplex analysis of HFE mutations. A. Principle of heteroduplex formation. A PCR product from the test specimen is mixed with the corresponding wild type PCR product, heated to denature the DNA, and then cooled to allow reannealing. The reannealed double stranded molecules will contain a mixture of homo- and heteroduplexes. B. A PCR product containing the aa 282 coding sequence of the *HFE* gene was analyzed either alone or after mixing with wild type (WT) PCR product on a Transgenomic Wave dHPLC instrument. The heteroduplex, detected in the heterozygous mutant by itself and both the heterozygous and homozygous mutants after mixing with the normal (WT) product, appears as a distinct peak eluting later than the homoduplex DNA species. Reprinted with permission from [37].

of the procedure, and Fig. 2.5B shows results obtained with normal and HH heterozygous and homozygous specimens. The PCR products are mixed, heat denatured, and allowed to reanneal. The reannealed products are loaded onto the dHPLC instrument. Approximately half the reannealed product will consist of a strand derived from the normal specimen hybridized to a strand from the test specimen, a structure termed a heteroduplex. If the test specimen contains a point mutation (or small deletion or insertion) anywhere along its length, there will be a region of the heteroduplex that tends to remain single stranded. As a result, the melting temperature around the mismatch will be lowered relative to the rest of the molecule. At some elevated temperature, the DNA strands around the mismatch separate while the perfectly matched normal duplex remains double stranded. This structural change alters the chemical properties of the heteroduplex, resulting in reduced affinity to the column matrix. Elution with an acetonitrile gradient allows good resolution between the normal molecule and the heteroduplex, with the heteroduplex eluting first from the column. Eluted DNA is detected by UV absorbance. The readout for a positive result is the presence of a peak not present with the normal specimen. Available models of the Transgenomic WAVE instrument can process 96- or 384-well plates automatically, making the system appropriate for high throughput applications.

Duchenne's muscular dystrophy provides a good example of a situation in which mutation scanning might be the preferred methodology for a diagnostic assay [97]. This degenerative muscle disease results from the presence of mutations in the X chromosome gene *DMD*, the product of which has been named dystrophin (see refs 98–100 for reviews). Mutations may be insertions, deletions, or point mutations, and may occur anywhere over a range of thousands of nucleotides of gene sequence. In such a case, screening for a panel of specific individual mutations would lack sensitivity; most mutations would be missed. One approach to this situation would be exon-by-exon PCR amplification followed by DNA sequencing. However, this approach would require the evaluation of many thousands of nucleotides of DNA sequence for each test, a process that would be prohibitively expensive. Other examples of diseases where heteroduplex analysis by dHPLC has been applied to detect a broad spectrum of possible mutations include CF [101], Fanconi anemia [102], and a genetic form of ocular hypertension [103]. Applications of dHPLC have also been described for detecting the two major hemochromatosis mutations [104,105] as well as for scanning the entire gene for mutations [106].

2.3.3.4 *Arrays of allele-specific oligonucleotides (ASOs)* One of the newest areas of technology in the nucleic acid testing field has been the development of methods for fabricating dense arrays of DNA species bound to a solid support – the 'DNA chip' or 'microarray.' Two methods have been widely used. One, developed by Affymetrix, involves the use of a technology for synthesis of oligonucleotides *in situ* in high density on the chip [107,108]. The method permits the generation of arrays containing thousands of distinct oligonucleotides per cm$^2$ on a chip. A second technology, developed in the laboratory of Patrick Brown, involves the deposition of prepared DNA species at the desired density on a glass slide [109]. The technology has found its most extensive use in studies of global gene expression patterns. In such studies, the probe consists of reverse transcribed and fluorescently labeled RNA purified from a tissue or cell line of interest. After incubation with the array, the intensity of fluorescence at each site corresponds to the level of expression of the corresponding gene. Characterization of global gene expression patterns has begun to provide a new classification scheme for cancers, particularly lymphomas [110] and breast cancer [111]. It is hoped that this emerging classification scheme will lead to the development of effective therapies targeted to the specific molecular features of the individual tumor.

To date, arrays of high density have not been widely applied in molecular diagnostics, although recent research findings suggest that this technology can classify tumors according to prognosis, and do a better job of this than traditional histopathological analysis. However, lower density arrays of oligonucleotides have found utility in molecular diagnostics in assays that screen for a defined panel of mutations or polymorphisms. In this context, the oligonucleotides are designed to bind to either the normal or mutant form of the target molecule, and are thus *allele-specific oligonucleotides* (ASOs).

Commercial products of this sort usually consist of a 1 × N array, with the oligonucleotides laid out along the length of a thin plastic strip. Since hybridization to any specific oligo on the strip gives rise to a line across the strip, these assays are called

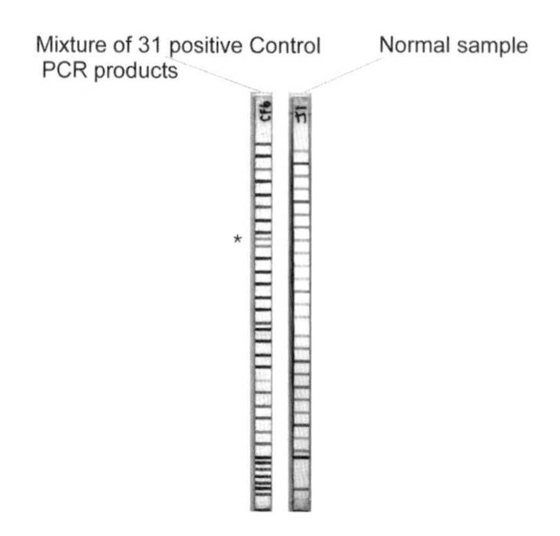


**Fig. 2.6** Roche CF Gold linear array panels (LAPs). The strips are manufactured with normal and mutant oligonucleotides bound corresponding to the 25 mutations recommended for CF testing by the ACMG/ACOG panel, plus six additional oligonucleotides that are only read in specific situations. After development, the appearance of a blue line at a specific position across the width of the strip corresponds to the presence of either the normal or mutant allele of one of the 25 mutations. Mutant alleles are identified by aligning the strip with a key provided by the manufacturer. Shown are a LAP that had been hybridized with PCR products from a normal DNA sample and a LAP that had been hybridized with PCR products from a mixture of synthetic DNAs representing all the mutants and six additional sites. Note that one mutant DNA cross-reacted with the normal probe (asterisk) in this assay.

'line probe assays' (Innogenetics) or 'linear array panels' (Roche). As an example, the Roche CF Gold linear array panel consists of 56 ASOs bound to a thin plastic strip. The panel of ASOs consists of normal and mutant sequences for 25 specific sites within the gene that codes for the CFTR protein, plus sequences for six polymorphisms that have clinical relevance in special situations. Performance of the assay involves several steps, including (1) DNA isolation, (2) PCR amplification, (3) hybridization to the linear array panel followed by washing off the unhybridized probe, and (4) development of the signal, in this case a blue color, where the probe has bound to the strip. Figure 2.6 shows the appearance of two strips after color development, one hybridized to PCR products from a normal specimen, the other to a mixture of synthetic DNAs containing all of the mutant sequences.

Although the assay is fairly tedious, two aspects of the assay may be considered 'hyphenated.' First, the PCR reaction simultaneously amplifies 14 separate segments of the *CFTR* gene, and second, the use of the array takes the place of 56 separate hybridizations. Similar technologies have been developed for use in determining the subtype of clinically important viruses, especially HCV and the HIV that causes AIDS [112,113].

2.3.3.5 *ELISA type detection systems* One strategy to increase the efficiency of molecular diagnostics methods makes use of a technology developed for the detection of an antigen by using an antibody. Termed the enzyme linked immunosorbent assay (ELISA), such assays involve binding the antigen to the plastic wall of a reaction vessel, which most commonly is a well of a 96-well plate. Depending on the application, the plastic may be pretreated in some manner to facilitate the interaction. For example, the wells may be coated with avidin to bind a biotinylated target, or with an antibody to bind a specific antigen. After the wells are washed, a solution is added containing an antibody to the specific target conjugated to an enzyme, generally either horseradish peroxidase (HRP) or alkaline phosphatase. The antibody–enzyme conjugate will bind only if the specific target molecule is present. After an appropriate incubation, the wells are washed and a solution containing a chromogenic substrate of the indicator enzyme is added. An absorbance reading above a predefined threshold constitutes a positive result.

As applied to molecular diagnostics, this type of technology is often used in assays for the detection of infectious agents. Although the ELISA methodology is applicable to the detection of protein antigens, the type of applications discussed here detect nucleic acids having a specific sequence. The high volume tests in this category include those for the common sexually transmitted agents *Neisseria gonorrhoeae*, *Chlamydia trachomatis*, and the human papillomaviruses (HPVs). US FDA approved PCR based tests are available for *N. gonorrhoeae* and *C. trachomatis* with signal detection in an ELISA format. The commonly used test for HPV relies exclusively on signal amplification, bypassing the PCR step and therefore avoiding problems of contamination control while at the same time reducing the labor required. This test, the Hybrid Capture II (HCII) assay manufactured by Digene, will be described here in detail as an example of the application of this technology in a clinical setting.

The sample for HCII consists of cells and mucus collected with a special brush designed for this purpose from the uterine cervix during a gynecological examination. The brush is placed into a tube of alcohol based fixative, which is transported to the laboratory. A portion of the specimen is taken to be processed for microscopic examination by a pathologist, similar to the traditional Pap smear. The cellular material in what remains is pelleted by centrifugation and resuspended in an alkaline buffer that lyses the cells and denatures the released DNA. A proprietary probe mixture, containing RNA species homologous to the large group of HPV types detected by the assay, is added in a buffer that neutralizes the pH and allows the probe to hybridize to any HPV DNA present in the specimen. The annealed mixture is then added to tubes coated with a proprietary antibody directed against DNA/RNA hybrids. Such hybrids form only when HPV DNA is present in the sample. After an incubation to allow the hybrids to bind, the reagents remaining in solution are removed, the tubes are washed, and a solution added containing the same DNA/RNA hybrid antibody conjugated to alkaline phosphatase. Another incubation and wash step follow, after which the substrate solution is added. The tubes are incubated to allow the reaction to proceed and the absorbance read in a spectrophotometer. The assay is scored qualitatively as positive or negative. This assay has been validated in very large clinical studies, and has become accepted as a standard of care for clinical follow-up of atypical cytology results. The aspects of the assay that may be considered hyphenated are the virtually simultaneous performance of the isolation–amplification–detection steps, with the elimination of complicated isolation procedures, PCR, and electrophoresis.

2.3.3.6 *Mass spectrometry* Matrix assisted laser desorption ionization–time of flight (MALDI-TOF) mass spectrometry has become a powerful analytical tool since its development in the late 1980s. In comparison with general laser desorption/ionization

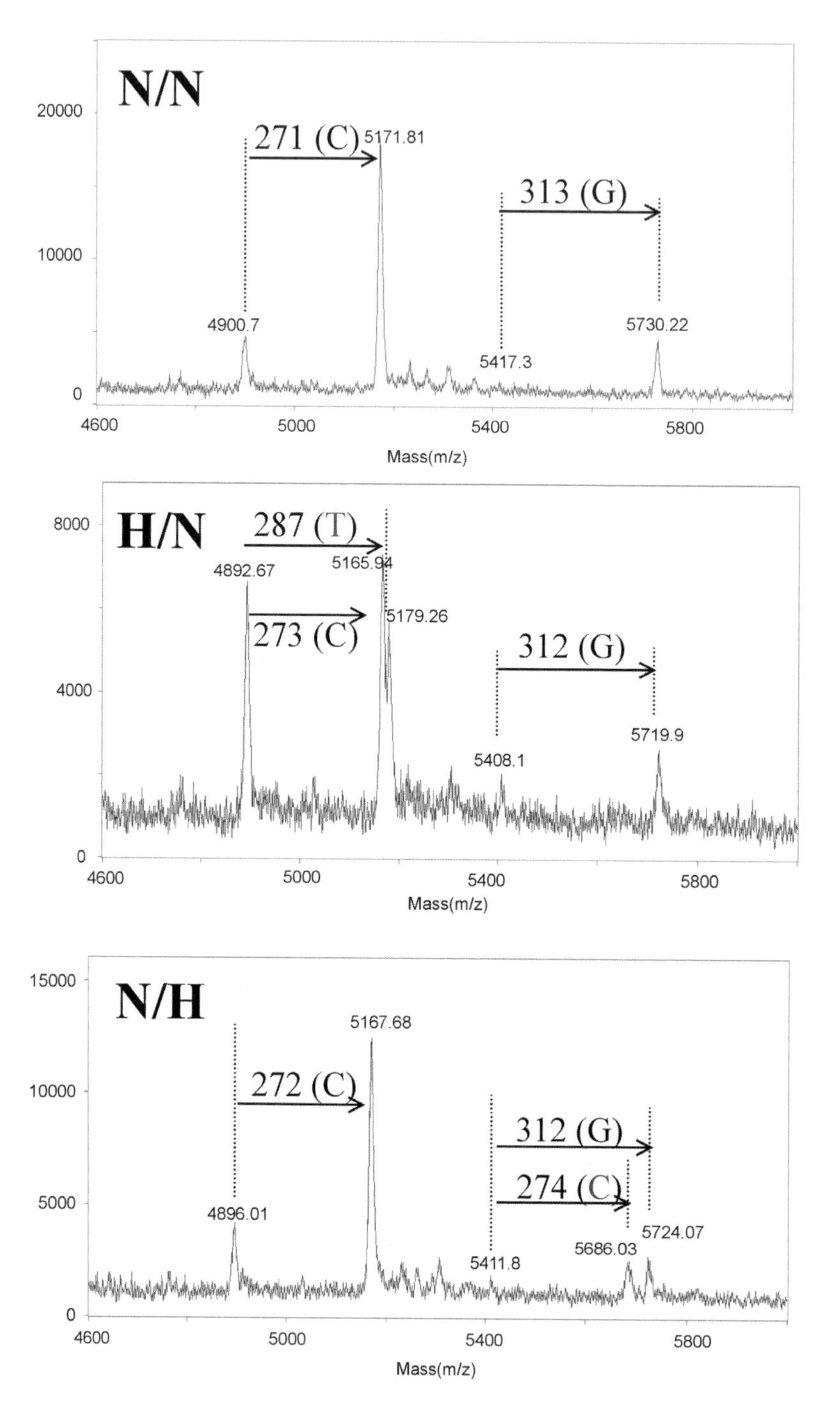


**Fig. 2.7** MALDI-TOF mass spectrometry. *HFE* PCR products were subjected to the SNE reaction as described in Fig. 2.4. The SNE products were analyzed by MALDI-TOF mass spectrometry. Reprinted with permission from [37].

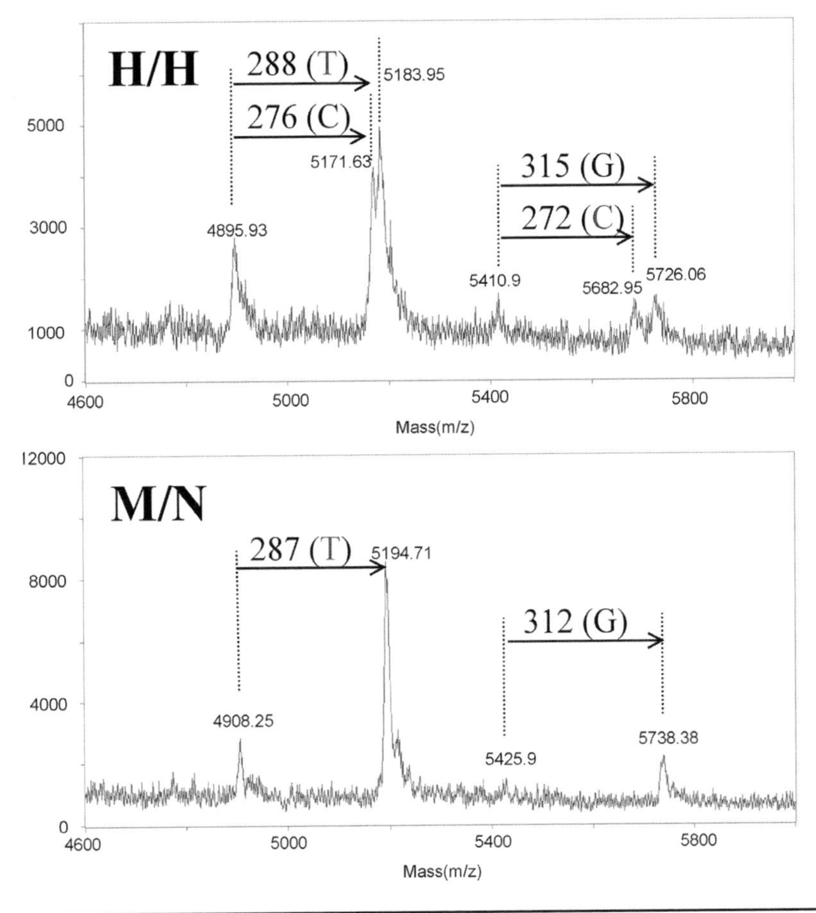


**Molecular Mass**

| | 282-site | 63-site |
|---|---|---|
| Normal | 272 (C) | 312 (G) |
| Mutant | 287 (T) | 272 (C) |

**Fig. 2.7** *Contd*

MS, which permits measurement of small molecules up to several thousand daltons, MALDI MS can generate ions of several hundred thousand daltons. The ability to detect larger molecules allows the analysis of DNA fragments up to 100 nt in length. The utility of MALDI MS depends on the generation of a suitable composite of matrix and analyte [114]. Therefore an important part of sample preparation involves finding a matrix and buffer system that will form homogeneous matrix/analyte co-crystals as the mixture dries. Sample analysis initiates when a brief pulse from a UV laser irradiates the matrix/analyte co-crystal. The matrix absorbs most of the laser energy, thereby preventing fragmentation of the biomolecules and aiding in the volatilization and ionization of the analytes. The ionized molecules are accelerated in an electric field and separated by time of flight (TOF) to the detector according to their mass/charge ($m/z$) ratio.

As a tool for genotyping DNA, MS methods have been developed with the ability to detect up to twelve mutations simultaneously with high sensitivity and specificity [115]. In contrast to capillary electrophoresis, the method does not require molecular weight standards or fluorescent tags on the DNA, and detection takes only a few seconds per sample. The major difficulty to be overcome is sample preparation, in particular the choice of the optimal matrix material.

As an example of applying MALDI-TOF analysis to genetic disease diagnosis, we again return to the two mutations associated with hereditary hemochromatosis (Fig. 2.7). The first steps in the procedure involved DNA isolation and amplification by PCR. For detection of the HH mutations, an SNE reaction was performed by mixing two primers, designed to bind adjacent to the sites of the two mutations, with the PCR products, ddNTPs and Taq polymerase.

To prepare the products of this reaction for MALDI-TOF analysis, they were desalted by loading the reaction mixtures onto Millipore ZipTip$C_{18}$ columns and eluted with 2.5 μl elution buffer (9 parts 50 mg/ml 3-hydroxypicolinic acid in 50% acetonitrile/water and 1 part 50 mg/ml ammonium citrate in water). An aliquot of eluate (1.5 μl) was spotted onto a stainless sample plate and air dried. Mass spectra were obtained on a linear MALDI-TOF mass spectrometer (Voyager DE, PerSeptive Biosystems). As shown in Fig. 2.7, the molecular mass determinations permitted the determination of the genotype at the two positions. Although the sample preparation steps used in this example were somewhat cumbersome for implementation as a high throughput assay, the rapid analysis possible by this approach would make it attractive for clinical use if the earlier steps in the assay could be performed on an automated platform.

## 2.4 More extensive hyphenation of molecular diagnostics assays

### 2.4.1 *Dedicated robotic methods*

2.4.1.1 *Cobas Amplicor* Perhaps the highest degree of hyphenation to date is realized in the Cobas Amplicor automated PCR instrument manufactured by Roche Diagnostics [116,117]. This instrument automates all steps in the assay procedure after sample preparation. The instrument automates several assays also sold in a manual format as Amplicor assays. These assays all involve PCR amplification followed by a detection procedure consisting of denaturation of the PCR product, hybridization in solution to a probe, a washing step to remove unhybridized probe, and finally a colorimetric detection step. The instrument is designed for use with US FDA approved assays for the quantitation of infectious agents. The available tests include the Amplicor HIV-1 monitor, a quantitative assay for detecting HIV in plasma. This assay has come into widespread use as a means for following the response to treatment and emergence of drug resistant strains of virus in AIDS patients. The other available test kits qualitatively detect hepatitis B virus (HBV), *C. trachomatis, N. gonorrhoeae*, and *Mycobacterium tuberculosis*. The automation, in combination with standardized kits and protocols, allows these assays to be performed in routine clinical laboratories as opposed to specialized molecular diagnostics sections.

Automatic progression from amplification through the detection steps on the Cobas Amplicor occurs with the physical movement of specimens from one station on the instrument to the next via a robotic gripper arm. All steps in the procedure after DNA isolation are performed by the instrument. For blood specimens, the Magna Pure LC instrument, as mentioned above, will distribute DNA samples into the reaction tubes used on the Cobas Amplicor, so that the only manual step is transfer of the reaction tubes from one instrument to the other. For the plasma specimens used for viral quantitation assays, sample preparation involves treatment with a chaotropic agent to lyse viral particles followed by alcohol precipitation of the RNA or DNA. The nucleic acids are mixed with a master mix and placed in the sample wells of the instrument. All subsequent steps of the assay are performed automatically, beginning with the thermocycling steps of PCR. In subsequent steps of the procedure, samples of the PCR reaction are transferred from one stage of the instrument to the next by a pipeting device attached to a robotic arm. In sequence, the PCR products are denatured, a biotinylated probe is added, and the sample is incubated at an optimal temperature for annealing of probe. Free probe is removed and the bound probed detected colorimetrically after addition of a chromogenic substrate.

2.4.1.2 *Programmable robotic liquid handlers* The most versatile, but possibly least user friendly, of the methods that bring hyphenation to the field of molecular diagnostics, is the programmable robotic liquid handler. These devices consist of a robotic arm and a platform on which specimens, reagents, pipet tips, and the full gamut of reaction vessels, from single tubes through high density multiwell plates, can be mounted. The robotic arm attaches to any of several different tools for liquid transfer, usually including single and multiple channel pipeting tools and a suction device for aspirating and discarding liquid from samples. A gripper tool may be available to allow the instrument to move samples in the common 96-well tray from one position on the platform to another. To make the instruments maximally adaptable to user defined procedures, they may have available modules such as a site on the platform that will apply a vacuum to a set of, for example, DNA purification columns arranged in a 96-well format. Thermocycler units exist that can be incorporated into such instruments as an add-on module, such that the robotic arm can set up a tray of PCR reactions, and then the gripper tool can place the tray into the thermocycler. Although the design of the software permits great flexibility in the development of user defined protocols, the complexities involved in creating the fully hyphenated application possible in theory, incorporating isolation–amplification–detection, make it impractical to do so for most users. Manufacturers of products, particularly for isolation and amplification steps, have dealt with this problem by generating applications suitable for implementation on the more common robotic instruments available.

## 2.5 Conclusion

The field of testing for genetic disease has undergone rapid evolution in the past few years resulting from the development of pieces of instrumentation intended for the clinical laboratory that automate each step in the assay procedure. Molecular diagnostic testing is no longer confined to specialized research laboratories, but can be performed in clinical genetics and molecular diagnostics laboratories, and, to a rapidly increasing extent, in routine clinical laboratories.

## References

1. Pawlotsky, J.M. (2003) Use and interpretation of hepatitis C virus diagnostic assays. *Clin. Liver Dis.*, **7**(1), 127–137.
2. Mass, R. (2000) The role of HER-2 expression in predicting response to therapy in breast cancer. *Semin. Oncol.*, **27**(6, Suppl 11), 46–52; discussion 92–100.
3. Bartlett, J., Mallon, E. & Cooke, T. (2003) The clinical evaluation of HER-2 status: which test to use? *J. Pathol.*, **199**(4), 411–417.
4. Di Leo, A., Dowsett, M., Horten, B. & Penault-Llorca, F. (2002) Current status of HER2 testing. *Oncology*, **63** (Suppl 1), 25–32.
5. Ravdin, P. (2000) The use of HER2 testing in the management of breast cancer. *Semin. Oncol.*, **27**(5, Suppl 9), 33–42.
6. Pegram, M.D., Pauletti, G. & Slamon, D.J. (1998) HER-2/Neu as a predictive marker of response to breast cancer therapy. *Breast Cancer Res. Treat.*, **52**(1-3), 65–77.
7. Swerdlow, S.H. & Williams, M.E. (2002) From centrocytic to mantle cell lymphoma: a clinicopathologic and molecular review of 3 decades. *Hum. Pathol.*, **33**(1), 7–20.
8. Barista, I., Romaguera, J.E. & Cabanillas, F. (2001) Mantle-cell lymphoma. *Lancet Oncol.*, **2**(3), 141–148.
9. Bijwaard, K.E., Aguilera, N.S., Monczak, Y., Trudel, M., Taubenberger, J.K. & Lichy, J.H. (2001) Quantitative real-time reverse transcription-PCR assay for cyclin D1 expression: utility in the diagnosis of mantle cell lymphoma. *Clin. Chem.*, **47**(2), 195–201.
10. Aguilera, N.S., Bijwaard, K.E., Duncan, B., Krafft, A.E., Chu, W.S., Abbondanzo, S.L. & Lichy, J.H. (1998) Differential expression of cyclin D1 in mantle cell lymphoma and other non-Hodgkin's lymphomas. *Am. J. Pathol.*, **153**(6), 1969–1976.
11. Weiss, L.M., Warnke, R.A., Sklar, J. & Cleary, M.L. (1987) Molecular analysis of the t(14;18) chromosomal translocation in malignant lymphomas. *N. Engl. J. Med.*, **317**(19), 1185–1189.
12. Lee, M.S., Chang, K.S., Cabanillas, F., Freireich, E.J., Trujillo, J.M. & Stass, S.A. (1987) Detection of minimal residual cells carrying the t(14;18) by DNA sequence amplification. *Science*, **237**, 175–178.
13. Dobrovic, A., Trainor, K.J. & Morley, A.A. (1988) Detection of the molecular abnormality in chronic myeloid leukemia by use of the polymerase chain reaction. *Blood*, **72**(6), 2063–2065.
14. Barnes, D.J. & Melo, J.V. (2002) Cytogenetic and molecular genetic aspects of chronic myeloid leukaemia. *Acta Haematol.*, **108**(4), 180–202.
15. Kantarjian, H.M., Talpaz, M., Cortes, J., O'Brien, S., Faderl, S., Thomas, D., Giles, F., Rios, M.B., Shan, J. & Arlinghaus, R. (2003) Quantitative polymerase chain reaction monitoring of Bcr-Abl during therapy with imatinib mesylate (Sti571; Gleevec) in chronic-phase chronic myelogenous leukemia. *Clin. Cancer Res.*, **9**(1), 160–166.
16. Downing, J.R., Head, D.R., Parham, D.M., Douglass, E.C., Hulshof, M.G., Link, M.P., Motroni, T.A., Grier, H.E., Curcio-Brint, A.M. & Shapiro, D.N. (1993) Detection of the t(11;22)(q24;q12) translocation of Ewing's sarcoma and peripheral neuroectodermal tumor by reverse transcription polymerase chain reaction. *Am. J. Pathol.*, **143**(5), 1294–1300.

17. Sorensen, P.H., Liu, X.F., Delattre, O., Rowland, J.M., Biggs, C.A., Thomas, G. & Triche, T.J. (1993) Reverse transcriptase PCR amplification of EWS/Fli-1 fusion transcripts as a diagnostic test for peripheral primitive neuroectodermal tumors of childhood. *Diagn. Mol. Pathol.*, **2**(3), 147–157.
18. Galili, N., Davis, R.J., Fredericks, W.J., Mukhopadhyay, S., Rauscher, F.J., 3rd, Emanuel, B.S., Rovera, G. & Barr, F.G. (1993) Fusion of a fork head domain gene to Pax3 in the solid tumour alveolar rhabdomyosarcoma. *Nat. Genet.*, **5**(3), 230–235.
19. Barr, F.G., Holick, J., Nycum, L., Biegel, J.A., & Emanuel, B.S. (1992) Localization of the t(2;13) breakpoint of alveolar rhabdomyosarcoma on a physical map of chromosome 2. *Genomics*, **13**(4), 1150–1156.
20. Barr, F.G., Chatten, J., D'Cruz, C.M., Wilson, A.E., Nauta, L.E., Nycum, L.M., Biegel, J.A. & Womer, R.B. (1995) Molecular assays for chromosomal translocations in the diagnosis of pediatric soft tissue sarcomas. *J. Am. Med. Assoc.*, **273**(7), 553–557.
21. Bijwaard, K.E., Fetsch, J.F., Przygodzki, R., Taubenberger, J.K. & Lichy, J.H. (2002) Detection of SYT-SSX fusion transcripts in archival synovial sarcomas by real-time reverse transcriptase-polymerase chain reaction. *J. Mol. Diagn.*, **4**(1), 59–64.
22. Argani, P., Zakowski, M.F., Klimstra, D.S., Rosai, J. & Ladanyi, M. (1998) Detection of the SYT-SSX chimeric RNA of synovial sarcoma in paraffin-embedded tissue and its application in problematic cases. *Mod. Pathol.*, **11**(1), 65–71.
23. Antonarakis, S.E., Kazazian, H.H., & Tuddenham, E.G. (1995) Molecular etiology of factor VIII deficiency in hemophilia *A. Hum. Mutat.*, **5**(1), 1–22.
24. Lakich, D., Kazazian, H.H., Jr., Antonarakis, S.E., & Gitschier, J. (1993) Inversions disrupting the factor VIII gene are a common cause of severe haemophilia *A. Nat. Genet.*, **5**(3), 236–241.
25. Weedn, V.W. & Roby, R.K. (1993) Forensic DNA testing. *Arch. Pathol. Lab. Med.* **117**(5), 486–491.
26. Waye, J.S. (1993) Forensic identity testing using highly polymorphic DNA markers: current status and emerging technologies. *Transfus. Med. Rev.*, **7**(3), 193–205.
27. Pena, S.D. & Chakraborty, R. (1994) Paternity testing in the DNA era. *Trends Genet.* **10**(6), 204–209.
28. Sajantila, A. & Budowle, B. (1991) Identification of individuals with DNA testing. *Ann. Med.*, **23**(6), 637–642.
29. Collins, F.S., Brooks, L.D., & Chakravarti, A. (1998) A DNA polymorphism discovery resource for research on human genetic variation. *Genome Res.*, **8**(12), 1229–1231.
30. Burke, W., Thomson, E., Khoury, M.J., McDonnell, S.M., Press, N., Adams, P.C., Barton, J.C., Beutler, E., Brittenham, G., Buchanan, A. & 8 other authors (1998) Hereditary hemochromatosis: gene discovery and its implications for population-based screening. *J. Am. Med. Assoc.*, **280**(2), 172–178.
31. Tripodi, A. & Mannucci, P.M. (2001) Laboratory investigation of thrombophilia. *Clin. Chem.*, **47**(9), 1597–1606.
32. Federman, D.G. & Kirsner, R.S. (2001) An update on hypercoagulable disorders. *Arch. Intern. Med.*, **161**(8), 1051–1056.
33. Press, R.D. (2001) Laboratory standards and guidelines for population-based cystic fibrosis carrier screening. *Mol. Diagn.*, **6**(3), 212–213.
34. Vajo, Z., Francomano, C.A. & Wilkin, D.J. (2000) The molecular and genetic basis of fibroblast growth factor receptor 3 disorders: the achondroplasia family of skeletal dysplasias, Muenke craniosynostosis, and Crouzon syndrome with acanthosis nigricans. *Endocr. Rev.*, **21**(1), 23–39.
35. Wenstrom, K.D. (2000) Fragile X and other trinucleotide repeat diseases. *Obstet. Gynecol. Clin. North Am.*, **29**(2), 367–388.
36. Hanson, E.H., Imperatore, G. & Burke, W. (2001) HFE gene and hereditary hemochromatosis: a huge review. Human genome epidemiology. *Am. J. Epidemiol.*, **154**(3), 193–206.
37. Liang, Q. & Lichy, J.H. (2002) Molecular testing for hereditary hemochromatosis. *Expert Rev. Mol. Diagn.*, **2**(1), 49–59.
38. Powell, L.W., George, D.K., McDonnell, S.M. & Kowdley, K.V. (1998) Diagnosis of hemochromatosis. *Ann. Intern. Med.*, **129**(11), 925–931.
39. Feder, J.N., Gnirke, A., Thomas, W., Tsuchihashi, Z., Ruddy, D.A., Basava, A., Dormishian, F., Domingo, R., Jr, Ellis, M.C., Fullan, A. & 23 other authors (1996) A novel MHC class I-like gene is mutated in patients with hereditary haemochromatosis. *Nat. Genet.*, **13**(4), 399–408.
40. Waalen, J., Felitti, V., Gelbart, T., Ho, N.J. & Beutler, E. (2002) Prevalence of hemochromatosis-related symptoms among individuals with mutations in the HFE gene. *Mayo Clin. Proc.*, **77**(6), 522–530.
41. Richards, C.S., Bradley, L.A., Amos, J., Allitto, B., Grody, W.W., Maddalena, A., McGinnis, M.J., Prior, T.W., Popovich, B.W., Watson, M.S. & Palomaki, G.E. (2002) Standards and guidelines for CFTR mutation testing. *Genet. Med.*, **4**(5), 379–391.
42. Lyon, E. & Miller, C. (2003) Current challenges in cystic fibrosis screening. *Arch. Pathol. Lab. Med.*, **127**(9), 1133–1139.
43. Maniatis, T., Fritsch, E.F. & Sambrook, J. (1982) *Molecular Cloning: A Laboratory Manual.* Cold Spring Harbor Laboratory, Cold Spring Harbor, NY.

44. Mullis, K.B. (1990) The unusual origin of the polymerase chain reaction. *Sci. Am.*, **262**(4), 56–61, 64–55.
45. Mullis, K.B. (1990) Target amplification for DNA analysis by the polymerase chain reaction. *Ann. Biol. Clin.* (Paris), **48**(8), 579–582.
46. Saiki, R.K.; Gelfand, D.H.; Stoffel, S., Scharf, S.J., Higuchi, R., Horn, G.T., Mullis, K.B. & Erlich, H.A. (1988) Primer-directed enzymatic amplification of DNA with a thermostable DNA polymerase. *Science*, **239**, 487–491.
47. Mullis, K.B. & Faloona, F.A. (1987) Specific synthesis of DNA in vitro via a polymerase-catalyzed chain reaction. *Methods Enzymol.*, **155**, 335–350.
48. Mullis, K., Faloona, F., Scharf, S., Saiki, R., Horn, G. & Erlich, H. (1986) Specific enzymatic amplification of DNA in vitro: the polymerase chain reaction. *Cold Spring Harb. Symp. Quant. Biol.*, **51**(1), 263–273.
49. Krafft, A.E., Duncan, B.W., Bijwaard, K.E., Taubenberger, J.K. & Lichy, J.H. (1997) Optimization of the isolation and amplification of RNA from formalin-fixed, paraffin-embedded tissue: the Armed Forces Institute of Pathology experience and literature review. *Mol. Diagn.*, **2**(3), 217–230.
50. Jackson, D.P., Lewis, F.A., Taylor, G.R., Boylston, A.W. & Quirke, P. (1990) Tissue extraction of DNA and RNA and analysis by the polymerase chain reaction. *J. Clin. Pathol.*, **43**(6), 499–504.
51. Wu, A.M., Ben-Ezra, J., Winberg, C., Colombero, A.M. & Rappaport, H. (1990) Analysis of antigen receptor gene rearrangements in ethanol and formaldehyde-fixed, paraffin-embedded specimens. *Lab. Invest.*, **63**(1), 107–114.
52. Farkas, D.H., Kaul, K.L., Wiedbrauk, D.L. & Kiechle, F.L. (1996) Specimen collection and storage for diagnostic molecular pathology investigation. *Arch. Pathol. Lab. Med.*, **120**(6), 591–596.
53. Farkas, D.H., Drevon, A.M., Kiechle, F.L., Di Carlo, R.G., Heath, E.M. & Crisan, D. (1996) Specimen stability for DNA-based diagnostic testing. *Diagn. Mol. Pathol.*, **5**(4), 227–235.
54. Yokota, M., Tatsumi, N., Nathalang, O., Yamada, T. & Tsuda, I. (1999) Effects of heparin on polymerase chain reaction for blood white cells. *J. Clin. Lab. Anal.*, **13**(3), 133–140.
55. Chomczynski, P. & Sacchi, N. (1987) Single-step method of RNA isolation by acid guanidinium thiocyanate–phenol–chloroform extraction. *Anal. Biochem.*, **162**(1), 156–159.
56. Kessler, H.H., Muhlbauer, G., Stelzl, E., Daghofer, E., Santner, B.I. & Marth, E. (2001) Fully automated nucleic acid extraction: Magna Pure LC. *Clin. Chem.*, **47**(6), 1124–1126.
57. Germer, J.J., Lins, M.M., Jensen, M.E., Harmsen, W.S., Ilstrup, D.M., Mitchell, P.S., Cockerill, F.R., 3rd & Patel, R. (2003) Evaluation of the Magna Pure LC instrument for extraction of hepatitis C virus RNA for the Cobas Amplicor hepatitis C virus test, version 2.0. *J. Clin. Microbiol.*, **41**(8), 3503–3508.
58. Roche Diagnostics (2003) Magna Pure LC. www.roche-applied-science.com/sis/magnapure.
59. Applied Biosystems (2000) Publication 117BR01-02: ABI Prism 6700 Automated Nucleic Acid Workstation. Applied Biosystems, Foster City, CA.
60. Applied Biosystems (2001). Publication 117PB04-01: ABI Prism 6100 Nucleic Acid Prepstation. Applied Biosystems, Foster City, CA.
61. Gentra Systems (2003) Autopure LS Product Information. Gentra Systems, Minneapolis, Minn. www.gentra.com.
62. Southern, E.M. (2000) Blotting at 25. *Trends Biochem. Sci.*, **25**(12), 585–588.
63. Southern, E.M. (1975) Detection of specific sequences among DNA fragments separated by gel electrophoresis. *J. Mol. Biol.*, **98**(3), 503–517.
64. Sarrazin, C. (2002) Highly sensitive hepatitis C virus RNA detection methods: molecular backgrounds and clinical significance. *J. Clin. Virol.*, **25** (Suppl 3), S23–S29.
65. Kubo, S., Hirohashi, K., Tanaka, H., Shuto, T., Takemura, S., Yamamoto, T., Uenishi, T., Kinoshita, H. & Nishiguchi, S. (2003) Usefulness of viral concentration measurement by transcription-mediated amplification and hybridization protection as a prognostic factor for recurrence after resection of hepatitis B virus-related hepatocellular carcinoma. *Hepatol. Res.*, **25**(1), 71–77.
66. Candotti, D., Richetin, A., Cant, B., Temple, J., Sims, C., Reeves, I., Barbara, J.A. & Allain, J.P. (2003) Evaluation of a transcription-mediated amplification-based HCV and HIV-1 RNA duplex assay for screening individual blood donations: a comparison with a minipool testing system. *transfusion*, **43**(2), 215–225.
67. Gorrin, G., Friesenhahn, M., Lin, P., Sanders, M., Pollner, R., Eguchi, B., Pham, J., Roma, G., Spidle, J., Nicol. S. & 3 other authors (2003) Performance evaluation of the versant HCV RNA qualitative assay by using transcription-mediated amplification. *J. Clin. Microbiol.*, **41**(1), 310–317.
68. Killeen, A.A., Breneman, J.W., 3rd, Carillo, A.R., Liu, J. & Hixson, C.S. (2003) Linked linear amplification for simultaneous analysis of the two most common hemochromatosis mutations. *Clin. Chem.*, **49**(7), 1050–1057.
69. Heid, C.A., Stevens, J., Livak, K.J., & Williams, P.M. (1996) Real time quantitative PCR. *Genome Res.*, **6**(10), 986–994.
70. Medhurst, A.D., Harrison, D.C., Read, S.J., Campbell, C.A., Robbins, M.J. & Pangalos, M.N. (2000) The use of Taqman RT-PCR assays for semiquantitative analysis of gene expression in CNS tissues and disease models. *J. Neurosci. Methods*, **98**(1), 9–20.

71. Wittwer, C.T., Ririe, K.M., Andrew, R.V., David, D.A., Gundry, R.A. & Balis, U.J. (1997) The Lightcycler: a microvolume multisample fluorimeter with rapid temperature control. *Biotechniques*, **22**(1), 176–181.
72. Piatek, A.S., Tyagi, S., Pol, A.C., Telenti, A., Miller, L.P., Kramer, F.R. & Alland, D. (1998) Molecular beacon sequence analysis for detecting drug resistance in *Mycobacterium tuberculosis. Nature Biotechnol.*, **16**(4), 359–363.
73. Leone, G., van Schijndel, H., van Gemen, B., Kramer, F.R. & Schoen, C.D. (1998) Molecular beacon probes combined with amplification by NASBA enable homogeneous, real-time detection of RNA. *Nucleic Acids Res.*, **26**(9), 2150–2155.
74. Marras, S.A., Kramer, F.R. & Tyagi, S. (1999) Multiplex detection of single-nucleotide variations using molecular beacons. *Genet. Anal.*, **14**(5-6), 151–156.
75. Bieche, I., Olivi, M., Champeme, M.H., Vidaud, D., Lidereau, R. & Vidaud, M. (1998) Novel approach to quantitative polymerase chain reaction using real-time detection: application to the detection of gene amplification in breast cancer. *Int. J. Cancer.*, **78**(5), 661–666.
76. Cao, Y., Ho, D.D., Todd, J., Kokka, R., Urdea, M., Lifson, J.D., Piatak, M., Jr, Chen, S., Hahn, B.H., Saag, M.S., & other authors (1995) Clinical evaluation of branched DNA signal amplification for quantifying HIV type 1 in human plasma. *AIDS Res. Hum. Retroviruses*, **11**(3), 353–361.
77. Caliendo, A.M. (1995) Laboratory methods for quantitating HIV RNA. *AIDS Clin. Care*, **7**(11), 89–91, 93, 96.
78. Ross, R.S., Viazov, S., Sarr, S., Hoffman, S., Kramer, A. & Roggendorf, M. (2002) Quantitation of hepatitis C virus RNA by third generation branched DNA-based signal amplification assay. *J. Virol. Methods*, **101**(1-2), 159–168.
79. Anastassopoulou, C.G., Touloumi, G., Katsoulidou, A., Hatzitheodorou, H., Pappa, M., Paraskevis, D., Lazanas, M., Gargalianos, P. & Hatzakis, A. (2001) Comparative evaluation of the Quantiplex HIV-1 RNA 2.0 and 3.0 (BDNA) assays and the Amplicor HIV-1 Monitor V1.5 test for the quantitation of human immunodeficiency virus type 1 RNA in plasma. *J. Virol. Methods*, **91**(1), 67–74.
80. Ryan, D., Nuccie, B. & Arvan, D. (1999) Non-PCR-dependent detection of the factor V Leiden mutation from genomic DNA using a homogeneous invader microtiter plate assay. *Mol. Diagn.*, **4**(2), 135–144.
81. Fors, L., Lieder, K.W., Vavra, S.H. & Kwiatkowski, R.W. (2000) Large-scale SNP scoring from unamplified genomic DNA. *Pharmacogenomics*, **1**(2), 219–229.
82. Hessner, M.J., Budish, M.A. & Friedman, K.D. (2000) Genotyping of factor V G1691A (Leiden) without the use of PCR by invasive cleavage of oligonucleotide probes. *Clin. Chem.*, **46**(8, Pt 1), 1051–1056.
83. Kwiatkowski, R.W., Lyamichev, V., de Arruda, M. & Neri, B. (1999) Clinical, genetic, and pharmacogenetic applications of the Invader assay. *Mol. Diagn.*, **4**(4), 353–364.
84. von Heeren, F. and Thormann, W. (1997) Capillary electrophoresis in clinical and forensic analysis. *Electrophoresis*, **18**(12-13), 2415–2426.
85. Thormann, W., Wey, A.B., Lurie, I.S., Gerber, H., Byland, C., Malik, N., Hochmeister, M. & Gehrig, C. (1999) Capillary electrophoresis in clinical and forensic analysis: recent advances and breakthrough to routine applications. *Electrophoresis*, **20**(15-16), 3203–3236.
86. Thormann, W., Lurie, I.S., McCord, B., Marti, U., Cenni, B. & Malik, N. (2001) Advances of capillary electrophoresis in clinical and forensic analysis (1999–2000). *Electrophoresis*, **22**(19), 4216–4243.
87. Petersen, J.R., Okorodudu, A.O., Mohammad, A. & Payne, D.A. (2003) Capillary electrophoresis and its application in the clinical laboratory. *Clin. Chim. Acta*, **330**(1-2), 1–30.
88. Watts, D. & MacBeath, J.R. (2001) Automated fluorescent DNA sequencing on the ABI Prism 310 genetic analyzer. *Methods Mol. Biol.* **167**, 153–170.
89. Liang, Q.; Davis, P.A.; Simpson, J.T., Thompson, B.H., Devaney, J.M. & Girard, J. (2000) Detection of hemochromatosis through the analysis of single-nucleotide extension products by capillary electrophoresis. *J. Biomol. Tech.*, **11**, 67–73.
90. Eggerding, F.A., Iovannisci, D.M., Brinson, E., Grossman, P. & Winn-Deen, E.S. (1995) Fluorescence-based oligonucleotide ligation assay for analysis of cystic fibrosis transmembrane conductance regulator gene mutations. *Hum. Mutat.*, **5**(2), 153–165.
91. Grossman, P.D., Bloch, W., Brinson, E., Chang, C.C., Eggerding, F.A., Fung, S., Iovannisci, D.M., Woo, S., Winn-Deen, E.S. & Iovannisci, D.A. (1994) High-density multiplex detection of nucleic acid sequences: oligonucleotide ligation assay and sequence-coded separation. *Nucleic Acids Res.*, **22**(21), 4527–4534.
92. Jarvius, J., Nilsson, M., & Landegren, U. (2003) Oligonucleotide ligation assay. *Methods Mol. Biol.*, **212**, 215–228.
93. Brinson, E.C., Adriano, T., Bloch, W., Brown, C.L., Chang, C.C., Chen, J., Eggerding, F.A., Grossman, P.D., Iovannisci, D.M., Madonik, A.M. & 6 other authors (1997) Introduction to PCR/OLA/SCS, a multiplex DNA test, and its application to cystic fibrosis. *Genet. Test.*, **1**(1), 61–68.
94. Frueh, F.W. & Noyer-Weidner, M. (2003) The use of denaturing high-performance liquid chromatography (dHPLC) for the analysis of genetic variations: impact for diagnostics and pharmacogenetics. *Clin. Chem. Lab. Med.*, **41**(4), 452–461.

95. Kuklin, A., Munson, K., Gjerde, D., Haefele, R. & Taylor, P. (1997) Detection of single-nucleotide polymorphisms with the Wave DNA fragment analysis system. *Genet. Test.*, **1**(3), 201–206.
96. Hecker, K.H., Taylor, P.D., & Gjerde, D.T. (1999) Mutation detection by denaturing DNA chromatography using fluorescently labeled polymerase chain reaction products. *Anal. Biochem.*, **272**(2), 156–164.
97. Bennett, R.R., den Dunnen, J., O'Brien, K.F., Darras, B.T. & Kunkel, L.M. (2001) Detection of mutations in the dystrophin gene via automated dHPLC screening and direct sequencing. *BMC Genet.*, **2**(1), 17.
98. Emery, A.E. (2002) The muscular dystrophies. *Lancet*, **359**, 687–695.
99. Matsuo, M. (2002) Duchenne and Becker muscular dystrophy: from gene diagnosis to molecular therapy. *IUBMB Life*, **53**(3), 147–152.
100. Biggar, W.D., Klamut, H.J., Demacio, P.C., Stevens, D.J. & Ray, P.N. (2002) Duchenne muscular dystrophy: current knowledge, treatment, and future prospects. *Clin. Orthop.*, **1**, 88–106.
101. Ravnik-Glavac, M., Atkinson, A., Glavac, D. & Dean, M. (2002) DHPLC Screening of cystic fibrosis gene mutations. *Hum. Mutat.*, **19**(4), 374–383.
102. Rischewski, J. & Schneppenheim, R. (2001) Screening strategies for a highly polymorphic gene: DHPLC Analysis of the Fanconi anemia group A gene. *J. Biochem. Biophys. Methods.*, **47**(1-2), 53–64.
103. Cobb, C.J., Scott, G., Swingler, R.J., Wilson, S., Ellis, J., MacEwen, C.J. & McLean, W.H. (2002) Rapid mutation detection by the transgenomic wave analyser DHPLC identifies MyoC mutations in patients with ocular hypertension and/or open angle glaucoma. *Br. J. Ophthalmol.*, **86**(2), 191–195.
104. Kaler, S.G., Devaney, J.M., Pettit, E.L., Kirshman, R. & Marino, M.A. (2000) Novel method for molecular detection of the two common hereditary hemochromatosis mutations. *Genet. Test.*, **4**(2), 125–129.
105. Devaney, J.M., Pettit, E.L., Kaler, S.G., Vallone, P.M., Butler, J.M. & Marino, M.A. (2001) Genotyping of two mutations in the HFE gene using single-base extension and high-performance liquid chromatography. *Anal. Chem.*, **73**(3), 620–624.
106. Le Gac, G., Mura, C. & Ferec, C. (2001) Complete scanning of the hereditary hemochromatosis gene (HFE) by use of denaturing HPLC. *Clin. Chem.*, **47**(9), 1633–1640.
107. McGall, G., Labadie, J., Brock, P., Wallraff, G., Nguyen, T. & Hinsberg, W. (1996) Light-directed synthesis of high-density oligonucleotide arrays using semiconductor photoresists. *Proc. Natl Acad. Sci. USA*, **93**(24), 13555–13560.
108. Yap, G. (2002) Affymetrix, Inc. *Pharmacogenomics*, **3**(5), 709–711.
109. Schena, M., Shalon, D., Heller, R., Chai, A., Brown, P.O. & Davis, R.W. (1996) Parallel human genome analysis: microarray-based expression monitoring of 1000 genes. *Proc. Natl Acad. Sci. USA*, **93**(20), 10614–10619.
110. Alizadeh, A., Eisen, M., Botstein, D., Brown, P.O. & Staudt, L.M. (1998) Probing lymphocyte biology by genomic-scale gene expression analysis. *J. Clin. Immunol.*, **18**(6), 373–379.
111. Sorlie, T., Tibshirani, R., Parker, J., Hastie, T., Marron, J.S., Nobel, A., Deng, S., Johnsen, H., Pesich, R., Geisler, S. & 6 other authors (2003) Repeated observation of breast tumor subtypes in independent gene expression data sets. *Proc. Natl Acad. Sci. USA*, **100**(14), 8418–8423.
112. Stuyver, L., Wyseur, A., Rombout, A., Louwagie, J., Scarcez, T., Verhofstede, C., Rimland, D., Schinazi, R.F. & Rossau, R. (1997) Line probe assay for rapid detection of drug-selected mutations in the human immunodeficiency virus type 1 reverse transcriptase gene. *Antimicrob. Agents Chemother.*, **41**(2), 284–291.
113. Nolte, F.S., Green, A.M., Fiebelkorn, K.R., Caliendo, A.M., Sturchio, C., Grunwald, A. & Healy, M. (2003) Clinical evaluation of two methods for genotyping hepatitis C virus based on analysis of the 5′ noncoding region. *J. Clin. Microbiol.*, 41(4), 1558–1564.
114. Smirnov, I.P., Hall, L.R., Ross, P.L., & Haff, L.A. (2001) Application of DNA-binding polymers for preparation of DNA for analysis by matrix-assisted laser desorption/ionization mass spectrometry. *Rapid Commun. Mass Spectrom.*, **15**(16), 1427–1432.
115. Ross, P., Hall, L., Smirnov, I., & Haff, L. (1998) High level multiplex genotyping by MALDI-TOF mass spectrometry. *Nature Biotechnol.*, **16**(13), 1347–1351.
116. Levidiotou, S., Vrioni, G., Galanakis, E., Gesouli, E., Pappa, C. & Stefanou, D. (2003) Four-year experience of use of the Cobas Amplicor system for rapid detection of *Mycobacterium tuberculosis* complex in respiratory and nonrespiratory specimens in Greece. *Eur. J. Clin. Microbiol. Infect. Dis.*, **22**(6), 349–356.
117. Van der Pol, B. (2002) Cobas Amplicor: an automated PCR system for detection of *C. trachomatis and N. gonorrhoeae*. *Expert Rev. Mol. Diagn.*, **2**(4), 379–389.

# 3 Measurement of oxidative DNA damage by gas chromatography–mass spectrometry and liquid chromatography–mass spectrometry

MIRAL DIZDAROGLU

## 3.1 Introduction

Oxidative DNA damage results from reactions of oxygen-derived species including free radicals, most notably hydroxyl radical, with the four heterocyclic bases and sugar moiety in DNA (reviewed in [1,2]). Reactions of free radicals result in a plethora of DNA products that comprise modified bases and sugars, DNA–protein crosslinks, strand breaks, clustered lesions and tandem lesions such as 8,5′-cyclopurine 2′-deoxynucleosides. Many of these products are genotoxic, leading to cell death and mutations (reviewed in [3]). Hydroxyl radicals can be generated in cells by ionizing radiation from cellular water or by reactions of superoxide radical and $H_2O_2$ with transition metal ions. Hydroxyl radicals react with DNA components at or near diffusion-controlled reaction rates by addition to double bonds of heterocyclic DNA bases and by abstraction of an H atom from the methyl group of thymine and from each of the C—H bonds of 2′-deoxyribose [4]. Subsequent reactions of thus-formed base and sugar radicals result in numerous products. Figure 3.1 illustrates the structures of the major modified DNA bases and 8,5′-cyclopurine 2′-deoxynucleosides. Unless repaired by complex DNA repair mechanisms in cells (reviewed in [5]), oxidative DNA damage can lead to a broad range of pathophysiological processes including carcinogenesis, neurological disorders and aging (reviewed in [3,6,7]). Accurate measurements are essential for the understanding of mechanisms and cellular repair of oxidative DNA damage, and its role in disease processes.

## 3.2 Measurement using mass spectrometric techniques

The measurement of oxidative DNA damage is an enormous technical challenge because of the low levels of modifications in cellular DNA that generally consist of a few lesions among several millions of intact DNA bases. A number of analytical techniques with their own advantages and drawbacks have been used for this purpose (reviewed in [8,9]). Some of these techniques generally measure a single compound and provide no spectroscopic evidence for identification. For example, HPLC with electrochemical detection (HPLC-EC) has been used for the measurement of the nucleoside form of 8-hydroxyguanine (8-OH-Gua), i.e. 8-hydroxy-2′-deoxyguanosine (8-OH-dGuo), following enzymic hydrolysis of DNA (reviewed in [8]). Most studies in the field of oxidative stress concentrated on this product as a biomarker for oxidative DNA damage because of its easy measurement by HPLC-EC and its mutagenic properties (reviewed in [10]). However, the measurement of a single product among numerous DNA products

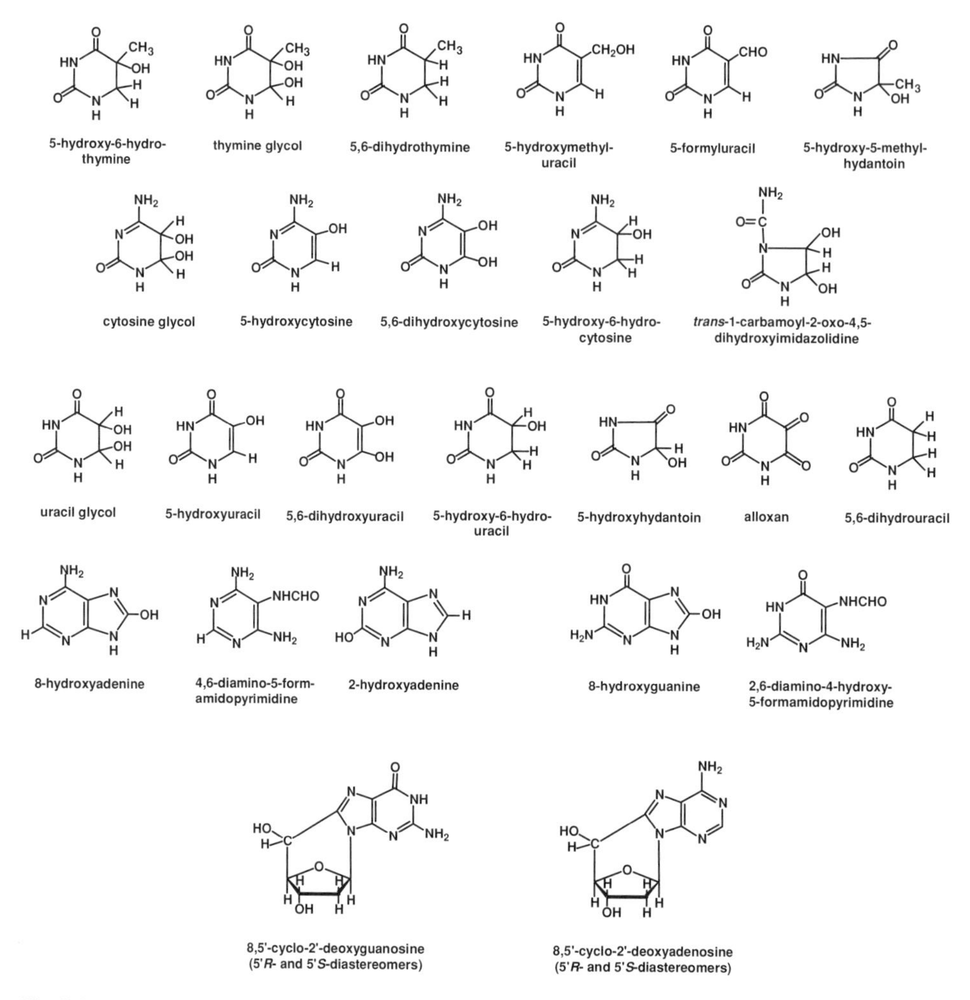


**Fig. 3.1** Structures of major oxidatively modified DNA bases and 8,5′-cyclopurine 2′-deoxynucleosides.

as a biomarker might be misleading, because relative yields of DNA products depend on experimental conditions, the redox status of cells and the availability of metal ions [2,7,11]. Furthermore, a single product might not necessarily reflect the overall rate of DNA damage. The advantage of simultaneously measuring multiple DNA products has been demonstrated [9,12,13]. Studies by the European Standards Committee for Oxidative DNA Damage (ESCODD) strongly suggested that the measurement of 8-OH-Gua or 8-OH-dGuo may depend on the laboratory and the technique used, and that no consensus exists between various laboratories involved in the measurement of this compound [14–16]. Mass spectrometry coupled with gas chromatography or liquid chromatography achieves unequivocal identification with structural evidence and quantification of oxidative DNA damage, and simultaneously measures a variety of DNA lesions (reviewed in [1]). The application of an isotope-dilution technique with the use of stable isotope-labeled analogues of DNA lesions as internal standards assures accurate quantification. In this chapter, we describe the methodologies for the measurement of oxidative DNA base damage by gas chromatography/mass spectrometry (GC/MS) and liquid chromatography/mass spectrometry (LC/MS).

## 3.3 Gas chromatography–mass spectrometry

Unlike other techniques, GC/MS is suited for the measurement of base products resulting from all four DNA bases, tandem lesions such as 8,5′-cyclopurine 2′-deoxynucleosides, products of the sugar moiety and DNA–protein crosslinks (reviewed in [1]). For GC/MS analysis, DNA must first be hydrolyzed to intact bases and modified bases by an acid, such as formic acid, or to intact and modified nucleosides by a mixture of endo- and exonucleases and phosphatase. Instead of hydrolysis with acid, DNA repair enzymes such as *Escherichia coli* formamidopyrimidine *N*-glycosylase (Fpg) and endonuclease III (Nth) are also used to release certain modified bases from DNA. Hydrolysates are dried and derivatized by trimethylsilylation to obtain volatile trimethylsilyl (TMS) derivatives of bases and nucleosides, and then analyzed by GC/MS. Electron-ionization mass spectra of these derivatives provide characteristic mass spectra for unequivocal identification [17–19]. Quantification is achieved using stable isotope-labeled analogues of modified bases and nucleosides as internal standards that are added to DNA samples before hydrolysis [20].

### 3.3.1 *Materials*

Most reference compounds are available commercially or can be synthesized [21]. A number of stable isotope-labeled analogues of modified bases that incorporate $^{13}C$, $^{15}N$, $^{18}O$ and $^{2}H$ are offered by Cambridge Isotope Laboratories (Cambridge, MA, USA) or by the National Institute of Standards and Technology (Gaithersburg, MD, USA) as a 'Standard Reference Material' (SRM). These compounds are 4,6-diamino-5-formamidopyrimidine-$^{13}C,^{15}N_2$ (FapyAde-$^{13}C,^{15}N_2$), 2,6-diamino-4-hydroxy-5-formamidopyrimidine-$^{13}C,^{15}N_2$ (FapyGua-$^{13}C,^{15}N_2$), 8-hydroxyadenine-$^{13}C,^{15}N_2$ (8-OH-Ade-$^{13}C,^{15}N_2$), 8-hydroxyguanine-$^{13}C,^{15}N_2$ (8-OH-Gua-$^{13}C,^{15}N_2$), 5-hydroxycytosine-$^{13}C,^{15}N_2$ (5-OH-Cyt-$^{13}C,^{15}N_2$), 5-hydroxyuracil-$^{13}C_4,^{15}N_2$ (5-OH-Ura-$^{13}C_4,^{15}N_2$), 5-(hydroxymethyl)uracil-$^{13}C_2,^{2}H_2$ (5-OHMeUra-$^{13}C_2,^{2}H_2$), thymine glycol-$^{2}H_4$ (ThyGly-$^{2}H_4$), 5-hydroxy-5-methylhydantoin-$^{13}C,^{15}N_2$ (5-OH-5-MeHyd-$^{13}C,^{15}N_2$), isodialuric acid-$^{13}C,^{15}N_2$ [5,6-dihydroxyuracil-$^{13}C,^{15}N_2$ (5,6-diOH-Ura-$^{13}C,^{15}N_2$)], dialuric acid-$^{13}C_2,^{15}N_2$ and 2′-deoxyguanosine-$^{15}N_5$ (dGuo-$^{15}N_5$). Except for 8-OH-Gua-$^{13}C,^{15}N_2$, all compounds are soluble in water. 8-OH-Gua-$^{13}C,^{15}N_2$ is dissolved in 10 mM NaOH; however, it is not stable and degrades gradually. Thus the UV-absorption spectrum of the solution must be measured before each use to determine the concentration using the absorption coefficient 11 900 $M^{-1}$ $cm^{-1}$ at 283 nm [22]. Dialuric acid-$^{13}C_2,^{15}N_2$ is readily oxidized in aqueous solution to alloxan-$^{13}C_2,^{15}N_2$ [20,23].

### 3.3.2 *Sample preparation*

First, the concentration of DNA in aqueous DNA samples is determined by UV spectroscopy (an absorbance unit ≡ 0.05 mg/ml of DNA). Aliquots of stable isotope-labeled internal standards are added to DNA samples (usually 0.05–0.1 mg of DNA). The samples are dried under vacuum (e.g. in a SpeedVac).

3.3.2.1 *Acidic hydrolysis* Acidic hydrolysis cleaves the glycosidic bond between the sugar moiety and the bases in DNA, releasing intact and modified bases; formic acid is used for this purpose. Stable isotope-labeled analogues of modified DNA bases and 2′-deoxyguanosine-$^{15}N_5$ are added to DNA samples as internal standards before acidic hydrolysis. The use of dGuo-$^{15}N_5$ permits the determination of the amount of DNA by GC/MS with the isotope-dilution technique. Upon hydrolysis, this compound yields guanine-$^{15}N_5$, which is used as an internal standard for the determination of the amount of guanine in DNA, and consequently the amount of DNA [24]. The results of DNA determination by this method and by UV absorbance measurements should correlate well with each other. DNA samples (e.g. 0.05–0.1 mg) are dissolved in 0.5 ml of 60% formic acid and heated at 140°C for 30 min in evacuated and sealed tubes. Hydrolysates are frozen in glass vials placed in liquid nitrogen and then lyophilized to dryness for 18 h. Most compounds are stable under these conditions except for cytosine-derived lesions. Cytosine glycol is quantitatively converted to 5-OH-Cyt and 5-OH-Ura, the former by dehydration and the latter by deamination and dehydration [25]. 5,6-Dihydroxycytosine and 5,6-dihydrocytosine deaminate to give rise to 5,6-diOH-Ura and 5,6-dihydrouracil. Upon acid-induced decarboxylation, alloxan quantitatively yields 5-hydroxyhydantoin (5-OH-Hyd) [20].

3.3.2.2 *Enzymic hydrolysis* Hydrolysis of DNA to nucleosides is achieved by a mixture of endo- and exonucleases and alkaline phosphatase [26]. Various versions of enzymic hydrolysis of DNA to nucleosides have been reported in the literature. The following procedure is used in our laboratory. An aliquot of DNA (e.g. 0.05–0.1 mg) is dissolved in 0.1 ml of 10 mM Tris-HCl buffer (pH 7.5). The pH is adjusted to 6.0 by adding 2.5 μl of 1 M sodium acetate (pH 4.5). DNA samples are incubated with nuclease P1 (10 units) at 37°C for 2 h and then with phosphodiesterase I (0.0045 units) and alkaline phosphatase (5 units) for another 2 h. Subsequently, hydrolysates are filtered using ultrafiltration membranes with molecular mass cut-off of 5 kDa by centrifugation at 15 000 *g* for 30 min, and then lyophilized for derivatization followed by GC/MS analysis.

Hydrolysis with the DNA repair enzyme Fpg is suitable for the measurement in DNA of its three substrates, namely 8-OH-Gua, FapyGua and FapyAde [27,28]. An aliquot of DNA (e.g. 0.05 mg) is dissolved in 0.1 ml of an incubation buffer consisting of 50 mM $Na_2HPO_4$ (pH 7.4), 100 mM KCl, 1 mM EDTA, and 0.1 mM dithiothreitol. Samples are incubated with 4 μg of Fpg for 30 min. After incubation, 0.25 ml of cold ethanol (−20°C) is added to each sample to stop the reaction and to precipitate DNA; this is followed by addition of aliquots of stable isotope-labeled analogues of 8-OH-Gua, FapyGua and FapyAde as internal standards. The samples are kept at −20°C for 2 h and then centrifuged at 15 000 *g* for 30 min at 4°C. DNA pellets and supernatant fractions are separated. Ethanol is removed from supernatant fractions under vacuum in a SpeedVac. Aqueous supernatant fractions are frozen in liquid nitrogen and lyophilized to dryness for 18 h. Hydrolysis with the DNA repair enzyme Nth is carried out in the same manner by replacing Fpg by Nth and by using stable isotope-labeled analogues of pyrimidine-derived modified bases (see above) as internal standards.

3.3.2.3 *Derivatization* DNA bases and nucleoside are not sufficiently volatile for GC/MS analysis and must be derivatized to obtain stable derivatives. Trimethylsilylation is the mode of derivatization most suitable for this purpose [29]. An aliquot (0.1 ml) of a mixture of nitrogen-bubbled bis(trimethylsilyl)trifluoroacetic acid (containing 1% trimethylchlorosilane) and dry pyridine (1/1, v/v) is added to lyophilized acid- or enzyme-hydrolysates of DNA samples. The amount of the reagents can be modified according to the amount of DNA. Each sample is purged with ultra high-purity nitrogen and then tightly sealed under nitrogen with Teflon-coated septa. Acid hydrolysates are derivatized at 23°C for 2 h by vigorously shaking the vials, whereas enzyme hydrolysates are heated at 120°C for 30 min and then cooled to room temperature. Aliquots (2–4 μl) of derivatized samples are directly injected into the injection port of the gas chromatograph.

### 3.3.3 *Instrumentation*

A GC/MS system with a capillary column equipped with a computer workstation serves best the purpose. In the past, most measurements have been performed using commercial quadrupole mass spectrometers. Fused silica capillary columns coated with 5% crosslinked phenyl methylsilicone stationary phase appear to be the best for analysis of trimethylsilylated DNA bases and nucleosides [17,18]. A 12.5 m long column with 0.2 mm internal diameter and 0.33 μm stationary phase film thickness is generally used. Ultra high-purity helium is the carrier gas of choice. The split mode of injection is preferred to avoid overloading the column. The injection port, and both GC/MS interface and ion source, are kept at 250°C and 280°C, respectively. The oven temperature of the gas chromatograph can be programmed from 130°C to 280°C at a rate of 8°C/min or any other way depending on desired experimental conditions. The column head pressure is generally kept at 50–100 kPa. The glass liner in the injection port of the gas chromatograph is filled with silanized glass wool that permits homogenous vaporization of injected samples. Electron-ionization (EI) mode at 70 eV is used for ionization of compounds entering the ion source of the mass spectrometer.

### 3.3.4 *Measurement*

3.3.4.1 *Separation and identification* The separation of trimethylsilyl (TMS) derivatives of a large number of modified DNA bases from one another and from four intact DNA bases can be achieved by GC in a single analysis [17]. TMS derivatives eluting from the gas chromatograph are ionized in the ion source of the mass spectrometer and analyzed by the mass analyzer. Electron-ionization mass spectra of TMS derivatives of modified DNA bases provide structural details that facilitate unequivocal identification. Their fragmentation patterns follow those of the intact DNA bases [30]. Generally, the mass spectra consist of a molecular ion ($M^{+\bullet}$), an $(M-15)^{+}$ ion, which results from loss of a methyl radical from $M^{+\bullet}$, and other characteristic ions with high relative intensity [17,18,31]. In some cases, an $(M-1)^{+}$ ion, which is formed by loss of an H atom from $M^{+\bullet}$ is also formed. The fragmentation patterns of TMS derivatives of modified nucleosides are similar to those of intact DNA nucleosides [32]. The $(\text{base}+\text{H})^{+}$ ion $[(\text{B}+1)^{+}$ ion] and the $(\text{base}+\text{H}-\text{CH}_3)^{+}$ ion dominate the mass spectra, but $M^{+\bullet}$ and $(M-15)^{+}$ ion are of low intensity [19,33]. The $(\text{B}+1)^{+}$ ion appears as the most intense ion in the mass spectra of TMS derivatives of 8-OH-dGuo and 8-hydroxy-2′-deoxyadenosine (8-OH-dAdo) due to stabilization by an electron donating substituent at C-8 of the purine ring. In contrast, TMS derivatives of 8,5′-cyclopurine 2′-deoxynucleosides provide intense $M^{+\bullet}$ and $(M-15)^{+}$ ions, most likely due to stabilization by the increased number of rings in the molecule [31,34,35]. A fragment ion characteristic of these compounds uniquely results from simultaneous cleavage of the glycosidic bond and the bond between the 5′-carbon and 4′-carbon. The resulting ions consist of the base and the HC5′-OH portion of the sugar moiety and appear as the most prominent ion in the mass spectra.

In GC/MS, selected-ion monitoring (SIM) by the mass spectrometer is used for identification and quantification of low levels of analytes in a complex mixture such as hydrolyzed DNA samples [36]. The use of this mode requires the knowledge of characteristic ions, their relative intensities and the GC-retention time of an analyte. After these parameters have been established, a number of characteristic ions of the analyte are recorded by the mass spectrometer during the time-period in which the analyte elutes from the GC column and enters the ion source. Subsequently, the analyte is identified by the signals of the monitored characteristic ions with typical relative intensities at its retention time. With capillary GC, retention times can be measured with great accuracy and precision, and thus facilitate unequivocal identification, in addition to the simultaneous measurement of mass/charge ($m/z$) ratios.

3.3.4.2 *Quantification* The quantification of modified DNA bases of nucleosides by GC/MS is achieved by the use of their stable isotope-labeled analogues as internal standards. Known amounts of internal standards are added to aliquots of DNA samples before hydrolysis by acid or enzymes. This procedure, called isotope-dilution MS, permits compensation for possible losses of an analyte during sample preparation or GC/MS analysis, because the analyte and its stable isotope-labeled analogue possess essentially the same physical and chemical properties [36]. Fragmentation patterns of stable isotope-labeled analogues are identical to those of the corresponding unlabeled compounds. However, the masses of most ions of the labeled analogues are shifted to higher values according to the number of isotopes in the molecule. TMS derivatives of modified bases or nucleosides co-elute with those of their $^{13}$C- and $^{15}$N-labeled analogues. In contrast, an analogue labeled with several $^{2}$H atoms elutes slightly earlier than its unlabeled counterpart. During GC/MS analysis, characteristic ions of the TMS derivatives of a modified DNA base or nucleoside and its stable isotope-labeled analogue are simultaneously recorded in the SIM mode. The measured areas of the ion-current profiles of the monitored ions are used to determine the level of modified base or nucleoside in a given DNA sample.

3.3.4.3 *Selectivity and sensitivity* GC/MS with SIM provides high selectivity because a few selected characteristic ions of an analyte are monitored at a time window, where the analyte elutes from the GC column. Sensitivities of approximately one fmol of a compound applied to the GC column are usually achieved (reviewed in [1]). However, the level of sensitivity might depend on the GC/MS instrument and the column used.

## 3.4 Liquid chromatography–mass spectrometry

In recent years, MS coupled to LC has been applied to the measurement of oxidatively modified nucleosides in DNA. The first application was the use of a quadrupole mass spectrometer with flow injection and the positive-ion electrospray-ionization mode (ESI) to record the mass spectra of 8-OH-dGuo and 8-OH-dAdo [37]. Subsequently, LC/tandem MS (LC/MS/MS) with ESI and with or without isotope-dilution technique has been used for the measurement of 8-OH-dGuo and other nucleosides such as 8-OH-dAdo in enzyme-hydrolysates of DNA or in urine [38–46]. In our laboratory, we recently utilized LC/MS with atmospheric pressure ionization–electrospray (ASI-ES) process in the positive ionization mode for the measurement of several modified nucleosides in DNA, which included 8-OH-dGuo, (5′*R*)-8,5′-cyclo-2′-deoxyadenosine [(5′*R*)-8,5′-cdAdo], (5′*S*)-8,5′-cyclo-2′-deoxyadenosine [(5′*S*)-8,5′-cdAdo)], 8-OH-dAdo, (5′*R*)-cyclo-2′-deoxyguanosine [(5′*R*)-cdGuo] and (5′*S*)-8,5′-cyclo-2′-deoxyguanosine [(5′*S*)-8,5′-cdGuo)] [47–50]. LC/MS is different from LC/MS/MS in that it has one quadrupole mass spectrometer only, and the cost of the instrument is substantially lower than that of an LC/MS/MS instrument. The following part describes the use of LC/MS for the measurement of some oxidatively modified nucleosides in DNA.

### 3.4.1 *Materials*

8-Hydroxy-2′-deoxyguanosine is available from Sigma (MO, USA). Other modified nucleosides and their stable isotope-labeled analogues are not available commercially. Some of these compounds have been synthesized in individual laboratories that used LC/MS/MS [41], or LC/MS [48,50,51], and the procedures can be found in the published papers. 8-Hydroxy-2′-deoxyguanosine-$^{18}$O (8-OH-dGuo-$^{18}$O) is available as a component of an SRM from the National Institute of Standards and Technology (Gaithersburg, MD, USA).

### 3.4.2 *Sample preparation*

Preparation of DNA samples for LC/MS analysis was described above in terms of the determination of the DNA amount and hydrolysis of DNA to nucleosides by endo- and exonucleases, and alkaline phosphatase. Following filtration, DNA hydrolysates are directly used for LC/MS analysis. No lyophilization and derivatization are necessary in this case.

### 3.4.3 *Instrumentation*

A commercially available liquid chromatograph–quadrupole mass selective detector with API-ES process serves best the purpose. The positive ionization mode is used for LC/MS analysis. The liquid chromatograph should be equipped with a UV spectrophotometer and an automatic sampler. The following conditions are used. Flow and temperature of the drying gas (nitrogen), 10 l/min and 350°C, respectively; nebulizing gas pressure, 172 kPa; capillary, fragmentor and electron multiplier potentials, 4000 V, 100 V and 2600 V, respectively; quadrupole temperature, 99°C. $C_{18}$-reversed-phase columns (15 cm × 2.1 mm i.d., 5 μm particle size) are generally used for separations. A guard column packed with the same stationary phase (1 cm × 2.1 mm i.d.) should be used to protect the analytical column. The following conditions for LC separations can be used: solvent A, a mixture of water and acetonitrile (98/2, v/v); solvent B, acetonitrile; gradient, 0.5% of solvent B/min; flow rate, 0.2 ml/min; column temperature, 30°C. Aliquots of up to 30 μl of filtered exzyme-hydrolysates of DNA samples can be injected on the column for analysis.

### 3.4.4 *Measurement*

3.4.4.1 *Separation and identification* Numerous modified nucleosides can be separated from one another and intact nucleosides by LC in a single analysis. Compounds eluting from the LC column enter the ion source of the mass spectrometer and are ionized by API-ES in the positive ionization mode. Formation mechanisms of the ions of modified DNA nucleosides are similar to those of intact RNA or DNA nucleosides [52]. In general, the ions consist of the protonated molecular ion ($MH^+$), the protonated free base ion ($BH_2^+$) and the Na-adduct ion ($MNa^+$) [37–42,45–50]. The relative abundances of these ions vary widely among modified nucleosides. Similar to their TMS derivatives (see above), 8,5′-cyclopurine 2′-deoxynucleosides undergo a unique fragmentation resulting from simultaneous cleavage of the glycosidic bond and the bond between the 5′-carbon and 4′-carbon, giving rise to intense ions that consist of the base and the HC5′-OH portion of the sugar moiety [48,50]. As in GC/MS, SIM is used in LC/MS for identification and quantification of low levels of modified nucleosides in enzyme hydrolysates of DNA. Again, this requires the knowledge of the mass spectra of the compounds and their time of elution from the LC column.

3.4.4.2 *Quantification* This procedure is also similar to that used in GC/MS. Stable isotope-labeled analogues of modified nucleosides are used as internal standards. These compounds, mostly labeled with $^{13}C$ and/or $^{15}N$, exhibit fragmentation patterns and retention times identical to their unlabeled counterparts. Known amounts of internal standards are added to aliquots of DNA samples before enzymic hydrolysis. Typical ions of modified nucleosides and internal standards are simultaneously recorded during LC/MS analysis in the SIM mode. Areas of ions signals are used to quantify modified nucleosides. Recently, LC/MS with an isotope-dilution technique has been used to identify and quantify 8-OH-dGuo and 8-OH-dAdo in DNA of living cells [51,53,54]. As an example, Figure 3.2 illustrates the ion-current profiles of the $BH_2^+$ ions of

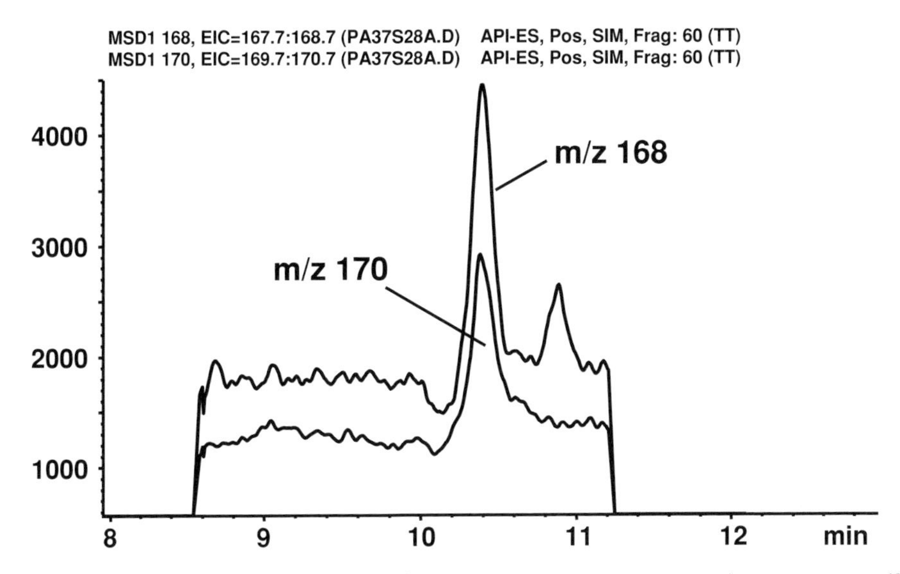


**Fig. 3.2** Ion-current profiles at $m/z$ 168 ($BH_2^+$) of 8-OH-dGuo and $m/z$ 170 ($BH_2^+$) of 8-OH-dGuo-$^{18}$O recorded during LC/MS analysis of the enzymic hydrolysate of a DNA sample isolated from human cells (from ref [53]).

8-OH-dGuo and 8-OH-dGuo-$^{18}$O recorded during LC/MS analysis of the enzymic hydrolysate of a DNA sample isolated from human cells [53].

3.4.4.3 *Selectivity and sensitivity* LC/MS with SIM facilitates selective and sensitive measurement of modified nucleosides in DNA. In contrast to GC/MS, which measures TMS derivatives with comparable sensitivity, the sensitivity of LC/MS varies from compound to compound. The sensitivity level achieved by LC/MS-SIM is similar to that by LC/MS/MS in the multiple-reaction monitoring mode (reviewed in [1]). For 8,5′-cdAdo, 8-OH-dAdo, 8,5′-cdGuo and 8-OH-dGuo, sensitivity levels of 2–35 fmol have been achieved. This means that detectable levels of these nucleosides in DNA are in the range of 6–10 lesions/$10^6$ DNA nucleosides per μg of DNA [1]. In our laboratory, we routinely use 30 μg of hydrolyzed DNA per injection that facilitates detection levels of 0.02–0.3 lesions/$10^6$ DNA nucleosides (or 0.2–3 lesions/$10^7$ DNA nucleosides).

### 3.5 Artifacts

When measuring oxidative DNA damage by mass spectrometric techniques, artifacts can be formed, as in any experiment, during numerous steps in sample preparation including isolation of DNA from living cells. Thus, guanine is the most prone to oxidative damage among DNA bases [55,56], and can readily be oxidized to give 8-OH-Gua. Various methods for DNA isolation have been used in the past (see, for example, refs [16,57] for an overview of the methods). A method that uses NaI has been suggested to provide the lowest level of 8-OH-Gua when measured by HPLC-EC in its nucleoside form 8-OH-dGuo [57,58]. However, the lowest level may not necessarily reflect the 'true' level of 8-OH-Gua in cells. Most work on artifacts in the past has concentrated on this compound. Trials by ESCODD suggested a significant variability between methods, techniques and laboratories in the measurement of 8-OH-Gua or 8-OH-dGuo [14–16]. In the case of GC/MS, the most likely step that can lead to artifacts is derivatization at high temperature because of the presence of intact guanine or other DNA bases in acid hydrolysates (reviewed in [59]). However, artifacts can be prevented if proper experimental conditions are used. Various methods of derivatization have been used to prevent artifact formation, such as derivatization at room temperature and/or addition of certain compounds to derivatization mixtures (reviewed in [1,16]). To avoid the presence of intact guanine in hydrolysates, Fpg instead of acid or endo- and exonucleases has been used to release 8-OH-Gua from DNA before analysis by GC/MS [28,60], or by HPLC-EC [61]. Fpg specifically excises 8-OH-Gua and formamidopyrimidines from DNA, and does not release intact guanine [27,62], thus avoiding the oxidation of guanine in hydrolysates. In the case of GC/MS, a comparative study showed that similar results on the level of 8-OH-Gua can be achieved when using Fpg or acid for hydrolysis of DNA [28]. This observation suggested that the derivatization of acid hydrolysates at room temperature does not generate 8-OH-Gua as an artifact. Subsequently, this fact has been confirmed by comparing the measurement of 8-OH-Gua by GC/MS after Fpg hydrolysis or acid hydrolysis, and the measurement of 8-OH-dGuo by LC/MS in enzyme hydrolysates of DNA [47]. Furthermore, this study provided evidence that GC/MS and LC/MS can provide similar results in terms of the measurement of 8-OH-Gua in DNA. Figure 3.3 illustrates the levels of this compound or its nucleoside form measured by GC/MS, LC/MS or LC/MS/MS under a variety of experimental conditions. Taken together, recent studies suggest that LC/MS and LC/MS/MS may be the techniques of choice to measure 8-OH-dGuo and other modified nucleosides at background levels, if

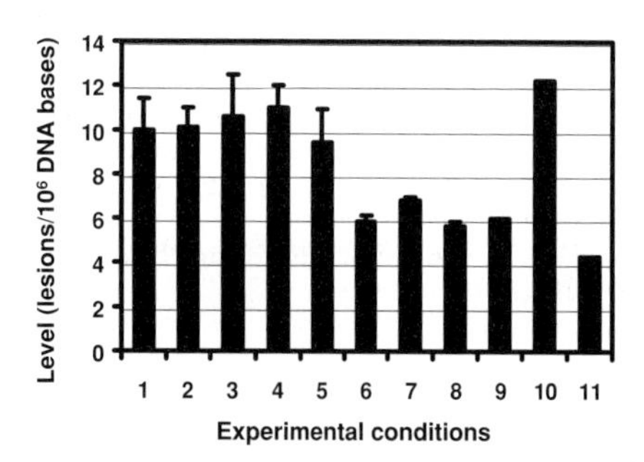


**Fig. 3.3** Level of 8-OH-dGuo measured by LC/MS and level of 8-OH-Gua measured by GC/MS in DNA under various conditions (from [47]). Columns 1–3: in calf thymus DNA, measured by LC/MS with enzymic hydrolysis, GC/MS with acidic hydrolysis and by GC/MS with Fpg hypdrolysis, respectively [47]. Columns 4 and 5: in calf thymus DNA, measured by GC/MS with acidic hydrolysis and by GC/MS with Fpg hydrolysis, respectively [28]. Columns 6–8: in DNA isoloated from cultured HeLa cells, measured by LC/MS with enzymic hydrolysis, GC/MS with acidic hydrolysis and by GC/MS with Fpg hydrolysis by Fpg protein [47]. All values in columns 1–8 represent the average (± standard deviation) of 3–6 independent measurements. Columns 9 and 10: in calf thymus DNA measured by LC/MS/MS [15,39]. Column 11: in rat liver measured by LC/MS/MS [38]. No standard deviation or standard error was given for the values represented by columns 9–11.

one wishes to avoid derivatization. Furthermore, when using GC/MS, derivatization of acid hydrolysates at room temperature or Fpg hydrolysis instead of acidic hydrolysis should be used for the measurement of background levels of 8-OH-Gua in DNA.

### 3.6 Disclaimer

Certain commercial equipment or materials are identified in this paper in order to specify adequately the experimental procedure. Such identification does not imply recommendation or endorsement by the National Institute of Standards and Technology, nor does it imply that the materials or equipment identified are necessarily the best available for the purpose.

## References


1. Dizdaroglu, M., Jaruga, P., Birincioglu, M. & Rodriguez, H. (2002) Free radical-induced damage to DNA: mechanisms and measurement. *Free Radic. Biol. Med.*, **32**, 1102–1115.
2. Breen, A.P. & Murphy, J.A. (1995) Reactions of oxyl radicals with DNA. *Free Radic. Biol. Med.*, **18**, 1033–1077.
3. Wallace, S.S., (2002) Biological consequences of free radical-damaged DNA bases. *Free Radic. Biol. Med.*, **33**, 1–14.
4. von Sonntag, C. (1987) *The Chemical Basis of Radiation Biology*, Taylor and Francis, New York.
5. Friedberg, E.C., Walker, G.C. & Siede, W. (1995) *DNA Repair and Mutagenesis*, ASM Press, Washingon DC.
6. Bohr, V.A. (2002) DNA damage and its processing. Relation to human disease. *J. Inherit. Metab. Dis.*, **25**, 215–222.
7. Halliwell, B. & Gutteridge, J.M.C. (1999) *Free Radicals in Biology and Medicine*, Oxford Science, Oxford.
8. Collins, A., Cadet, J., Epe, B. & Gedik, C. (1997) Problems in the measurement of 8-oxoguanine in human DNA. Report of a workshop, DNA oxidation, held in Aberdeen, UK, 19–21 January, 1997. *Carcinogenesis*, **18**, 1833–1836.
9. Dizdaroglu, M. (1998) Mechanisms of free radical damage to DNA. In: *DNA and Free Radicals: Techniques, Mechanisms and Applications*. (e.d. O.I. Auoma & B. Halliwell), pp. 3–26. OICA International, Saint Lucia.
10. Grollman, A.P. & Moiya, M. (1993) Mutagenesis by 8-oxoguanine: an enemy within. *Trends Genet.*, **9**, 246–249.
11. Dizdaroglu, M. (1992) Oxidative damage to DNA in mammalian chromatin. *Mutat. Res.*, **275**, 331–342.
12. Rehman, A., Collis, C.S., Yang, M., Kelly, M., Diplock, A.T., Halliwell, B. & Rice-Evans, C. (1998) The effects of iron and vitamin C co-supplementation on oxidative damage to DNA in healthy volunteers. *Biochem. Biophys. Res. Commun.*, **246**, 293–298.
13. Podmore, I.D., Griffiths, H.R., Herbert, K.E., Mistry, N., Mistry, P. & Lunec, J. (1998) Vitamin C exhibits pro-oxidant properties. *Nature*, **392**, 559.
14. Lunec, J. ESCODD: European Standards Committee on Oxidative DNA Damage. *Free Radic. Res.*, **29**, 601–608.
15. ESCODD (2000) Comparison of different methods of measuring 8-oxoguanine as a marker of oxidative DNA damage. *Free Radic. Res.*, **32**, 333–341.
16. ESCODD (2003) Measurement of DNA oxidation in human cells by chromatographic and enzymic methods. *Free Radic. Biol. Med.*, **34**, 1089–1099.
17. Dizdaroglu, M. (1985) Application of capillary gas chromatography–mass spectrometry to chemical characterization of radiation-induced base damage of DNA; implications for assessing DNA repair processes. *Anal. Biochem.*, **144**, 593–603.
18. Dizdaroglu, M. (1984) The use of capillary gas chromatography–mass spectrometry for identification of radiation-induced DNA base damage and DNA base–amino acid crosslinks. *J. Chromatogr.*, **295**, 103–121.
19. Dizdaroglu, M. (1986) Characterization of free radical-induced base damage in DNA by the combined use of enzymatic hydrolysis and gas chromatography–mass spectrometry. *J. Chromatogr.*, **367**, 357–366.
20. Dizdaroglu, M. (1993) Quantitative determination of oxidative base damage in DNA by stable isotope-dilution mass spectrometry. *FEBS Lett.*, **315**, 1–6.

21. Dizdaroglu, M. (1994) Chemical determination of oxidative DNA damage by gas chromatography–mass spectrometry. *Methods Enzymol.*, **234**, 3–16.
22. Hamberg, M. & Zhang, L.-Y. (1995) Quantitative determination of 8-hydroxyguaine and guanine by isotope dilution mass spectrometry. *Anal. Biochem.*, **229**, 336–344.
23. Behrend, R. & Roosen, O. (1889) Synthese der Harnsäure. *Justus Liebigs Ann. Chem.*, **251**, 235–256.
24. Sentürker, S. & Dizdaroglu, M. (1999) The effect of experimental conditions on the levels of oxidatively modified bases in DNA as measured by gas chromatography–mass spectrometry: how many modified bases are involved? Prepurification or not? *Free Radic. Biol. Med.*, **27**, 370–380.
25. Dizdaroglu, M., Holwitt, E., Hagan, M.P. & Blakely, W.F. (1986) Formation of cytosine glycol and 5,6-dihydroxycytosine in deoxyribonucleic acid on treatment with osmium tetroxide. *Biochem. J.*, **235**, 531–536.
26. Crain, P.F. (1990) Preparation and enzymatic hydrolysis of DNA and RNA for mass spectrometry. *Methods Enzymol.*, **193**, 782–790.
27. Boiteux, S., Gajewski, E., Laval, J. & Dizdaroglu, M. (1992) Substrate specificity of the *Escherichia coli* Fpg protein (formamidopyrimidine–DNA glycosylase): excision of purine lesions in DNA produced by ionizing radiation or photosensitization. *Biochemistry.*, **31**, 106–110.
28. Rodriguez, H., Jurado, J., Laval, J. & Dizdaroglu, M. (2000) Comparison of the levels of 8-hydroxyguanine in DNA as measured by gas chromatography mass spectrometry following hydrolysis of DNA by *Escherichia coli* Fpg protein or formic acid. *Nucleic Acids Res.*, **28**, E75.
29. Schram, K.H. (1990) Preparation of trimethylsilyl derivatives of nucleic acid components for analysis by mass spectrometry. *Methods Enzymol.*, **193**, 791–796.
30. White, E., Krueger, V.P. & McCloskey, J.A. (1972) Mass spectra of trimethylsilyl derivatives of pyrimidine and purine bases. *J. Org. Chem.*, **37**, 430–438.
31. Dizdaroglu, M. (1990) Gas chromatography–mass spectrometry of free radical-induced products of pyrimidines and purines in DNA. *Methods Enzymol.*, **193**, 842–857.
32. McCloskey, J.A. (1990) Electron ionization mass spectra of trimethylsilyl derivatives of nucleosides. *Methods Enzymol.*, **193**, 825–842.
33. Dizdaroglu, M. (1985) Formation of an 8-hydroxyguanine moiety in deoxyribonucleic acid on gamma-irradiation in aqueous solution. *Biochemistry.*, **24**, 4476–4481.
34. Dizdaroglu, M. (1986) Free-radical-induced formation of an 8,5′-cyclo-2′-deoxyguanosine moiety in deoxyribonucleic acid. *Biochem. J.*, **238**, 247–254.
35. Dirksen, M.L., Blakely, W.F., Holwitt, E. & Dizdaroglu, M. (1988) Effect of DNA conformation on the hydroxyl radical-induced formation of 8,5′-cyclopurine-2′-deoxyribonucleoside residues in DNA. *Int. J. Radiat. Biol.*, **54**, 195–204.
36. Watson, J.T. (1990) Selected-ion measurements. *Methods Enzymol.*, **193**, 86–106.
37. Reddy, D.M. & Iden, C.R. (1993) Analysis of modified deoxynucleosides by electrospray ionization mass spectrometry. *Nucleosides Nucleotides*, **12**, 815–826.
38. Serrano, J., Palmeira, C.M., Wallace, K.B. & Kuehl, D. W. (1996) Determination of 8-hydroxy-deoxyguanosine in biological tissue by liquid chromatography/electrospray ionization–mass spectrometry/mass spectrometry. *Rapid Commun. Mass Spectrom.*, **10**, 1789–1791.
39. Ravanat, J.L., Duretz, B., Guiller, A., Douki, T. & Cadet, J. (1998) Isotope dilution high-performance liquid chromatography–electrospray tandem mass spectrometry assay for the measurement of 8-oxo-7,8-dihydro-2′-deoxyguanisine in biological samples. *J.Chromatogr. B Biomed. Sci. Appl.*, **715**, 349–356.
40. Hua, Y., Wainhaus, S.B., Yang, T., Shen, N., Yiong, Y., Xu, X., Zhang, Z., Bolton, J.L. & van Breemen, R.B. (2000) Comparison of negative and positive ion electrospray tandem mass spectrometry for the liquid chromatography tandem mass spectrometry analysis of oxidized deoxynucleosides. *J. Am. Soc. Mass Spectrom.*, **12**, 80–87.
41. Frelon S., Douki, T., Ravanat, J.L., Pouget, J.P., Tornabene, C. & Cadet, J. (2000) High-performance liquid chromatography–tandem mass spectrometry measurement of radiation-induced base damage to isolated and cellular DNA. *Chem. Res. Toxicol.*, **13**, 1002–1010.
42. Podmore, I.D., Cooper, D., Evans, M.D., Wood, M, & Lunec, J. (2000) Simultaneous measurement of 8-oxo-2′-deoxyguanosine and 8-oxo-2′-deoxyadenosine by HPLC-MS/MS. *Biochem. Biophys. Res. Commun.*, **277**, 764–770.
43. Weimann, A., Belling, D. & Poulsen, H.E. (2001) Measurement of 8-oxo-2′-deoxyguanosine and 8-oxo-2′-deoxyadenosine in DNA and human urine by high performance liquid chromatography–electrospray tandem mass spectrometry. *Free Radic. Biol. Med.*, **30**, 757–764.
44. Weimann, A., Belling, D. & Poulsen, H. (2002) Quantification of 8-oxo-guanine and guanine as the nucleobase, nucleoside and deoxynucleoside forms in human urine by high-performance liquid chromatography–electrospray tandem mass spectrometry. *Nucleic Acids Res.*, **30**, E7.
45. Pouget, J.P., Frelon, S., Ravanat, J.L., Testard, I., Odin, F. & Cadet, J. (2002) Formation of modified DNA bases in cells exposed either to gamma radiation or to high-LET particles. *Radiat. Res.*, **157**, 589–595.

46. Cadet, J., Douki, T., Frelon, S., Sauvaio, S., Pouget, J.P. & Ravanat, J.L. (2002) Assessment of oxidative base damage to isolated and cellular DNA by HPLC-MS/MS measurement. *Free Radic. Biol. Med.*, **33**, 441–449.
47. Dizdaroglu, M., Jaruga, P. & Rodriguez, H. (2001) Measurement of 8-hydroxy-2′-deoxyguanosine in DNA by high-performance liquid chromatography–mass spectroemetry: comparison with measurement by gas chromatography–mass spectrometry. *Nucleic Acids Res.*, **29**, E12.
48. Dizdaroglu, M., Jaruga, P. & Rodriguez., H. (2001) Identification and qualification of 8,5′-cyclo-2′-deoxyadenosine in DNA by liquid chromatography/mass spectrometry. *Free Radic. Biol. Med.*, **30**, 774–784.
49. Jaruga, P., Rodriguez, H. & Dizdaroglu, M. (2001) Measurement of 8-hydroxy-2′-deoxyadenosine in DNA by liquid chromatography/mass spectrometry. *Free Radic. Biol. Med.*, **31**, 336–344.
50. Jaruga, P., Birincioglu, M., Rodriguez, H. & Dizdaroglu, M. (2002) Mass spectrometric assays for the tandem lesion 8,5′-cyclo-2′-deoxyguanosine in mammalian DNA. *Biochemistry.*, **41**, 3703–3711.
51. Tuo, J., Jaruga, P., Rodriguez, H., Dizdaroglu, M. & Bohr, V.A. (2002) The Cockayne syndrome group B gene product is involved in cellular repair of 8-hydroxyadenine in DNA. *J. Biol. Chem.*, **277**, 30832–30837.
52. Pomerantz, S.C. & McCloskey, J.A. (1990) Analysis of RNA hydrolyzates by liquid chromatography–mass spectrometry. *Methods Enzymol.*, **193**, 796–824.
53. Tuo, J., Muftuoglu, M., Chen C., Jaruga, P., Selzer, R.R., Brosh, R.M., Jr., Rodriguez, H., Dizdaroglu, M. & Bohr, V.A. (2001) The Cockayne syndrome group B gene product is involved in general genome base excision repair of 8-hydroxyguanine in DNA. *J. Biol. Chem.*, **276**, 45772–45779.
54. Tuo, J., Jaruga, P., Rodriguez, H., Bohr, V.A. & Dizdaroglu, M. (2003) Primary fibroblasts of Cockayne syndrome patients are defective in cellular repair of 8-hydroxyguanine and 8-hydroxyadenine resulting from oxidative stress. *FASEB J.*, **17**, 668–674.
55. Steenken, S, and Jovanovic, S.V. (1997) How easily oxidizable is DNA? One- electron reduction potentials of adenosine and guanosine radicals in aqueous solution. *J. Am. Chem. Soc.*, **119**, 617–618.
56. Candeias, L.P. & Steenken, S. (2000) Reaction of $HO^{\cdot}$ with guanine derivatives in aqueous solution: formation of two different redox-active OH-adduct radicals and their unimolecular transformation reactions. Properties of $G(-H)^{\cdot}$. *Chem. Eur. J.*, **6**, 475–484.
57. Helbock, H.J., Beckman, K.B., Shigenaga, M.K., Walter, P.B., Woodall, A.A., Yeo. H.C. & Ames, B.N. (1998) DNA oxidation matters: The HPLC-electrochemical detection assay of 8-oxo-deoxyguanosine and 8-oxo-guanine. *Proc. Natl Acad. Sci. USA.*, **95**, 288–293.
58. Nakae, D., Mizumoto, Y., Kobayashi, E., Noguchi, O. & Konishi, Y. (1995) Improved genomic/nuclear DNA extraction for 8-hydroxydeoxyguanosine analysis of small amounts of rat liver tissue. *Cancer Lett.*, **97**, 233–239.
59. Cadet, J., Douki, T. & Ravanat, J.L. (1997) Artifacts associated with the measurement of oxidized DNA bases. *Environ. Health Perspect.*, **105**, 1034–1039.
60. Jaruga, P., Speina, E., Gackowski, D., Tudek, B. & Olinski, R. (2000) Endogenous oxidative DNA base modifications analysed with repair enzymes and GC/MS technique. *Nucleic Acids Res.*, **28**, E16.
61. Beckman, K.B., Saljoughi, S., Mashiyma, S.T. & Ames, B.N. (2000) A simpler, more robust method for the analysis of 8-oxoguanine in DNA. *Free Radic. Biol. Med.*, **29**, 357–367.
62. Tchou, J., Kasai, H., Shibutani, S., Chung, M.H., Laval, J., Grollman, A.P. & Nishimura, S. (1991) 8-Oxoguanine (8-hydroxyguanine) DNA glycosylase and its substrate specificity. *Proc. Natl Acad. Sci. U.S.A.*, **88**, 4690–4694.

# 4 Utility of chemical derivatization schemes for peptide mass fingerprinting

JONATHAN A. KARTY, TAE-YOUNG KIM and JAMES P. REILLY

## 4.1 Introduction

Discoveries associated with the structure of DNA, the genetic code, the polymerase chain reaction (PCR), recombinant technologies and nucleic acid sequencing have firmly established genomics at the center of molecular and cellular biology. Nevertheless, it is universally appreciated that while DNA provides the blueprint for assembling cells, proteins actually do the work. Intriguingly, a cell's genome is essentially static but its proteome, i.e. its complement of expressed proteins, is dynamic. In general, a cellular proteome may be modulated by disease, environmental stress induced by changes in available nutrients, drugs, and developmental processes such as cell differentiation, aging, etc. To the extent that the field of medicine is moving toward a molecular understanding of disease, proteomes must be monitored. The high current interest in methods to identify and quantitate cellular protein distributions is therefore easy to understand.

It is fortunate that the information embedded in a cell's proteome is so valuable, because extracting it is a formidable task. Several characteristics amplify the technological challenge of proteomics relative to that of genomics. The first involves dynamic range. While a cell has one copy of its genome, its proteins will typically vary in abundance by five to six orders of magnitude. The second problem is associated with chemical variability. The *four* nucleotides that act as DNA building blocks are chemically quite similar, while the *twenty* amino acids that constitute proteins are chemically divergent. Some are polar, some non-polar; some have basic side-chains, some acidic, some neutral. Some amino acids are hydrophobic, while some are hydrophilic. Thus, while one analytical separation technology may work well for all DNA oligomers, no single method appears to be appropriate for all peptides and proteins. Third, PCR works only with nucleic acids. While an insignificant amount of DNA can be amplified to analytically detectable quantities, no equivalent method exists for proteins. Thus, analytical sensitivity is often an issue in proteomics. Fourth, there is no analogue of DNA chips or cDNA microarrays. While the expression of large numbers of genes can be readily and simultaneously monitored, proteins corresponding to these genes cannot be so easily detected and identified. Fifth, post-translational modifications occur. Proteins, particularly in higher organisms, are often modified following their synthesis in ways that we cannot currently predict. To the extent that these modifications reflect an organism's response to disease or an applied stress, it is important that they be identified and interpreted. Finally, because of the dynamic nature of the proteome it is naturally of interest to study how it evolves under some set of conditions, usually over a period of time. Thus *multiple* experiments involving a single subject are often necessary, enhancing the value of *high throughput* methodologies. In summary, these challenges have motivated and will continue to motivate the development of strategies for separating, detecting and identifying cellular proteins.

Emerging methods in biological mass spectrometry (MS) and separations science are enabling the accurate, rapid, and sensitive characterization of proteins in complex biological mixtures [1–3]. A popular technique used to identify proteins involves the gas phase fragmentation of one or more of their enzymatic or chemical digestion products, followed by correlation of MS/MS data with that stored in theoretical databases [4,5]. This general approach has been utilized with both electrospray ionization (ESI) and matrix assisted laser desorption/ionization (MALDI) using low- and high-energy ion activation methods and various types of mass analyzers. Although low-energy collision-induced dissociation (CID) has probably been the prevailing activation method [2], high-energy CID [6], surface induced dissociation (SID) [7], photodissociation (PD) [8], electron-capture dissociation (ECD) [9], and infrared multi-photon dissociation (IRMPD) [10] have also been investigated. Complementary sequence-specific information can sometimes be derived using a combination of these methods. In any case, because a mass spectrometer can only handle a limited number of molecules at any one time, all proteomic samples must be thoroughly fractionated before their introduction into the ion source. This can be accomplished in several ways. One popular approach is to digest proteins and separate the resulting tryptic peptides by any number of stages of liquid chromatography. Alternatively, intact proteins can be separated by chromatography or electrophoresis and the separated proteins digested.

Each of these approaches has advantages and disadvantages. For example, membrane and hydrophobic proteins tend to precipitate out of solution and not elute from chromatography columns. If these types of proteins are digested before the separation, at least some characteristic peptides will remain soluble and usually be detected, but, because an average protein produces 30–40 tryptic peptides, the number of components in a proteolytic digest is 30–40 times higher than in the undigested mixture. If proteins are digested after they are separated, all peptides derived from a given protein remain together and can be mass spectrometrically analyzed in conjunction. This facilitates their interpretation. However, as noted above, some proteins are lost during the separation process and are never detected.

While two-dimensional liquid chromatography (2D-LC) of proteins is an area of increasing interest [11–15], historically the most popular method of separating proteins was two-dimensional (2D) gel electrophoresis [16,17]. This technology is often criticized because it is slow, labor-intensive, has reproducibility problems, has rather limited dynamic range, and performs poorly with very large, very small and most hydrophobic proteins. Nevertheless, life scientists are comfortable with the method, and recent advances in automation have improved its user-friendliness. Most important, 2D gels offer considerable peak capacity and they present an attractive, simple and visually recognizable overview of the protein distribution in a complex sample. These advantages guarantee that 2D gels will continue to be used until a better protein separation technology is developed. Not surprisingly, over the past decade, the identification of proteins separated by 2D gels has been a major preoccupation of mass spectrometrists.

In 1993, five research groups simultaneously and independently developed the method of peptide mass fingerprinting for identifying proteins [18–22]. Since that time, the technique has become widely used in proteomics. Applications exploiting peptide mass mapping increase as the number of sequenced genomes expands. One of the major advantages of peptide mass mapping using MALDI-time of flight (TOF) MS is its speed. Mass spectra can be acquired in seconds, and database searches can be performed in near real-time, making MALDI-TOF peptide mass mapping amenable to high-throughput applications [23,24]. Peptide mass mapping is based on matching a limited number of experimentally measured ion masses to a set of theoretically predicted proteolytic fragment masses [25]. Because it works best with samples that contain a single or at most a few proteins, 2D gel electrophoresis commonly precedes peptide mass mapping analyses of complex samples.

Unfortunately, not all proteins isolated by 2D gel electrophoresis can be identified by MALDI-TOF peptide mass mapping. In some cases, slower techniques such as micro-Edman degradation or nano-electrospray ionization MS/MS are used to generate sequence information enabling the identification of the components in these samples [25]. Often, the inability to identify a protein in a MALDI-TOF peptide mass mapping experiment is due to the observation of an insufficient number of interpretable mass spectral peaks [26]. Unfortunately MALDI-TOF MS only provides the masses of proteolytic fragments. Rapid techniques that can derive additional sequence information from MALDI-TOF mass spectra would facilitate a wide variety of proteomics experiments.

It is generally recognized that MALDI-TOF MS displays a sensitivity bias towards arginine-terminated peptides that can affect peptide mass mapping of tryptic protein digests. For example, Krause *et al.* [27] described a 4–15-fold decrease in signal intensity for peptides whose C-terminal arginine residues were exchanged for lysines. Along similar lines, four other groups demonstrated that guanidination can increase the sensitivity of MALDI-TOF MS to lysine-terminated tryptic peptides [28–31]. In addition to increasing peak intensities, guanidination also provides information about the lysine content of peptides. Mass shifts between corresponding peaks in unguanidinated and guanidinated mass spectra of a digested protein immediately establish the number of lysines in peptides. We have previously demonstrated that the lysine content can be exploited to refine the database searches performed during peptide mass mapping experiments [32].

One can easily envision an analogous esterification-based experiment that enables one to establish the number of carboxylic-acid-containing residues in a peptide. Because the total number of aspartic and glutamic acid residues is substantially higher than the number of lysines in most proteins, peptides will typically display greater variability in the former number. Consequently, one would expect that establishing the number of acidic residues in a peptide would be an even more useful constraint in peptide mass mapping than knowing the number of lysines. However, while guanidination has been shown to enhance MALDI ion yields for peptides, the effect of esterification on these yields is less certain. In this chapter, this type of experiment is described and the utility of guanidination and esterification in peptide mass mapping of electrophoretically-separated proteins is discussed.

The source of the proteins studied was *Caulobacter crescentus,* an aquatic bacterium with a dimorphic life cycle that represents an attractive prototype system for studying cellular differentiation [33]. The complete genome of *C. crescentus* has been published [34]; translation of this genome enables the prediction of the proteome and a complete set of tryptic fragment masses. The membrane protein study described below presented an opportunity to probe the efficacy of two different chemical derivatizations to facilitate the identification of proteins present in a number of gel spots. In an earlier investigation of the *C. crescentus* life cycle, approximately 70% of the 2D gel spots analyzed were identified [35]. In this work, membrane proteins were initially isolated from this organism. Because they represent just a small subset of the proteins synthesized by *C. crescentus* and because they are very hydrophobic, we employed 1D polyacrylamide gel electrophoresis (PAGE) to separate the proteins. Procedures used to handle 1D and 2D gel bands are similar, and the data interpretation issues that arise are essentially identical. In this work, the combination of guanidination and esterification derivatizations is exploited to facilitate peptide mass mapping. Mass spectral data derived from several bands are discussed and interpreted in order to demonstrate the strengths and shortcomings of peptide mass mapping. The logic followed in interpreting the data and drawing conclusions is presented, and the contribution to the process that chemical derivatizations can make is elucidated. Based on previous statistical modeling [32], absolute confidence limits associated with the protein identifications are derivable. An example of one case in which the peptide mass fingerprint is inconclusive is presented. MALDI-TOF/TOF experiments provide crucial supplementary information.

## 4.2 Experimental methodology

### 4.2.1 *Band destaining and tryptic digestion prior to derivatizations*

Intact pieces of sodium dodecyl sulfate (SDS)-PAGE gel bands are typically excised from complete gels, destained and then introduced to a digestion solution [32,36]. We often cut the gel band into smaller pieces to accelerate the diffusion of reagents. The pieces are incubated twice in a solution of 50% acetonitrile and 50% $NH_4HCO_3$ buffer for 20 min under mild agitation. It is assumed that the proteins remain entrained in the gel during this procedure. The gel pieces are allowed to stand in water for 15 min, washed one more time with pure water, then soaked in 100 μl of acetonitrile for 5 min and dried in a SpeedVac.

To digest the proteins, we employ a solution of methylated porcine trypsin and L-1-tosylamide-2-phenylethyl chloromethyl ketone (TPCK) treated, non-methylated bovine trypsin in ammonium bicarbonatebuffer. Non-methylated bovine trypsin provides autolysis peptides that are useful for internal calibration. The digestion is performed overnight at 37°C, then stopped by addition of trifluoroacetic acid. Peptides are extracted by soaking the gel in water/acetonitrile solutions in Eppendorff tubes that are immersed an ultrasonic bath.

### 4.2.2 *Guanidination*

This reaction protocol has recently been described [29,32,37]. Basically, we combine the peptide solution with 8 mol/kg *O*-methylisourea hemisulfate and 7 M ammonia and incubate for 20 min at 65°C. The ammonia is pumped away, although not to complete dryness. Solid phase micro-extraction onto cartridges packed with Zorbax

$C_{18}$ is necessary to remove the rather concentrated reagents. Peptides can be eluted directly with α-cyano-4-hydroxycinnamic acid matrix solution.

#### 4.2.3 *Esterification*

Carboxylic groups of digested peptides can be esterified with a solution of anhydrous ethanol and acetyl chloride [38,39]. The solution is incubated for 4 h at 37°C. Esterified peptides are extracted and prepared for sample spotting as described above.

#### 4.2.4 *MALDI*

MALDI spots from the derivatized or underivatized samples are made by depositing a mixture of peptide and α-cyano-4-hydroxycinnamic acid matrix in a solution that contains water, acetonitrile and a small amount of trifluoroacetic acid. Our peptide mass spectra are recorded with a Bruker Reflex III reflectron time-of-flight mass spectrometer. Spectra are internally calibrated using various tryptic autolysis peaks (842.510, 2163.059, and 2289.155 Da for unguanidinated samples; 990.549, 1195.596, and 2205.081 Da for guanidinated samples; 870.541, 1073.596, 2303.213, and 2351.262 Da for esterified samples).

Unfortunately in many esterified mass spectra, it is difficult to find more than one or two calibrant peaks, because the intensity of the autolysis peptide signals appears to be reduced upon esterification. In these cases, an external calibration can be applied at first. Then, once a protein is putatively identified (based on any of the three mass spectra we record for each gel band), the esterified mass spectrum can be calibrated internally using peptide masses from the assigned protein. MALDI-TOF/TOF data were recorded on an ABI 4700 Proteomics Analyzer.

#### 4.2.5 *Database searching algorithms*

Although a few commercial software packages for interpreting peptide mass mapping data are available, when analyzing data involving both derivatized and underivatized samples, we find it fastest to employ our own software package called PRODIGIES (*Pro*tein *Dig*est *I*dentification and *E*lucidation *S*oftware) [32]. The program reads the ASCII formatted peak reports generated by the mass spectrometer and creates a list of masses for database comparison. The experimental masses are compared with theoretical tryptic peptide masses that the program calculates based on the translated *C. crescentus* genome [34]. For these analyses, we typically use a conservative peptide mass error window of ±0.15 Da. For guanidinated samples, complete conversion of lysine to homoarginine is assumed [37]. Likewise, for esterified samples, complete ethyl esterification of acidic groups is expected by the software (although, as discussed below, this assumption is often invalid). We assume that trypsin will miss up to two cleavage sites, that methionines may or may not be oxidized, that each protein's N-terminal methionine residue may or may not have been cleaved off, that all cysteines are totally alkylated by iodoacetamide before the electrophoresis separation and that peptide N-terminal glutamine residues may or may not have been converted to pyroglutamic acid.

Naturally, there is a trade-off associated with considering all of these sample modifications. On the one hand, it enables more matches to be made between experimental and theoretical peptide masses; on the other hand, taking into account partial reactions can significantly increase the size of the peptide database against which experimental data are matched. In the present case, the modifications increased the database from about 130 000 (assuming no modifications or missed cleavages) to 600 000 tryptic fragment entries. The expanded database yields more random matches and a corresponding increase in the number of false positive assignments made. About 80% of our peptide mass matches are associated with theoretical peptides having no missed cleavages. Part of the reason this number is high is that the proteolytic digestion was allowed to proceed for a full 16 h. Deciding how to consider missed cleavages represents a typical example of how proteomics studies often involve compromises. Allowing more missed cleavages improves total protein sequence coverage but this gain is achieved at the expense of statistical confidence of assignments.

PRODIGIES treats mass spectrometric data in two ways. In the first mode of operation, it compares the data from a single mass spectrum to the masses generated by the *in silico* tryptic digest of the entire proteome and it creates a list of open reading frame (ORF) products that could be present in the sample. Results are summarized in a 'master hit array' (MHA). In the MHA, ORFs are ranked by the number of observed masses that match theoretical tryptic peptide masses of a particular ORF, and the errors between the observed and theoretical values are displayed. Many publicly available peptide mass mapping computer programs function similarly [5,21,40–43].

In its second, more advanced, mode PRODIGIES directly exploits the sequence information obtained by comparing derivatized and underivatized data. Use of this mode in conjunction with the guanidination derivatization will initially be described. In this case, data from both guanidinated and unguanidinated mass spectra are used to generate three MHAs. PRODIGIES creates an MHA for the unguanidinated data as described above. It performs a second *in silico* tryptic digest assuming that all lysine residues have been converted to homoarginines. This new list is compared against the peak report from the guanidinated mass spectrum to produce a second MHA. Often, only one ORF will be common to both the unguanidinated and guanidinated MHAs, and it can be inferred that the product of this ORF is present in the gel band. PRODIGIES then tabulates those features that appear either at the same mass or shifted by an integral multiple of 42.02 Da in the two MHAs. These are referred to as 'consistent' masses. The number of 42.02 Da shifts corresponds to the number of lysine residues in a particular peptide. Each mass in this new table is again compared to the unguanidinated *in silico* digest; a match is indicated only when a theoretical peptide mass matches an observed mass *and* its sequence contains the correct number of lysines. This new consistent MHA lists those ORFs with the largest numbers of matches based on the criteria just described. The consistent MHA is a powerful tool for assigning proteins to gel band tryptic digests. It reduces the number of false positives that arise when multiple peptides with different sequences share the same mass. Random matches occur because there are hundreds of thousands of predicted tryptic peptides compressed into a relatively narrow mass range. For example, using the digest conditions described above, there are 162 predicted *C. crescentus* peptides having masses within 0.15 Da of 1061.6 Da. Lysine content information excludes many of these peptides from appearing in the consistent

MHA since it only lists those peptides with the correct mass *and* number of lysines. For this particular example, 82 of the 162 peptides contain no lysines, 67 have one, and only 13 peptides have two lysines in their sequences. PRODIGIES requires less than 1 s to search a pair of mass spectra against the *C. crescentus* proteome on a 700 MHz computer.

Esterfication data are treated analogously, except that the mass shift associated with ethyl esterfication is 28.03 Da per acidic site (which includes aspartate, glutamate, and the C terminus of the peptide). While all peptides have a C terminus, they vary considerably in terms of their abundance of acidic sites. Thus, the mass shifts associated with esterfication represent a more diverse distribution than the shifts due to guanidination.

### 4.2.6 *Random spectra generation and statistical analysis*

The confidence with which one makes a protein identification is as important as the identification itself. We previously reported a statistical analysis of several million database searches with simulated data that allowed the assignment of absolute confidence levels to protein identifications [32]. The question that these simulations addressed is rather simple: given a mass spectrum containing a certain number of peaks, how many of these must match peptides associated with a single ORF for the identification to be certified with a specific confidence level? A simulation proceeds as follows. One of the 3762 *C. crescentus* ORFs is selected at random. A specified number of 'real' tryptic peptides are chosen at random from the list of predicted tryptic peptides for that particular ORF. Likewise, a specified number of 'random' tryptic peptides are selected from the list of all predicted tryptic peptides from all ORFs in the *C. crescentus* proteome. The combination of these sets of masses defines a theoretical mass spectrum. PRODIGIES shifts the masses of all 'real' and 'random' peptides based on the number of lysines in their sequences, generating a corresponding theoretical guanidinated mass spectrum. It then analyzes the theoretical mass spectra just as it handles experimental data. For each simulated spectrum, the 'winning' ORF is determined. The 'winning' ORF is defined as the ORF with the most matches in an MHA. By repeating the process many times, PRODIGIES determines the percentage of theoretical mass spectra (containing varying numbers of 'real' and 'random' masses) that are properly interpreted, i.e. the 'winning' ORF is the one from which the real peptides were selected.

## 4.3 Results

The gel electrophoresis separated our mixture of *C. crescentus* membrane proteins into 34 resolved bands. These were arbitrarily numbered and will be referred to by these numbers. Based on their varying shades of darkness, protein abundances probably varied by 1–2 orders of magnitude. Mass spectral analysis of these bands led to identification of one or more proteins in 29 of the bands. Several representative cases will be discussed that exemplify our general experience with peptide mass mapping of gel band digests.

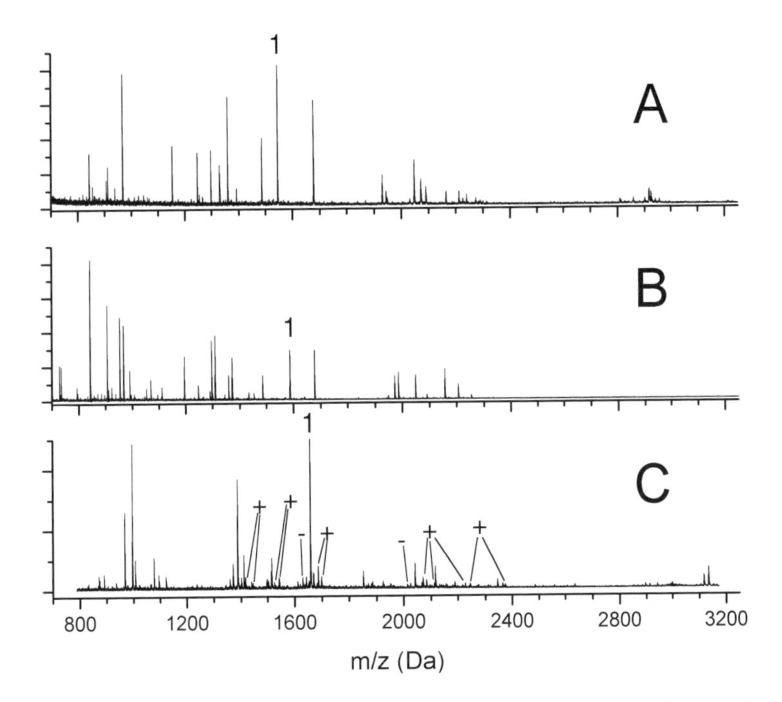


**Fig. 4.1** MALDI-TOF mass spectra from peptide extracts of gel band 33. (A) Unmodified sample (B) guanidinated sample (C) esterified mass spectra. +, solvolysis of N or Q to ethyl-esterified D or E; −, incomplete esterification of D or E.

### 4.3.1 *Identification of a strong gel band*

Band 33 was one of the most intensely stained bands on our gel. Figure 4.1 displays mass spectra obtained from the unmodified, guanidinated and esterified tryptic digests of this band. The signal to noise ratio in the spectra is excellent. 21, 30 and 63 peaks were noted in these spectra and their masses measured. Tables 4.1A–F contain six truncated MHAs for gel band 33. (Note that since master hit arrays can be arbitrarily large depending on the minimum number of matches that a user selects for listing an ORF, some method of limiting their size in a printed document is needed. In the present case, the portion of the master hit array to the right of the right-hand column is not displayed to save space, and in some cases the bottom rows of the MHA are deleted. These truncations do not affect data interpretation in practice and they facilitate inspection and interpretation of the master hit arrays.) The top row of an MHA lists the ORFs ranked by number of peptides matched. The left-hand column lists the experimental peptide masses in order of decreasing intensity. The fractional elements in the table are the differences, in Da, between the measured and theoretical masses. If an experimental mass does not match any peptide in a given ORF, this is indicated by an asterisk. Numbers in braces indicate the total number of matches in each column or row. (Note that some of these numbers may appear to be incorrect because part of the master hit array is not displayed.) The 'winning' ORF appears in the upper left-hand corner of the MHA and is the protein most likely to be present in the gel band. The preliminary interpretation of these data is

**Table 4.1** Master hit arrays (MHAs) derived from the mass spectra of gel band 33 peptide extracts. The MHAs have been truncated for space considerations. These have been limited to 25 lines in height and 11 columns in width. This truncation does not affect the analysis.

(A) Unmodified tryptic digest (21 masses submitted).

| | 163 {17} | 1594 {6} | 2898 {6} | hits |
|---|---|---|---|---|
| 1544.83 | −0.05 | −0.04 | * | {2} |
| 967.51 | −0.01 | 0.04 | −0.02 | {3} |
| 1358.77 | −0.04 | −0.07 | * | {2} |
| 1676.85 | −0.03 | * | * | {1} |
| 1152.54 | 0.02 | * | * | {1} |
| 1484.81 | −0.06 | * | * | {1} |
| 1297.59 | −0.02 | * | 0.02 | {2} |
| 1247.63 | 0.00 | * | 0.11 | {2} |
| 1328.72 | −0.03 | * | * | {1} |
| 2046.90 | * | 0.05 | * | {1} |
| 911.59 | −0.02 | * | * | {1} |
| 1929.07 | −0.10 | 0.03 | * | {2} |
| 938.46 | * | * | 0.05 | {1} |
| 2071.10 | −0.07 | * | −0.07 | {2} |
| 1390.77 | * | * | * | {0} |
| 2089.97 | −0.02 | * | * | {1} |
| 1942.95 | −0.01 | * | * | {1} |
| 2915.57 | −0.11 | −0.14 | * | {2} |
| 2045.98 | −0.11 | * | 0.05 | {2} |
| 2922.47 | −0.03 | * | * | {1} |
| 2807.48 | * | * | * | {0} |

(B) Guanidinated sample (30 masses submitted).

| | 163 {22} | 657 {8} | 1238 {8} | 42 {7} | 503 {7} | 789 {7} | 1580 {7} | ... | 3585 {7} | hits |
|---|---|---|---|---|---|---|---|---|---|---|
| 906.45 | −0.02 | * | −0.08 | * | 0.05 | * | 0.07 | ... | * | {8} |
| 953.62 | −0.03 | * | * | −0.04 | * | * | * | ... | * | {4} |
| 967.53 | −0.03 | * | * | * | * | * | * | ... | * | {6} |
| 1308.76 | −0.03 | −0.07 | * | * | * | * | * | ... | * | {5} |
| 1296.71 | −0.03 | * | 0.00 | −0.02 | * | * | * | ... | * | {5} |
| 1586.81 | −0.01 | * | 0.14 | * | * | * | 0.08 | ... | * | {6} |
| 1676.82 | 0.00 | * | * | * | * | * | * | ... | 0.08 | {7} |
| 1370.77 | −0.06 | 0.01 | −0.08 | −0.06 | * | * | * | ... | * | {6} |
| 728.30 | 0.04 | 0.14 | * | 0.11 | * | -0.03 | 0.00 | ... | * | {8} |
| 1194.58 | 0.01 | * | * | * | * | * | 0.04 | ... | * | {4} |
| 1358.78 | −0.05 | * | * | * | 0.00 | * | * | ... | * | {7} |
| 1984.99 | −0.03 | * | * | * | * | * | * | ... | * | {3} |
| 2047.86 | * | * | * | * | * | * | * | ... | * | {1} |
| 1068.51 | 0.00 | * | * | * | * | * | * | ... | * | {5} |
| 2155.08 | 0.00 | * | * | −0.09 | * | −0.04 | * | ... | −0.02 | {7} |
| 1484.81 | −0.06 | −0.04 | * | * | 0.01 | −0.07 | * | ... | 0.01 | {9} |
| 1971.02 | −0.03 | * | * | * | * | * | * | ... | * | {5} |
| 1247.64 | −0.01 | * | 0.07 | * | * | * | * | ... | * | {5} |
| 1110.53 | 0.00 | * | * | * | 0.13 | * | * | ... | 0.05 | {5} |
| 923.49 | −0.03 | 0.01 | 0.03 | * | 0.04 | 0.05 | * | ... | * | {7} |
| 1052.55 | −0.03 | * | * | * | * | * | 0.03 | ... | * | {3} |
| 911.49 | * | 0.10 | * | * | * | 0.02 | 0.01 | ... | * | {5} |
| 1289.68 | * | * | * | * | * | 0.03 | * | ... | −0.03 | {4} |

*Contd*

**Table 4.1** *Contd*

(C) Esterified sample (50 masses submitted).

| | 163 {20} | 545 {16} | 42 {12} | 373 {12} | 502 {11} | 1580 {11} | 2043 {11} | ... | 3758 {6} | hits |
|---|---|---|---|---|---|---|---|---|---|---|
| 1656.98 | −0.07 | 0.03 | * | −0.05 | * | * | * | ... | * | {19} |
| 995.49 | 0.04 | * | * | * | 0.09 | * | * | ... | * | {14} |
| 1386.75 | 0.01 | * | * | * | * | * | 0.06 | ... | * | {31} |
| 967.59 | 0.04 | * | * | * | * | * | −0.09 | ... | * | {26} |
| 994.46 | * | 0.14 | * | 0.12 | * | * | * | ... | * | {14} |
| 1409.67 | 0.02 | * | * | * | * | * | * | ... | * | {15} |
| 1076.59 | * | * | * | * | * | * | * | ... | * | {22} |
| 1514.89 | −0.02 | * | −0.05 | −0.07 | * | −0.06 | −0.01 | ... | * | {31} |
| 1006.47 | * | * | * | 0.09 | * | * | * | ... | * | {18} |
| 1369.73 | * | * | * | * | * | * | * | ... | 0.05 | {28} |
| 1686.00 | * | * | −0.07 | * | −0.12 | * | * | ... | * | {24} |
| 2041.10 | 0.00 | * | * | * | * | 0.03 | * | ... | * | {19} |
| 2115.13 | * | 0.04 | * | * | * | * | * | ... | 0.06 | {22} |
| 1512.81 | −0.03 | 0.01 | * | 0.03 | * | * | 0.03 | ... | * | {30} |
| 1668.95 | −0.07 | −0.02 | * | * | * | −0.03 | −0.02 | ... | 0.09 | {24} |
| 890.50 | 0.02 | * | −0.02 | * | * | * | * | ... | * | {24} |
| 1094.54 | * | * | * | * | * | * | * | ... | * | {12} |
| 870.52 | * | 0.03 | * | * | * | * | * | ... | * | {27} |
| 1850.99 | 0.05 | * | * | 0.07 | * | * | * | ... | * | {32} |
| 1120.67 | * | * | * | * | * | * | * | ... | * | {18} |
| 1400.72 | * | * | * | * | * | * | * | ... | * | {23} |
| 1415.78 | * | * | * | * | * | 0.06 | * | ... | * | {28} |
| 1641.95 | * | * | * | * | −0.07 | * | * | ... | * | {27} |

(D) Consistent MHA using both unmodified and guanidinated data.

| | 163 {13} | 887 {3} | 1352 {3} | 1757 {3} | 2525 {3} | 3458 {3} | 3569 {3} | 3665 {3} | hits |
|---|---|---|---|---|---|---|---|---|---|
| 1544.83 | −0.05 | * | −0.03 | * | * | * | * | * | {2} |
| 967.51 | −0.01 | * | * | * | * | −0.01 | * | * | {2} |
| 1358.77 | −0.04 | * | * | 0.03 | −0.10 | −0.01 | −0.09 | −0.02 | {6} |
| 1676.85 | −0.03 | * | * | * | * | * | * | * | {1} |
| 1152.54 | 0.02 | 0.12 | * | * | * | * | * | * | {2} |
| 1484.81 | −0.06 | * | * | 0.04 | −0.13 | * | * | −0.05 | {4} |
| 1247.63 | 0.00 | * | * | * | * | * | 0.02 | * | {2} |
| 1328.72 | −0.03 | * | * | * | * | * | * | * | {1} |
| 911.59 | −0.02 | −0.08 | −0.05 | * | −0.10 | * | * | * | {4} |
| 1929.07 | −0.10 | * | * | * | * | * | * | * | {1} |
| 2071.10 | −0.07 | * | 0.04 | * | * | * | * | * | {2} |
| 1390.77 | * | * | * | −0.04 | * | * | * | 0.00 | {2} |
| 2089.97 | −0.02 | 0.14 | * | * | * | 0.13 | 0.13 | * | {4} |
| 1942.95 | −0.01 | * | * | * | * | * | * | * | {1} |

*Contd*

obvious. In Table 4.1A, 17 of the 21 observed masses from the unmodified digest match predicted tryptic peptides from ORF 163. *C. crescentus* genome annotation indicates that this is a 46 kDa 'hypothetical' protein whose function is unknown.

The next question that we can attempt to answer is whether any proteins other than ORF 163 are also present in this gel spot. Guanidinated data help to address this. Table 4.1B shows the MHA derived from the tryptic digest of band 33 after guanidination. Since the reaction is quantitative [37] complete conversion of lysine to

**Table 4.1** *Contd*

(E) Consistent MHA using both unmodified and esterified data.

| | 163 {8} | 139 {2} | 230 {2} | 465 {2} | 502 {2} | 506 {2} | 526 {2} | ... | 3568 {2} | hits |
|---|---|---|---|---|---|---|---|---|---|---|
| 1544.83 | −0.05 | * | * | −0.04 | * | −0.04 | −0.05 | ... | −0.13 | {8} |
| 967.51 | −0.01 | 0.07 | * | * | 0.03 | * | 0.08 | ... | * | {5} |
| 1358.77 | −0.04 | * | * | * | * | * | * | ... | * | {2} |
| 1484.81 | −0.06 | * | * | 0.03 | * | * | * | ... | * | {3} |
| 1297.59 | −0.02 | * | * | * | * | 0.00 | * | ... | * | {2} |
| 1247.63 | * | * | * | * | * | * | * | ... | * | {0} |
| 1328.72 | * | * | * | * | * | * | * | ... | * | {0} |
| 2046.90 | * | * | * | * | * | * | * | ... | * | {0} |
| 911.59 | −0.02 | * | * | * | * | * | * | ... | −0.05 | {4} |
| 1929.07 | −0.10 | 0.00 | * | * | * | * | * | ... | * | {7} |
| 938.46 | * | * | 0.00 | * | * | * | * | ... | * | {3} |
| 2071.10 | −0.07 | * | −0.04 | * | 0.09 | * | * | ... | * | {6} |
| 1942.95 | * | * | * | * | * | * | * | ... | * | {0} |

(F) MHA arising from the esterified sample after the 13 masses consistent with incomplete esterification and amide solvolysis were removed from the peak report.

| | 163 {20} | 545 {14} | 2043 {10} | 2751 {10} | 672 {9} | 1580 {9} | 1847 {9} | ... | 3707 {6} | hits |
|---|---|---|---|---|---|---|---|---|---|---|
| 1656.98 | −0.07 | 0.03 | * | −0.04 | * | * | * | ... | * | {9} |
| 995.49 | 0.04 | * | * | * | * | * | * | ... | * | {11} |
| 1386.75 | 0.01 | * | 0.06 | * | * | * | * | ... | * | {22} |
| 967.59 | 0.04 | * | −0.09 | * | * | * | * | ... | −0.09 | {14} |
| 994.46 | * | 0.14 | * | * | * | * | * | ... | * | {9} |
| 1409.67 | 0.02 | * | * | * | * | * | * | ... | | {6} |
| 1076.59 | * | * | * | * | * | * | 0.09 | ... | * | {11} |
| 1514.89 | −0.02 | * | −0.01 | −0.05 | * | −0.06 | * | ... | * | {20} |
| 1006.47 | * | * | * | 0.07 | 0.08 | * | 0.03 | ... | * | {12} |
| 1369.73 | * | * | * | 0.07 | * | * | * | ... | 0.09 | {16} |
| 2041.10 | 0.00 | * | * | * | * | 0.03 | * | ... | * | {14} |
| 2115.13 | * | 0.04 | * | * | * | * | * | ... | * | {12} |
| 1512.81 | −0.03 | 0.01 | 0.03 | * | * | * | * | ... | * | {15} |
| 1668.95 | −0.07 | −0.02 | −0.02 | * | * | −0.03 | * | ... | * | {17} |
| 890.50 | 0.02 | * | * | * | −0.02 | * | −0.02 | ... | * | {17} |
| 1094.54 | * | * | * | * | * | * | * | ... | * | {3} |
| 870.52 | * | 0.03 | * | * | 0.00 | * | * | ... | * | {16} |
| 1850.99 | 0.05 | * | * | * | * | * | 0.03 | ... | * | {15} |
| 1120.67 | * | * | * | * | * | * | * | ... | * | {6} |
| 1400.72 | * | * | * | * | * | * | * | ... | * | {13} |
| 1641.95 | * | * | * | * | 0.02 | * | * | ... | * | {11} |
| 3132.68 | 0.01 | 0.06 | * | * | * | * | * | ... | * | {9} |
| 1358.72 | 0.05 | 0.07 | * | 0.02 | * | * | * | ... | * | {18} |

homoarginine is assumed. The MHA in Table 4.1B compares observed guanidinated masses with theoretical masses from the guanidinated proteome. ORF163 once again has the largest number of matches, with 22 out of 30 masses submitted. More importantly, no other ORFs appear in both Tables 4.1A and 4.1B and thus no single ORF appears to be associated with the peptides not assigned to ORF 163. Likewise, the MHA in Table 4.1C compares observed esterified masses with theoretical masses from an esterified proteome. Twenty of the 63 submitted masses can be associated with the same

ORF 163. Once again, no other ORFs are common to Tables 4.1A, 4.1B and 4.1C, providing further confirmation of an unambiguous assignment of this single protein being present in the gel band.

As noted above, one of the previously demonstrated capabilities of our data analysis software PRODIGIES is the simultaneous interpretation of masses that appear in both unmodified and guanidinated digests of a protein. Consistent masses derived from a pair of spectra represent an abridged set of data whose interpretation leads to a reduction in the noise associated with random peptide matching [32]. Table 4.1D presents a consistent MHA generated from the unmodified and guanidinated digest spectra from gel band 33. In this case, it appears that 13 of the 14 consistent masses are associated with the single ORF 163. Only three of the consistent masses could be associated with any other ORF, again underlining the lack of significant evidence of any other protein being present in this gel band. Table 4.1E similarly presents a consistent MHA generated from analyzing unmodified and esterified digest spectra. Theoretical masses from ORF 163 match eight of the 13 consistent masses provided by these two spectra and no other ORF can match more than two of the consistent masses.

Upon closer examination of the data, it was found that two of the peaks in the esterified mass spectrum had masses that were 28 Da lower than predicted masses for ORF 163 peptides. These are interpretable as incompletely esterified ORF 163 tryptic peptides [39]. In both cases, the fully esterified peaks were much more intense than their partially esterified counterparts. The latter are marked with a minus sign in Fig. 4.1. More troubling, 11 peaks in the esterified gel band 33 mass spectrum were 29 Da higher than predicted ORF 163 tryptic peptides. Such a shift occurs when an amide-containing residue (N or Q) is converted to the ethyl ester of the corresponding acidic group (D or E) by solvolysis [44]. In ten of these instances, the correctly esterified peak appeared to be more intense than its incorrectly modified counterpart. Peaks consistent with amide solvolysis are marked by a plus sign in Fig. 4.1C. The 13 masses resulting from either incomplete esterification or amide solvolysis were removed from the esterified sample peak report and it was reanalyzed by PRODIGIES. These results are presented in Table 4.1F. Notice that no matches to ORF 163 were lost. Furthermore, because the false positives arising from random matches are removed, one finds that ORF 163 now stands out more from the other suggested identifications in the MHA.

In summary, the analysis of consistent masses, whether they are derived from comparison of an unmodified digest spectrum with a guanidinated or an esterified spectrum, corroborates our interpretation derived from analyzing individual mass spectra. For a case as unambiguous as this particular gel band, the extra work associated with derivatizing the tryptic digest hardly seems necessary. However, the next example describes a more common situation in which the added value of chemical derivatizations is apparent.

### 4.3.2 *Identification of a weak gel band*

Gel band 18 is an example of one that is below average in stain intensity. Its unmodified, guanidinated and esterified mass spectra, containing 21, 14 and 32 peaks, respectively, do not lead to obvious interpretation. Table 4.2A displays an MHA associated with the unmodified spectrum. Only eight masses match ORF 1460, a flagellin protein, suggesting an ambiguous assignment. However, these eight are among the 14 most intense peaks in the mass spectrum, lending some credence to the interpretation. Unfortunately, ORF 76 matches the experimental data almost as well, albeit with somewhat inferior mass errors.

By itself, the guanidinated MHA in Table 4.2B provides a similarly encouraging but hardly unambiguous picture. The 'winning' ORF (1460) provides a slightly more convincing assignment than ORF 2879. Most importantly, however, this conclusion matches that derived from the unmodified data. No other ORF appears in both MHAs.

**Table 4.2** Truncated master hit arrays (MHAs) derived from the mass spectra of gel band 18 peptide extracts.

(A) Unmodified tryptic digest (21 masses submitted).

| | 1460 | 76 | 85 | 88 | 562 | 625 | 2533 | ... | 3325 | hits |
|---|---|---|---|---|---|---|---|---|---|---|
| | {8} | {7} | {5} | {5} | {5} | {5} | {5} | ... | {4} | |
| 1449.80 | −0.07 | * | −0.06 | 0.02 | * | −0.07 | * | ... | −0.06 | {11} |
| 1914.02 | −0.04 | −0.07 | 0.05 | * | 0.02 | * | * | ... | * | {10} |
| 2019.11 | −0.03 | −0.03 | * | * | * | * | * | ... | 0.04 | {8} |
| 837.40 | 0.01 | 0.07 | * | * | * | * | 0.04 | ... | * | {6} |
| 2103.13 | −0.01 | * | −0.12 | 0.03 | * | * | −0.14 | ... | * | {13} |
| 2597.54 | * | −0.15 | * | * | * | * | * | ... | −0.13 | {3} |
| 1048.63 | * | * | * | * | −0.11 | * | * | ... | * | {5} |
| 3256.75 | −0.09 | * | * | * | * | * | * | ... | * | {3} |
| 1112.48 | * | 0.09 | 0.06 | * | * | * | * | ... | * | {6} |
| 1339.69 | −0.04 | * | * | * | * | * | * | ... | * | {5} |
| 1760.01 | * | * | * | * | * | * | * | ... | * | {3} |
| 1374.86 | * | * | * | * | * | −0.13 | * | ... | −0.09 | {6} |
| 1124.47 | * | 0.10 | 0.09 | 0.08 | * | * | * | ... | * | {10} |
| 1794.99 | −0.07 | * | * | * | * | * | * | ... | * | {7} |
| 2199.22 | * | * | * | * | * | −0.03 | −0.11 | ... | * | {8} |
| 1095.44 | * | * | * | * | 0.13 | 0.02 | * | ... | * | {8} |
| 2580.49 | * | −0.12 | * | −0.11 | * | * | * | ... | * | {4} |
| 2154.12 | * | * | * | * | −0.08 | * | 0.06 | ... | * | {10} |
| 920.61 | * | * | * | −0.13 | −0.10 | * | * | ... | * | {3} |
| 1589.80 | * | * | * | * | * | * | * | ... | * | {8} |
| 1671.76 | * | * | * | * | * | 0.11 | 0.07 | ... | * | {8} |

(B) Guanidinated sample (14 masses submitted).

| | 1460 | 2879 | 42 | 615 | 690 | 739 | 1069 | ... | 3651 | hits |
|---|---|---|---|---|---|---|---|---|---|---|
| | {7} | {6} | {4} | {4} | {4} | {4} | {4} | ... | {4} | |
| 1491.76 | 0.00 | * | * | * | * | * | * | ... | * | {1} |
| 1956.00 | 0.00 | −0.03 | 0.08 | 0.14 | * | * | * | ... | * | {7} |
| 859.50 | 0.00 | 0.00 | * | −0.05 | 0.01 | * | * | ... | * | {5} |
| 837.41 | 0.00 | * | −0.03 | * | * | * | * | ... | 0.00 | {5} |
| 1048.61 | * | −0.04 | −0.09 | −0.09 | * | * | * | ... | * | {3} |
| 1155.50 | * | * | * | * | * | * | 0.11 | ... | * | {3} |
| 1381.70 | −0.03 | −0.03 | * | * | 0.07 | 0.04 | 0.10 | ... | * | {5} |
| 2061.10 | 0.00 | * | * | * | * | 0.02 | * | ... | −0.03 | {4} |
| 1137.49 | * | * | * | * | 0.14 | 0.11 | * | ... | * | {3} |
| 1837.01 | −0.06 | * | * | −0.11 | 0.11 | * | −0.01 | ... | * | {4} |
| 964.42 | * | 0.03 | * | * | * | 0.09 | * | ... | * | {3} |
| 1545.62 | * | 0.09 | 0.10 | * | * | * | 0.10 | ... | 0.15 | {4} |
| 1818.94 | * | * | * | * | * | * | * | ... | −0.14 | {2} |
| 3256.25 | * | * | * | * | * | * | * | ... | * | {0} |

*Contd*

**Table 4.2** *Contd*

(C) Esterified sample (34 masses submitted).

| | 503 {7} | 1493 {6} | 76 {5} | 562 {5} | 655 {5} | ... | 1460 {5} | ... | 3569 {5} | hits |
|---|---|---|---|---|---|---|---|---|---|---|
| 1561.85 | 0.03 | * | * | −0.01 | * | ... | 0.01 | ... | * | {4} |
| 1590.87 | * | −0.01 | * | * | −0.03 | ... | * | ... | * | {3} |
| 1533.79 | * | * | 0.06 | * | * | ... | * | ... | * | {4} |
| 948.59 | * | −0.05 | * | * | −0.04 | ... | * | ... | −0.07 | {6} |
| 1274.84 | * | * | * | * | * | ... | * | ... | * | {0} |
| 854.99 | * | * | * | * | * | ... | * | ... | * | {0} |
| 2075.05 | * | * | * | * | * | ... | 0.09 | ... | * | {2} |
| 993.94 | * | * | * | * | * | ... | * | ... | * | {0} |
| 1579.85 | * | * | 0.03 | 0.07 | * | ... | * | ... | * | {4} |
| 921.51 | * | * | −0.06 | * | * | ... | −0.01 | ... | * | {3} |
| 2054.14 | * | * | * | * | * | ... | 0.00 | ... | * | {2} |
| 2625.40 | * | * | * | * | * | ... | * | ... | * | {3} |
| 2654.39 | * | −0.03 | * | * | * | ... | * | ... | * | {1} |
| 1619.89 | −0.09 | * | 0.00 | −0.01 | * | ... | * | ... | * | {3} |
| 876.38 | 0.14 | 0.10 | * | * | * | ... | * | ... | * | {4} |
| 1402.82 | * | −0.01 | * | * | * | ... | * | ... | −0.04 | {4} |
| 1478.75 | * | * | * | * | * | ... | * | ... | * | {1} |
| 1515.85 | * | * | * | * | * | ... | * | ... | * | {1} |
| 2608.27 | * | * | * | * | 0.07 | ... | * | ... | * | {4} |
| 1505.75 | * | * | * | * | * | ... | * | ... | * | {0} |
| 898.50 | * | 0.07 | * | * | * | ... | * | ... | * | {2} |
| 2637.52 | * | * | * | * | * | ... | * | ... | * | {0} |
| 939.46 | 0.04 | * | * | * | * | ... | * | ... | * | {2} |

(D) Consistent MHA using unmodified and guandinated data.

| | 1460 {6} | 373 {3} | 2059 {3} | 42 {2} | 85 {2} | 118 {2} | 615 {2} | ... | 3758 {2} | hits |
|---|---|---|---|---|---|---|---|---|---|---|
| 1449.80 | −0.07 | * | * | * | −0.06 | * | * | ... | * | {3} |
| 1914.02 | −0.04 | * | * | * | 0.05 | * | * | ... | * | {8} |
| 2019.11 | −0.03 | −0.08 | * | * | * | 0.03 | * | ... | −0.05 | {5} |
| 837.40 | 0.01 | * | 0.05 | −0.02 | * | * | * | ... | * | {9} |
| 1048.63 | * | −0.08 | −0.15 | −0.11 | * | * | −0.11 | ... | * | {8} |
| 1339.69 | −0.04 | * | * | * | * | * | * | ... | * | {5} |
| 1794.99 | −0.07 | 0.00 | 0.12 | * | * | −0.11 | −0.12 | ... | * | {10} |
| 1095.44 | * | * | * | * | * | * | * | ... | 0.08 | {2} |

(E) Consistent MHA using unmodified and esterified data.

| | 1460 {5} | 85 {2} | 285 {2} | 745 {2} | 905 {2} | 1059 {2} | 1298 {2} | ... | 3722 {2} | hits |
|---|---|---|---|---|---|---|---|---|---|---|
| 1449.73 | 0.00 | 0.01 | * | −0.05 | * | * | * | ... | * | {7} |
| 842.51 | * | * | 0.01 | 0.00 | * | * | −0.12 | ... | 0.00 | {5} |
| 837.39 | 0.02 | * | * | * | 0.05 | * | * | ... | * | {2} |
| 1913.98 | 0.00 | 0.09 | * | * | * | * | 0.00 | ... | −0.05 | {6} |
| 1339.67 | * | * | * | * | * | * | * | ... | * | {0} |
| 1045.50 | * | * | * | * | * | * | * | ... | * | {0} |
| 1108.42 | * | * | * | * | * | * | * | ... | * | {1} |
| 2019.08 | 0.00 | * | * | * | * | * | * | ... | * | {2} |
| 1487.69 | * | * | * | * | 0.10 | * | * | ... | * | {2} |
| 2103.06 | 0.06 | * | −0.06 | * | * | 0.02 | * | ... | * | {4} |
| 1394.70 | * | * | * | * | * | 0.04 | * | ... | * | {4} |
| 1794.86 | * | * | * | * | * | * | * | ... | * | {0} |

The esterified data provide less convincing support for this assignment. ORF 1460 is one of several that match five of the 32 peptide masses, and ORFs 503 and 1493 actually provide more matches to the experimental data. Encouragingly, ORF 1460 is the only one that appears in all three MHAs. Much more convincing support for this assignment is offered by the consistent MHAs. The winning ORF in the guanidinated (Table 4.2D) and esterified (Table 4.2E) MHAs is clearly and unambiguously ORF 1460. Although the number of matches (six out of eight masses for the guanidinated, and five out of 12 for the esterified case) is not large, the reduction in the number of random matches is so significant that no other ORF provides a viable alternative assignment. In conclusion, at least two mass spectra, one from unmodified, one from derivatized peptides are required to convincingly assign an ORF to this gel band. Nevertheless, this identification is far less definitive than that for band 33, and a more quantitative measure of our confidence in this assignment is desirable. From our previous Monte Carlo simulations of unguanidinated and guanidinated peptide tryptic digests of *C. crescentus* proteins we learned that when six out of eight *consistent* masses are assignable to one protein, the protein is correctly identified in more than 99% of all cases.

### 4.3.3 *More complex gel bands*

While the previous examples exemplify the common situation of detecting and identifying a single protein in a gel band, on occasion, bands may contain two or more unresolved proteins. This situation certainly should be relatively common when complex samples are fractionated by 1D PAGE alone. In the present experiment, gel band 5 represents such a case. Tables 4.3A and 4.4B display MHAs derived from the analysis of unmodified and guanidinated tryptic digests of this gel band. Two ORFs, 995 and 171, which correspond to a hypothetical protein and a tonB dependent receptor, respectively, dominate the tables. Not only do they match considerably more experimental masses than any other ORF, between them they match all but two of the 23 most intense peaks in the spectrum. The striking alternating pattern of matches apparent in both MHAs suggests that the two proteins are probably present at comparable levels in the gel band. Also impressive is the small number of overlapping assignments: in Table 4.3A only a single experimental mass, 1443.78 Da, is ambiguously assigned to both ORFs. At first glance, one might assume that this peak corresponds to an ORF 171 peptide due to the smaller mass error associated with this assignment. However, the appearance of a peak at 1485.77 Da in the guanidinated spectrum indicates that the peptide corresponding to this mass contains one lysine; as displayed in the guanidinated consistent MHA of Table 4.3C this mass is uniquely matched to ORF 995. Also remarkable is the fact that ORFs 995 and 171 together match the 28 most intense masses that appear in the consistent MHA. The co-migration of these proteins is not surprising considering that their predicted masses are 99.5 kDa and 97.3 kDa, respectively.

### 4.3.4 *Homologous proteins*

Distinguishing homologous proteins is not straightforward for any proteomics technology, and this is certainly not a particular strength of peptide mass mapping. Gel band 17 provides a good example of the problems that arise. The guanidinated and esterified

**Table 4.3** Truncated master hit arrays (MHAs) derived from the mass spectra of gel band 5; these MHAs have been truncated at 30 lines in length.

(A) Unmodified tryptic digest (48 masses submitted).

| | 995 | 171 | 3569 | 373 | 623 | 786 | 1113 | ... | 3536 | hits |
|---|---|---|---|---|---|---|---|---|---|---|
| | {23} | {19} | {11} | {10} | {10} | {10} | {10} | ... | {8} | |
| 1587.87 | * | 0.00 | * | * | * | * | * | ... | −0.14 | {4} |
| 1423.68 | 0.00 | * | * | −0.02 | * | * | * | ... | * | {6} |
| 1448.65 | −0.02 | * | * | * | * | * | * | ... | * | {6} |
| 1376.71 | * | 0.00 | * | * | * | * | * | ... | * | {4} |
| 1759.85 | * | 0.01 | * | * | * | * | * | ... | * | {6} |
| 1410.76 | * | * | * | * | * | * | * | ... | * | {2} |
| 1891.93 | * | 0.00 | * | * | * | * | * | ... | * | {1} |
| 2642.26 | 0.00 | * | * | * | 0.12 | * | * | ... | * | {2} |
| 1134.63 | 0.00 | * | * | * | * | −0.08 | * | ... | * | {4} |
| 1457.74 | * | −0.06 | * | −0.05 | 0.05 | * | * | ... | −0.02 | {10} |
| 1443.78 | −0.08 | 0.01 | −0.01 | −0.06 | * | * | * | ... | 0.04 | {9} |
| 1230.53 | 0.05 | * | * | 0.08 | * | * | * | ... | 0.10 | {8} |
| 1142.64 | * | 0.02 | −0.01 | * | −0.01 | * | * | ... | * | {10} |
| 1686.87 | * | −0.04 | * | * | * | * | −0.02 | ... | * | {4} |
| 1327.69 | 0.01 | * | * | * | * | 0.07 | * | ... | * | {6} |
| 1400.68 | * | −0.01 | * | 0.03 | * | * | * | ... | 0.07 | {6} |
| 1184.64 | * | * | * | * | 0.02 | * | * | ... | * | {2} |
| 2101.13 | −0.04 | * | −0.03 | * | * | 0.07 | * | ... | * | {8} |
| 813.41 | 0.02 | * | 0.08 | * | 0.02 | 0.08 | * | ... | 0.04 | {8} |
| 1221.60 | 0.00 | * | 0.10 | * | 0.08 | * | * | ... | * | {6} |
| 1717.83 | 0.11 | * | * | * | * | * | * | ... | * | {3} |
| 1827.88 | 0.05 | * | * | * | * | * | * | ... | * | {5} |
| 1247.55 | 0.08 | * | 0.10 | −0.06 | 0.15 | * | 0.11 | ... | * | {9} |
| 1710.77 | 0.04 | * | * | * | 0.12 | * | * | ... | * | {5} |
| 1851.06 | * | 0.02 | * | * | * | * | * | ... | * | {3} |
| 1657.85 | 0.02 | * | 0.06 | * | * | * | * | ... | 0.05 | {6} |
| 2098.95 | 0.04 | * | * | * | * | * | * | ... | * | {5} |
| 1806.81 | 0.01 | 0.02 | * | * | * | * | 0.11 | ... | * | {6} |

(B) Guanidinated sample (51 masses submitted).

| | 995 | 171 | 3569 | 3605 | 2449 | 502 | 503 | ... | 3452 | hits |
|---|---|---|---|---|---|---|---|---|---|---|
| | {27} | {20} | {15} | {13} | {11} | {10} | {10} | ... | {8} | |
| 1111.57 | −0.03 | * | * | * | * | * | * | ... | * | {7} |
| 1442.71 | * | −0.02 | * | 0.03 | −0.04 | * | 0.12 | ... | * | {10} |
| 1376.71 | 0.08 | 0.00 | * | * | * | * | * | ... | * | {8} |
| 1423.69 | −0.01 | * | * | * | * | * | * | ... | 0.05 | {10} |
| 1225.58 | * | 0.04 | * | * | * | * | * | ... | 0.00 | {7} |
| 1448.64 | −0.01 | * | * | * | * | * | * | ... | * | {9} |
| 1242.59 | * | 0.04 | 0.13 | 0.09 | * | * | * | ... | 0.09 | {11} |
| 1587.84 | * | 0.03 | * | * | * | * | * | ... | * | {10} |
| 969.58 | −0.10 | −0.06 | * | −0.02 | * | −0.05 | * | ... | * | {7} |
| 1129.57 | −0.04 | * | * | * | * | * | * | ... | * | {9} |
| 813.44 | −0.01 | * | * | −0.02 | 0.01 | * | * | ... | * | {11} |
| 1289.63 | 0.02 | * | 0.05 | * | * | * | * | ... | 0.01 | {9} |
| 1329.70 | 0.05 | * | * | * | * | * | 0.04 | ... | * | {8} |
| 1457.68 | * | 0.00 | * | 0.06 | * | * | 0.12 | ... | * | {9} |
| 1134.62 | 0.01 | * | * | * | * | * | * | ... | * | {7} |
| 1327.66 | 0.04 | * | * | 0.08 | * | 0.14 | * | ... | * | {10} |
| 1269.49 | * | * | * | 0.13 | * | * | * | ... | * | {2} |
| 940.56 | * | −0.03 | * | * | * | * | −0.01 | ... | * | {5} |
| 2642.28 | 0.02 | * | * | * | * | * | * | ... | * | {4} |

*Contd*

**Table 4.3** *Contd*

| | | | | | | | | | | |
|---|---|---|---|---|---|---|---|---|---|---|
| 1520.88 | * | −0.04 | * | −0.08 | −0.12 | * | * | ... | * | {11} |
| 948.57 | * | −0.05 | * | * | −0.03 | * | * | ... | * | {8} |
| 1230.55 | 0.03 | * | * | 0.05 | 0.10 | * | * | ... | * | {8} |
| 841.42 | 0.01 | * | 0.04 | * | * | * | * | ... | * | {8} |
| 1539.79 | * | −0.05 | 0.11 | * | * | * | * | ... | 0.00 | {6} |
| 1891.93 | * | 0.00 | * | * | * | * | * | ... | * | {4} |
| 1759.91 | 0.05 | −0.05 | * | * | * | * | 0.06 | ... | * | {10} |
| 1752.79 | 0.04 | * | * | * | * | 0.14 | * | ... | * | {6} |
| 1485.77 | −0.05 | * | 0.04 | * | * | * | −0.13 | ... | * | {7} |

(C) Consistent MHA using unmodified and guandinated data.

| | 995 {17} | 171 {11} | 1185 {7} | 713 {6} | 373 {5} | 701 {5} | 2367 {5} | ... | 3697 {4} | hits |
|---|---|---|---|---|---|---|---|---|---|---|
| 1587.87 | * | 0.00 | * | * | * | * | * | ... | * | {4} |
| 1423.68 | 0.00 | * | * | * | −0.02 | 0.04 | * | ... | * | {5} |
| 1448.65 | −0.02 | * | * | 0.10 | * | * | 0.08 | ... | * | {6} |
| 1376.71 | * | 0.00 | * | −0.06 | * | * | 0.01 | ... | * | {9} |
| 1759.85 | * | 0.01 | * | * | * | 0.07 | * | ... | * | {6} |
| 1891.93 | * | 0.00 | * | * | * | * | * | ... | * | {5} |
| 2642.26 | 0.00 | * | * | * | * | * | * | ... | * | {3} |
| 1134.63 | 0.00 | * | * | * | * | * | * | ... | * | {5} |
| 1457.74 | * | −0.06 | 0.02 | * | −0.05 | * | * | ... | * | {7} |
| 1443.78 | −0.08 | * | 0.00 | * | * | * | * | ... | * | {3} |
| 1230.53 | 0.05 | * | 0.09 | * | 0.13 | * | * | ... | * | {6} |
| 1142.64 | * | 0.02 | 0.00 | * | * | * | −0.03 | ... | * | {6} |
| 1686.87 | * | −0.04 | * | * | * | * | * | ... | * | {4} |
| 1327.69 | 0.01 | * | * | * | * | * | * | ... | 0.03 | {7} |
| 1400.68 | * | −0.01 | * | * | 0.03 | * | * | ... | * | {2} |
| 2101.13 | −0.04 | * | −0.11 | * | * | −0.13 | * | ... | * | {8} |
| 813.41 | 0.02 | * | 0.01 | 0.03 | * | * | * | ... | * | {7} |
| 1221.60 | 0.00 | * | * | 0.01 | * | * | * | ... | * | {7} |
| 1717.83 | 0.11 | * | * | * | * | * | 0.04 | ... | * | {5} |
| 1827.88 | 0.05 | * | * | 0.07 | * | * | * | ... | * | {6} |
| 1247.55 | 0.08 | * | * | * | * | * | * | ... | * | {3} |
| 1710.77 | 0.04 | * | * | * | * | * | * | ... | * | {1} |
| 1657.85 | 0.02 | * | * | * | * | * | * | ... | −0.01 | {3} |

consistent MHAs appear in Tables 4.4A and 4.4B, respectively. While ORFs 792, 793 and 1461 appear to match substantially more masses than all others, their relative merits can hardly be distinguished. The similarity in the particular experimental masses that they match is obvious. Inspection of the characteristics of the proteins immediately reveals their remarkable homology. The three ORFs encode flagellin proteins FljM, FljN and FljK with masses of 27 927 Da, 27 926 Da, and 27 999 Da, respectively. The three exhibit between 79% and 82% sequence identity (evaluated by comparing them two at a time). While unique peaks matching ORFs 792 and 1461 appear in the two MHAs, and no unique peaks are assigned to ORF 793, it would be difficult to rule out any of these three proteins.

Although the data will not be presented, the use of guanidination and esterification is helpful in ruling out a fourth ORF from being assigned to this gel band. ORF 88 matches nine observed masses in the unmodified master hit array and eight of the nine are *not* shared by masses assigned to ORFs 792, 793, and 1461. From this observation alone,

**Table 4.4** Consistent master hit arrays (MHAs) derived from the mass spectra of gel band 17.

(A) Guanidinated consistent MHA.

| | 792 | 1461 | 793 | 2440 | 536 | 794 | 1460 | ... | 3549 | hits |
|---|---|---|---|---|---|---|---|---|---|---|
| | {10} | {10} | {7} | {4} | {3} | {3} | {3} | ... | {3} | |
| 1424.79 | −0.09 | * | −0.09 | * | * | * | * | ... | * | {3} |
| 1452.79 | * | −0.09 | * | * | * | * | * | ... | 0.00 | {3} |
| 2005.09 | −0.03 | * | −0.03 | * | * | −0.03 | * | ... | * | {6} |
| 1914.00 | −0.02 | −0.02 | −0.02 | * | * | * | −0.02 | ... | * | {4} |
| 2059.12 | −0.03 | −0.03 | −0.03 | * | * | −0.03 | * | ... | −0.12 | {7} |
| 2447.37 | −0.04 | −0.04 | * | 0.04 | 0.05 | * | * | ... | * | {6} |
| 2003.12 | * | −0.04 | * | * | * | * | * | ... | * | {1} |
| 839.40 | −0.01 | −0.01 | −0.01 | * | * | * | * | ... | * | {4} |
| 1224.67 | * | * | * | 0.00 | −0.07 | * | * | ... | * | {2} |
| 1609.91 | * | * | * | −0.02 | −0.11 | * | * | ... | * | {4} |
| 2430.36 | −0.06 | −0.06 | * | 0.02 | * | * | * | ... | −0.02 | {6} |
| 1308.65 | * | −0.03 | * | * | * | * | * | ... | * | {1} |
| 817.44 | 0.04 | 0.04 | 0.04 | * | * | 0.04 | 0.04 | ... | * | {7} |
| 1029.54 | * | * | * | * | * | * | * | ... | * | {2} |
| 1294.64 | −0.03 | * | * | * | * | * | * | ... | * | {2} |
| 1794.98 | −0.06 | −0.06 | −0.06 | * | * | * | −0.06 | ... | * | {6} |

(B) Esterified consistent MHA.

| | 792 | 793 | 1461 | 373 | 794 | 922 | 408 | ... | 3549 | hits |
|---|---|---|---|---|---|---|---|---|---|---|
| | {6} | {6} | {5} | {4} | {3} | {3} | {2} | ... | {2} | |
| 1424.70 | 0.00 | 0.00 | * | * | * | −0.02 | * | ... | * | {7} |
| 1452.67 | * | * | 0.03 | 0.11 | * | 0.14 | * | ... | * | {6} |
| 1913.99 | −0.01 | −0.01 | −0.01 | * | * | * | * | ... | * | {5} |
| 2005.05 | 0.01 | 0.01 | * | * | 0.01 | * | * | ... | * | {7} |
| 839.36 | 0.03 | 0.03 | 0.03 | * | * | * | * | ... | * | {4} |
| 2059.08 | 0.01 | 0.01 | 0.01 | * | 0.01 | * | * | ... | 0.02 | {6} |
| 817.47 | 0.01 | 0.01 | 0.01 | 0.01 | 0.01 | * | −0.04 | ... | −0.05 | {13} |
| 1033.44 | * | * | * | 0.08 | * | * | 0.10 | ... | * | {3} |
| 1936.02 | * | * | * | * | * | * | * | ... | * | {0} |
| 1294.62 | * | * | * | * | * | * | * | ... | * | {0} |
| 2154.09 | * | * | * | * | * | 0.06 | * | ... | * | {2} |
| 1895.93 | * | * | * | * | * | * | * | ... | * | {0} |
| 1179.66 | * | * | * | 0.01 | * | * | * | ... | * | {4} |
| 1577.80 | * | * | * | * | * | * | * | ... | * | {1} |
| 2126.20 | * | * | * | * | * | * | * | ... | * | {1} |

at one point it seemed quite likely that ORF 88 was also associated with this gel band. However, ORF 88 did not appear in either the guanidinated or esterified MHAs, nor in either consistent MHA. Since ORF 88 corresponds to a large 176 kDa protein, with 748 different predicted tryptic peptides (based on the cleavage rules and parameters noted above) it represents a great source of random mass matches that the consistent MHAs effectively weed out.

### 4.3.5 *Advantages of esterification*

While guanidination generally appears to be more effective for generating consistent masses than esterification, gel band 25 represents a fine counterexample. Guanidinated and esterfied consistent MHAs are displayed in Tables 4.5A and 4.5B, respectively.

**Table 4.5** Consistent master hit arrays (MHAs) derived from the mass spectra of gel band 25.

(A) Guandinated consistent MHA.

| | 362 {5} | 695 {4} | 3045 {4} | 12 {3} | 26 {3} | 88 {3} | 183 {3} | . . . | 3313 {3} | hits |
|---|---|---|---|---|---|---|---|---|---|---|
| 1410.83 | * | * | * | * | −0.07 | * | * | . . . | −0.08 | {5} |
| 2129.13 | −0.14 | * | * | * | * | * | * | . . . | * | {3} |
| 1938.21 | −0.13 | * | −0.13 | * | * | * | * | . . . | * | {2} |
| 1258.74 | * | * | * | * | * | * | −0.06 | . . . | * | {5} |
| 1529.84 | * | * | * | * | −0.03 | * | −0.02 | . . . | * | {4} |
| 1129.69 | * | * | * | * | −0.03 | * | * | . . . | * | {8} |
| 1908.07 | −0.14 | * | * | * | * | 0.03 | −0.02 | . . . | * | {5} |
| 1363.71 | −0.06 | * | 0.06 | * | * | * | * | . . . | * | {7} |
| 1329.76 | * | * | −0.04 | * | * | * | * | . . . | * | {8} |
| 1724.98 | * | * | * | * | * | * | * | . . . | * | {1} |
| 1373.85 | * | −0.12 | * | * | * | −0.12 | * | . . . | * | {4} |
| 1333.78 | * | * | −0.08 | −0.04 | * | * | * | . . . | −0.08 | {6} |
| 1900.06 | * | −0.12 | * | * | * | * | * | . . . | * | {1} |
| 1550.86 | * | −0.03 | * | 0.04 | * | * | * | . . . | * | {5} |
| 1187.54 | −0.02 | 0.11 | * | 0.11 | * | * | * | . . . | * | {3} |
| 1617.84 | * | * | * | * | * | −0.09 | * | . . . | 0.06 | {6} |

(B) Esterified consistent MHA.

| | 362 {8} | 1063 {3} | 2872 {3} | 21 {2} | 42 {2} | 110 {2} | 386 {2} | . . . | 3605 {2} | hits |
|---|---|---|---|---|---|---|---|---|---|---|
| 2129.04 | −0.05 | * | * | * | * | * | * | . . . | * | {3} |
| 1529.77 | * | * | * | * | * | 0.04 | * | . . . | * | {4} |
| 1938.10 | −0.02 | −0.07 | * | −0.15 | * | * | −0.04 | . . . | * | {5} |
| 1129.70 | * | −0.02 | * | * | * | * | −0.09 | . . . | * | {6} |
| 1363.65 | * | * | * | * | * | * | * | . . . | * | {0} |
| 986.58 | −0.02 | −0.05 | −0.03 | * | * | * | * | . . . | −0.02 | {9} |
| 862.42 | * | * | * | * | * | * | * | . . . | * | {3} |
| 832.34 | * | * | * | * | 0.06 | * | * | . . . | * | {1} |
| 1994.08 | * | * | * | * | * | * | * | . . . | * | {3} |
| 1201.67 | 0.00 | * | * | * | * | * | * | . . . | * | {6} |
| 2112.03 | * | * | * | * | * | * | * | . . . | * | {2} |
| 894.42 | * | * | 0.02 | * | * | * | * | . . . | * | {3} |
| 1117.56 | * | * | * | 0.09 | * | * | * | . . . | * | {3} |
| 2185.07 | * | * | * | * | * | * | * | . . . | * | {0} |
| 1079.55 | 0.02 | * | * | * | 0.03 | * | * | . . . | 0.03 | {5} |
| 974.50 | 0.03 | * | −0.01 | * | * | * | * | . . . | * | {8} |
| 958.50 | * | * | * | * | * | * | * | . . . | * | {1} |
| 1347.63 | 0.03 | * | * | * | * | 0.01 | * | . . . | * | {5} |
| 1106.57 | * | * | * | * | * | * | * | . . . | * | {3} |
| 2046.12 | −0.05 | * | * | * | * | * | * | . . . | * | {2} |

Five of the 16 guanidinated consistent masses match ORF 362, which corresponds to a phosphonate ABC transporter; however, this is hardly more than the four matches that ORFs 695 and 3045 provide. Based on our previous statistical analysis, the probability that this assignment is correct is about 95%. In contrast, eight of the 20 esterified consistent masses match ORF 362, while the next likeliest alternatives, ORFs 1063 and 2872, match only three experimental masses. Without a simulation of the esterification experiment we cannot specify the probability that this is correct. The fact that the two MHAs lead to the same conclusion instills the most confidence in this assignment.

**Table 4.6** Consistent master hit arrays (MHAs) derived from the mass spectra of gel band 4.

(A) Guanidianted consistent MHA.

| | 1007 | 3032 | 502 | 619 | 1385 | 1478 | 1812 | ... | 3569 | hits |
|---|---|---|---|---|---|---|---|---|---|---|
| | {4} | {4} | {3} | {3} | {3} | {3} | {3} | ... | {2} | |
| 1410.81 | −0.07 | −0.03 | * | * | * | 0.03 | * | ... | * | {16} |
| 764.35 | 0.02 | * | 0.08 | * | * | * | * | ... | * | {15} |
| 1899.97 | * | * | 0.12 | 0.09 | * | * | 0.12 | ... | * | {15} |
| 712.22 | * | * | * | * | * | * | * | ... | * | {7} |
| 1367.71 | * | * | * | 0.02 | 0.00 | * | 0.09 | ... | −0.08 | {18} |
| 1617.83 | 0.02 | 0.08 | * | * | * | * | * | ... | * | {9} |
| 1393.77 | −0.06 | −0.07 | * | * | * | −0.04 | −0.03 | ... | * | {18} |
| 1094.58 | * | −0.07 | 0.04 | −0.03 | 0.00 | * | * | ... | * | {21} |
| 1959.98 | * | * | * | * | * | * | * | ... | 0.06 | {9} |
| 1950.91 | * | * | * | * | 0.07 | 0.08 | * | ... | * | {8} |
| 1745.83 | * | * | * | * | * | * | * | ... | * | {8} |

(B) Esterified consistent MHA.

| | 1007 | 502 | 3032 | 26 | 543 | 562 | 568 | ... | 3597 | hits |
|---|---|---|---|---|---|---|---|---|---|---|
| | {4} | {3} | {3} | {2} | {2} | {2} | {2} | ... | {2} | |
| 1410.76 | −0.02 | * | 0.02 | 0.00 | 0.05 | * | * | ... | * | {8} |
| 1094.57 | * | 0.05 | * | −0.05 | * | * | * | ... | 0.07 | {6} |
| 1393.72 | −0.01 | * | −0.02 | * | * | 0.09 | 0.03 | ... | 0.06 | {9} |
| 1899.99 | * | 0.10 | * | * | * | * | * | ... | * | {3} |
| 764.37 | 0.00 | 0.06 | * | * | * | 0.04 | * | ... | * | {9} |
| 1617.87 | −0.02 | * | * | * | * | * | * | ... | * | {3} |
| 1676.79 | * | * | * | * | * | * | * | ... | * | {0} |
| 1106.55 | * | * | * | * | −0.02 | * | 0.03 | ... | * | {11} |
| 1465.72 | * | * | * | * | * | * | * | ... | * | {2} |
| 1437.72 | * | * | −0.01 | * | * | * | * | ... | * | {1} |
| 1449.71 | * | * | * | * | * | * | * | ... | * | {0} |
| 1922.01 | * | * | * | * | * | * | * | ... | * | {3} |
| 1037.48 | * | * | * | * | * | * | * | ... | * | {0} |
| 1545.87 | * | * | * | * | * | * | * | ... | * | {3} |

### 4.3.6 *Inadequacy of peptide mass mapping*

While the darkness of a gel band correlates fairly well with the ease with which the proteins in it can be identified, disappointing mass spectra occasionally arise from promising-looking gel bands. Gel band 4 provided an example of this. Though it is a very dark band, its rather unconvincing guanidinated and esterified consistent master hit arrays appear in Tables 4.6A and 4.6B. While ORF 1007, corresponding to S-layer protein RsaA, is the 'winning' ORF in both cases, ORFs 3032 and 502 are either tied or very close behind 1007 in *both* MHAs. Peptide mass mapping simply does not yield a definitive conclusion in this case.

The best way to resolve such an ambiguous situation is through a different type of experiment. MALDI-TOF/TOF provides the capability of fragmenting selected peptide ions to obtain unique structural information that can, in turn, lead to credible peptide identifications. The MALDI-TOF/TOF spectrum of the most intense peak listed in Tables 4.6A and 4.6B (1410.8 Da) appears in Figure 4.2. The spectrum is assigned to peptide QANIDYLTAFVR that is derived from ORF 1007. Interestingly, peptide ions having masses of 1393.7 and 1617.8 Da were also analyzed with the TOF/TOF

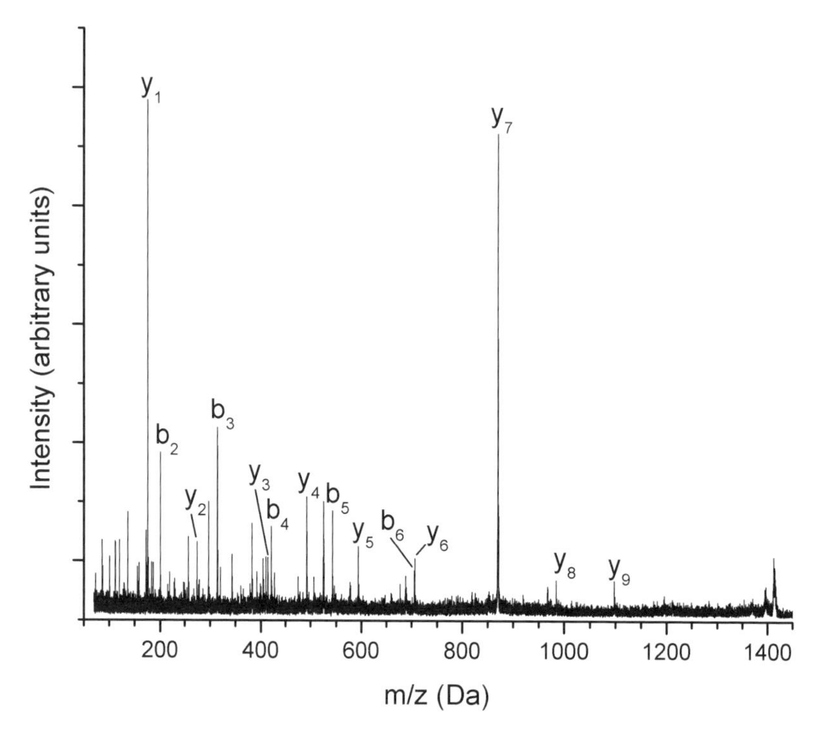


**Fig. 4.2** MALDI-TOF/TOF mass spectrum of 1410.76 Da peak from gel band 4. b and y notation corresponds to peptide QANIDYLTAFVR.

instrument. Nevertheless, both ion fragmentation spectra matched peptides from ORF 1007 rather than peptides from ORF 3032. Thus the ambiguity reflected in Table 4.6 is resolved. In conclusion, there are certainly cases where peptide mass mapping suggests a certain protein assignment for a gel band, but additional experiments are needed to provide authentication.

### 4.3.7 *Sample preparation issues*

A question that often arises in MS experiments on protein digests is the extent to which samples should be purified. A simple and popular approach for removing some undesirable impurities from digests is microcolumn extraction through $C_{18}$ resin, sold commercially as ZipTips. Figure 4.3 displays mass spectra of gel band 34 that were recorded following two sample handling procedures. In Fig. 4.3A the sample was not passed through a ZipTip but in Fig. 4.3B it was. The observed results are rather typical for this procedure: smaller peptides appear to be unaffected or enhanced by this procedure while larger peptides generally suffer a loss in intensity. Similar effects have been previously reported [21,37,45]. These intensity variations serve as a reminder that peak intensities in MALDI mass spectra are generally difficult to interpret. Intensities are usually related to sample concentration but they can be strongly

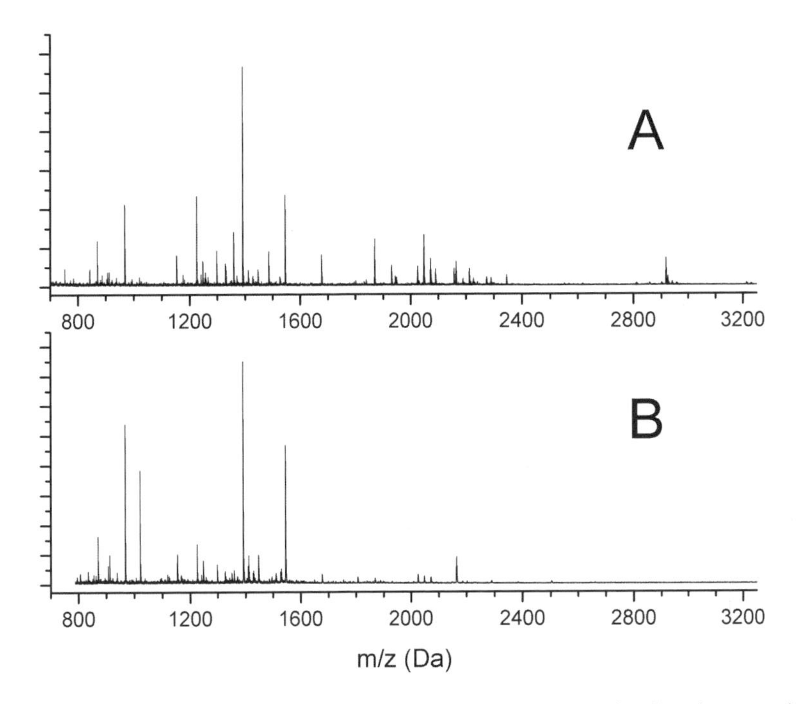


**Fig. 4.3** Mass spectra from gel band 34 peptide extracts (A) before and (B) after $C_{18}$ microextraction.

affected by sample preparation procedures and the ionizing propensities of different molecules.

### 4.3.8 *Correlation between theoretical and experimental data*

The correspondence between simulated and experimental data is as follows. The number of 'real' peptides in a simulated mass spectrum corresponds to the number of experimentally measured masses that match the 'winning' ORF. The number of 'random' masses in a simulated spectrum corresponds to the number of experimentally measured masses that do not match the 'winning' ORF. For these gel bands an average experimental unguanidinated mass spectrum contained 29 masses. For this number of masses, at least 11 matches (11 'real' peptides and 18 'random' peptides) to the 'winning' ORF are required to make an assignment with 95% confidence, and at least 13 matches are needed for 99% confidence. The average experimental guanidinated mass spectrum contained 32 masses; 11 matches are required to make an identification with 95% confidence, and 13 matches for 99% confidence. The average experimental consistent MHA obtained during the stalk gel analyses had 19 masses. Perusal of Figure 5 of Karty *et al.* [32] reveals that only six consistent matches are required for 95% confidence and eight for 99% confidence. Simulations demonstrated that spot identifications based on consistent mass spectral features are less affected by the presence of random data and they require fewer matches. This is especially helpful for analyzing faint gel spots that typically yield fewer mass spectral peaks.

## 4.4 Discussion

### 4.4.1 *Overview of peptide mass mapping data*

Following the analysis of 34 gel bands, 14 distinct proteins were identified. They include eight different *tonB* dependent receptors, one *ompA* family protein, three flagellins, and two proteins annotated as hypothetical. In these experiments, we observed 29 masses per unguanidinated tryptic digest spectrum. Of those 29 measured masses, on average, 15 matched the 'winning' ORF. For the guanidinated data, there were on average 16 'winning' ORF matches for 32 observed masses, while for esterified data these numbers were 13 and 57, respectively. When guanidinated and esterified consistent masses were considered, the percentage of those matched to the 'winning' ORF increased to 58% (11/19) and 33% (8/24), respectively. Note that these percentages are artificially slightly low because they are based on the assumption that each band can be correctly associated with only a single ORF, and this is clearly incorrect.

There is usually a strong correlation between the intensity of a gel band and our ability to identify a protein in it. Of the six spots in which no proteins were identified by MALDI-TOF MS, two were very faint. Most of the gel spots that were not identifiable had few (3–10) peaks in their digest mass spectra. Poor enzyme digestion efficiency at low substrate concentrations is a well known problem [23,45,46], and could be the cause of these uninformative mass spectra.

A total of 35 proteins (although only 14 different ones) were identified in the gel bands in this study. Of those 35 identifications, all but 13 could have been made at the 99% confidence level using *only* unguanidinated *or* guanidinated data. In 11 cases a 90% confidence identification was possible only after both unguanidinated and guanidinated data were interpreted using the PRODIGIES algorithm. There were two cases in which the consistent MHA was less definitive than either the unguanidinated or the guanidinated MHAs. In one of these, the guanidinated sample yielded no mass spectral peaks, suggesting a sample handling error. In the other case, the consistent MHA provided the same identification but with a slightly reduced confidence of 98%. This resulted because two proteins were present in the gel band, apparently leading to some signal discrimination in one of the spectra. The lesson evident from this is that if the signal to noise ratio of either the derivatized or the underivatized sample is inferior for any reason, the process of observing and interpreting consistent masses will suffer. Although esterification data contributed positively to the identification of a protein in gel band 25, lack of complete reaction together with the occurrence of amide solvolysis limited the effectiveness of this procedure and its ultimate potential is not encouraging. For this reason, a statistical simulation of the effect of esterification on peptide mass mapping has not been performed, and quantitative confidence levels associated with interpretations based on esterification data are not available.

### 4.4.2 *Problems with esterification for routine digest derivatizations*

Two of the more important characteristics of a peptide modification reaction involve its completeness and whether or not unintended side reactions occur. Ideally, a peptide modification reaction is quantitative and specific: all residues of a particular type are modified, and no other reactions occur. Beardsley and Reilly [37] have observed complete guanidination of lysine residues in as few as 5 min. Partial guanidination of peptide N-terminal glycine residues and deamidation of aspargines that are N-terminal to glycine residues are minor side reactions [37,47].

Unfortunately, the instances of incomplete esterification were not isolated to the gel band 33 analysis that was discussed above. In fact, evidence of incomplete esterification appeared in many mass spectra. In spectra of 35 gel bands, we found 24 such cases. Incomplete esterification creates two major problems. The first is a reduction in sensitivity because the intensity arising from a particular peptide is divided between two or three masses. The second arises during data analysis. An incompletely esterified peptide will cause an incorrect number of acidic groups to be inferred during the interpretation of consistent masses. This leads to false positives in the consistent MHA. As an example of this, peak 1 in Fig. 4.1A has a mass of 1544.83 Da. After esterification, peaks are observed at both 1628.55 Da (implying two acidic residues) and 1656.98 Da (implying three acidic residues). In fact, the ORF 163 peptide (ERPRPDYEAVGQK) contains three acidic residues. In the consistent MHA (Table 4.1E), ORF 465 is found also to have a consistent peptide at 1544.83 Da. Unfortunately, the sequence of the peptide (VLNTDGSEQRGGRR) contains only two acidic residues. Its appearance in the consistent MHA results from the inference of an incorrect number of acidic residues caused by the observation of incomplete esterification.

Under the acidic conditions used for the esterification derivatization, amide groups, not just carboxylic acids, can be replaced by esters in a process called solvolysis [44]. This serious problem was observed in several of the esterified mass spectra. Overall, in the analysis of 35 gel bands, we observed 72 cases of N or Q solvolysis and three cases of double solvolysis (solvolysis of two N/Q residues in the same peptide). This side reaction creates even more problems than incomplete esterification of acidic residues. First of all, solvolysis under acidic conditions proceeds by a different mechanism than under neutral or basic conditions, and the former is 2–3 orders of magnitude slower [48,49]. The process is not quantitative under the conditions that we used for esterification, leading to considerable spectral complexity and confusion in identifying and interpreting consistent masses. Second, the masses corresponding to esterified amides will *not* be included in the table of consistent masses (they are shifted by 29.02 Da, not 28.03 Da), thus confounding analysis. Conceivably, one could extend the reaction time significantly to allow all reactions to run to completion [44].

If a side reaction affects a minor component of a sample, it can often be ignored. Conversely, if a side reaction occurs quantitatively, it can be taken into account during data interpretation. Unfortunately, the esterification side reactions meet neither of these criteria. N and Q residues make up 5.6% of all amino acids in the translated proteome of *C. crescentus*. This suggests that at least one N or Q will be found in a majority of *C. crescentus* tryptic peptides. Since the reaction is not quantitative, it cannot be easily taken into account during automated data analysis. In fact, the problem with many side reactions is that they are easy to recognize only *after* the protein has been identified. In the case of unknown proteins, side reaction products just confound the analysis. Since esterification at acidic groups is not complete and can lead to the modification of amide-containing residues, it should not be used as a general technique for identifying unknown proteins in gel bands; however, it can be quite useful to confirm specific identifications or aid in sequencing particular peptides.

### 4.4.3 *Sources of chemical noise*

There are many sources of chemical noise (real mass spectral peaks that do not correspond to predicted tryptic fragments of *C. crescentus* proteins) in these peptide mass mapping experiments. By increasing the number of unmatched masses, chemical noise decreases the confidence of assignments. Chemical noise sources such as tryptic autolysis peaks or human keratin peptides are quite well known and can be removed from mass lists before analysis [50–53]. Clusters of alkali metals and matrix molecules are less familiar, but must also be taken into account or the background chemical noise will increase [53,54]. We routinely filter out most of our matrix cluster ions before they are considered by PRODIGIES by taking into account their unusual mass defects. $C_{18}$ microextraction effectively removes alkali ions and thereby reduces the contribution of these alkali/matrix clusters. Other sources of unmatched peaks include: errors in the proteome [55–57], non-tryptic cleavages [58–60], and undocumented post-translational modifications [48,49,61–79]. We recently presented a more detailed study of chemical noise sources in peptide mass mapping experiments, and a more complete interpretation of the unmatched masses that appear in these types of experiments [47].

### 4.4.4 *Multiple proteins in a gel spot*

When complex cellular lysates containing hundreds or thousands of proteins are separated on 2D gels, co-migrating proteins are occasionally problematic [23,26,80]. The situation is bound to become more serious when only 1D PAGE is used to separate proteins. As clearly demonstrated in the analysis of gel band 5, two proteins present in high abundance can easily be detected by peptide mass mapping. However, a more likely scenario would be having a plentiful protein present with one or more scarce proteins. Because the latter do not contribute a significant number of matches to the MHA, they are rarely identified. The number of peptide mass matches associated with the low-abundance proteins is usually comparable to the number of fortuitous matches with fragments from random proteins.

The inherent weakness of peptide mass mapping is that several unique peptide masses are required to identify a protein. When the number of experimental masses matched to any single ORF is inadequate, or the number of noise sources is too large, peptide mass mapping fails to be definitive. Scoring algorithms that weight the number of matches by the size of the protein (and effectively the number of potential tryptic fragments) can reduce this problem somewhat [21], but they do not eliminate it. Guanidination or esterification improves the situation by decreasing the number of matches required to make identifications and increasing peptide mass mapping's tolerance of chemical noise. Nevertheless, the best way to identify a protein when too few peptides are observed is through sequencing using some form of tandem mass spectrometry [25,51,57,79,81].

### 4.4.5 *Effect of guanidination and esterification on sensitivity*

Although we have previously demonstrated that guanidination improves MALDI ionization efficiencies, guanidinated spectra sometimes contain fewer peaks than their unguanidinated counterparts. We have attributed the loss of high-mass peptides following guanidination [29] to the purification step required to remove the *O*-methylisourea from the peptide sample. Stewart *et al.* [45] quantitated many forms of sample handling losses using $^{18}O$ labeling and isotope ratio mass spectrometry. They reported signal losses of 30–90% when 1 pmol to 50 fmol of an albumin digest were purified with a Millipore $\mu C_{18}$ ZipTip. Unfortunately, 8 mol/kg *O*-methylisourea is used to make guanidination complete and fast. Our laboratory is currently investigating other derivatization protocols that may not require post-derivatization purification.

The efficacy of guanidination for improving the detection of lysine-terminated tryptic peptides in MALDI experiments is well established [28–30,37]. These experiments provided another opportunity for observing this phenomenon. In the analysis of underivatized peptides, 29.1% of all matches involved lysine-terminated peptides. Likewise among the five most intense matched peaks in each mass spectrum, 22.5% of the peaks involved lysine-terminated peptides. After guanidination, 47.4% of all matches and 50.3% of the five most intense matched peaks in each spectrum corresponded to lysine-terminated peptides. Esterification was certainly not expected to affect the partitioning of signal between lysine- and arginine-terminated peptides. Indeed, we found that after esterification, 31.0% of all matches and 28.7% of the five most intense matched peaks in each spectrum corresponded to lysine-terminated peptides. A complication associated with derivatizations in general is that some previously non-observable peptides are observed after the reaction. More surprisingly, some are observed before and not after the reaction. Derivatizations, particularly guanidination, affect the ionizability of peptides, and because peptides seem to compete with each other for available protons in a MALDI plume, what is good for one peptide may be bad for another. The additional sample handling associated with derivatized peptides, especially when $C_{18}$ microextraction is required to clean samples up, can reduce the sensitivity for detecting guanidinated peptides. This effect appears to be particularly noticeable for large peptides [29,37].

In contrast with guanidination, the effect of esterification on peptide ionization efficiency is not obvious or consistent. Some peptides ionize better, some less well, after esterification [82]. This effect is quite reasonable, because esterification does not introduce a proton-attracting basic moiety. Rather, it changes a peptide's polarity and hydrophobicity. MALDI appears to be quite sensitive to these molecular characteristics, and it is not surprising that for some peptides the ionization yield increases while for others it decreases after esterification. Unfortunately, this characteristic is not optimal for peptide mass mapping especially in terms of 'consistent' mass production. Esterified peptides that do not ionize well may not be detected. Consequently the masses never appear in the consistent MHA and these peptides are effectively discriminated against just as if they derived from random chemical noise. The necessity to clean up esterified samples by $C_{18}$ microextraction undoubtedly has a negative impact on their MALDI signals. However, the weakest characteristic of the esterification experiment is the incomplete derivatization and the unexpected amide solvolysis discussed above.

### 4.4.6 *Peptide ion fragmentation*

As demonstrated for the 1410.8 Da peptide in gel band 4, MS/MS ion fragmentation data can be quite definitive when peptide mass mapping leads to ambiguous results. Unfortunately, these fragmentation experiments and their interpretation take considerable time. Desirable improvements in the technologies would include faster data acquisition and methods of data analysis that are more efficient than current database matching algorithms. *De novo* sequencing, for example, would be an attractive alternative. One promising way to accomplish this involves the derivatization of peptide N-termini using sulfonic acid labels [83]. With this approach, Keough *et al.* [84] have demonstrated the ability to fragment tryptic peptide ions and produce contiguous series of y-type fragment ions using a variety of different tandem mass spectrometers. The rationale behind this phenomenon is easily understood. A strongly acidic tag introduces a negative charge at the N-terminus. Creation of a positively charged peptide ion thus requires the binding of *two* protons. One of them is expected to be sequestered at the basic C-terminal arginine or lysine. The second is expected to be mobile and thus capable of attaching itself to any amide on the peptide backbone. Protonated amides lead to weakened peptide bonds, thereby facilitating the production of a contiguous series of y-type ions. Keough *et al.* [84] have demonstrated the value of these contiguous ion series for both the identification of proteins and the *de novo* sequencing of peptides.

### 4.4.7 *Comparison of* Prodigies *to other mass mapping algorithms*

Many of the current publicly available web-based peptide mass mapping programs are not configured to exploit fully the information gained from guanidination or esterification derivatizations. ProFound [40] (www.prowl.rockefeller.edu/cgi-bin/Profound), PeptIdent [43] (kr.expasy.org/tools/peptident.html), ms-fit [42] (prospector.ucsf.edu), and Mascot [5] (www.matrixscience.com) only allow comparison of a single set of mass spectral data to a publicly available proteome, such as SWISS-PROT (www.expasy.ch) or NCBI (www.ncbi.nih.gov). Most of these programs allow for covalent modification of specific types of residues, although PeptIdent does not. PepMapper5 [41] (wolf.bms.umist.ac.uk/mapper/mapper5.html) takes into account guanidination and allows for up to three datasets to be compared, but it does not consider esterification. PepMapper1 [41] (wolfs.bms.umist.ac.uk/mapper/mapper1.html) allows for methyl esterification but only analyzes one dataset; it also restricts interpretations to only publicly available proteomes. The proteomes of some organisms are not currently accessible to any of the above-mentioned free, web-based programs. Prodigies can use any FASTA formatted proteome file. The file can contain either a whole organism proteome, or a smaller subset of proteins of interest. Each of the programs offers some statistical ranking method associated with different interpretations, but it can be difficult to extract absolute confidence limits associated with assignments. The statistical simulations of Prodigies can be regenerated for any set of conditions. Based on the time spent for the simulation of numerous sets of 'real' and 'random' peptides, a simulation of 10 000 database searches takes about an hour. If the conditions of analysis are constant (same modifications, number of missed cleavages, mass error, etc.), then the statistical analysis may be run once and the results used as a reference table for future assignments.

## 4.5 Conclusions

In general, guanidination increases the efficacy of peptide mass mapping. Less often, esterification can play a similar role. By exploiting the lysine count obtained by comparing unguanidinated and guanidinated data as a search parameter in peptide mass mapping experiments, the number of sub-99% confidence ORF assignments can be cut by half. Simulations demonstrate that analysis of consistent data requires one-third fewer matches than analysis of unguanidinated or guanidinated data alone [32]. The contribution of esterification to the problem of identifying gel bands appears to be somewhat disappointing as it provides only a small increment to our capabilities. Because of this, theoretical simulations to establish statistical confidence of our esterification-based assignments have not yet been performed. In the membrane protein study described in this chapter, proteins were identified in 29 out of 34 gel bands examined, and 14% of those identifications were made possible by guanidination. Fourteen unique proteins were identified in the 34 gel bands including two hypothetical or conserved hypothetical proteins of unknown function, eight *tonB* dependent receptors, and one *ompA* family protein.

## Acknowledgements

Our proteomics methodology developments have been supported by grants from the National Institutes of Health (grant GM-61336) and the National Science Foundation (grant 0094579).

We wish to thank Marcia Ireland and Dr Yves Brun for providing the 1D gel-separated *Caulobacter crescentus* membrane protein sample that was mass spectrometrically analyzed.

## References

1. Aebersold, R. & Mann, M. (2003) Mass spectrometry-based proteomics, *Nature*, **422**, 198–207.
2. Mann, M., Hendrickson, R.C. & Pandey, A. (2001) Analysis of proteins and proteomes by mass spectrometry, *Annu. Rev. Biochem.*, **70**, 437–473.
3. Burlingame, A.L., Boyd, R.K. & Gaskell, S.J. (1998) Mass spectrometry, *Anal. Chem.*, **70**, 647R–717R.
4. Eng, J.K., McCormack, A.L. & Yates, J.R., III (1994) An approach to correlate tandem mass spectral data of peptides with amino acid sequences in a protein database, *J. Am. Soc. Mass Spectrom.*, **5**, 976–989.
5. Perkins, D.N., Pappin, D.J.C., Creasy, D.M. & Cottrell, J.S. (1999) Probability-based protein identification by searching databases using mass spectrometry data, *Electrophoresis*, **20**, 3551–3567.
6. Stimson, E., Truong, O., Richter, W.J., Waterfield, M.D. & Burlingame, A.L. (1997) Enhancement of charge remote fragmentation in protonated peptides by high-energy CID MALDI-TOF-MS using 'cold' matrices, *Int. J. Mass Spectrom. Ion Processes*, **169/170**, 231–240.
7. Schwartz, J.C., Kaiser, R.E., Jr., Cooks, R.G. & Savickas, P.J.A. (1990) Sector/ion trap hybrid mass spectrometer of BE/trap configuration, *Int. J. Mass Spectrom. Ion Processes*, **98**, 209–224.

8. Martin, S.A., Hill, J.A, Kittrell, C. & Biemann, K. (1990) Photon-induced dissociation with a four-sector tandem mass spectrometer, *J. Am. Soc. Mass Spectrom.*, **1**, 107–109.
9. Zubarev, R.A., Kelleher, N.L. & McLafferty, F.W. (1998) Electron capture dissociation of multiply charged protein cations, a nonergodic process, *J. Am. Chem. Soc.*, **120**, 3265–3266.
10. Little, D.P., Speir, J.P., Senko, M.W., O'Connor, P.B. & McLafferty, F.W. (1944) Infrared multiphoton dissociation of large multiply charged ions for biomolecule sequencing, *Anal. Chem.*, **66**, 2809–2815.
11. Opticek, G.J., Ramirez, S.M., Jorgenson, J.W. & Moseley, M.A., III, (1998) Comprehensive two-dimensional high-performance liquid chromatography for the isolation of overexpressed proteins and proteome mapping, *Anal. Biochem.*, **258**, 349–361.
12. Wagner, K., Miliotis, T., Marko-Varga, G., Bischoff, R. & Unger, K.K. (2002) An automated on-line multidimensional HPLC system for protein and peptide mapping with integrated sample preparation, *Anal. Chem.*, **74**, 809–820.
13. Wang, H., Kachman, M. T., Schwartz, D.R., Cho, K.R. & Lubman, D.M. (2002) A protein molecular weight map of ES2 clear cell ovarian carcinoma cells using a two-dimensional liquid separations/mass mapping technique, *Electrophoresis,* **23**, 3168–3181.
14. Lubman, D.M., Kachman, M.T., Wang, H., Gong, S., Yan, F., Hamler, R.L., O'Neil, K.A., Zhu, K., Buchanan, N.S. & Barder, T.J. (2002) Two-dimensional liquid separations – mass mapping of proteins from human cancer lysates, *J. Chromatog. B*, **782**, 183–196.
15. Liu, H., Berger, S. J., Chakraborty, A. B., Plumb, R. S. & Cohen, S.A. (2002) Multidimensional chromatography coupled to electrospray ionization time-of-flight mass spectrometry as an alternative to two-dimensional gels for the identification and analysis of complex mixture of intact proteins, *J. Chromatog. B*, **782**, 267–289.
16. O'Farrell, P.H. (1975) High resolution two-dimensional electrophoresis of proteins, *J. Biol. Chem.*, **250**, 4007–4021.
17. Jungblut, P. & Thiede, B. (1997) Protein identification from 2-DE gels by MALDI mass spectrometry, *Mass Spectrom. Rev.*, **16**, 145–162.
18. Henzel, W.J., Billeci, T.M. Stults, J.T., Wong, S.C, Grimley, C. & Watanabe, C. (1993) Identifying proteins from two-dimensional gels by molecular mass searching of peptide fragments in protein sequence databases, *Proc. Natl Acad. Sci. USA,* **90**, 5011–5015.
19. James, P., Quadroni, M., Carafoli, E. & Gonnet, G. (1993) Protein identification by mass profile fingerprinting, *Biophys. Biochem. Res. Commun.*, **195**, 58–64.
20. Mann, M., Hojrup, P. & Roepstorff, P. (1993) Use of mass spectrometric molecular weight information to identify proteins in sequence databases, *Biol. Mass Spectrom.*, **22**, 338–345.
21. Pappin, D.J.C., Hojrup, P. & Bleasby, A. J. (1993) Rapid indentification of proteins by peptide-mass fingerprinting, *Curr. Biol.*, **3**, 327–332.
22. Yates, J.R., Speicher, S., Griffin, P.R. & Hunkapiller, T. (1993) Peptide mass maps: a highly informative approach to protein identification, *Anal. Biochem.*, **214**, 397–408.
23. Quadroni, M. & James, P. (1999) Proteomics and automation, *Electrophoresis,* **20**, 664–677.
24. Nordhoff, E., Egelhofer, V., Giavalisco, P., Eickhoff, H., Horn, M., Przewieslik, T., Theiss, D., Schneider, U., Lehrach, H. & Gobom, J. (2001) Large-gel two-dimensional electrophoresis – matrix assisted laser desorption/ionization–time of flight-mass spectrometry: an analytical challenge for studying complex protein mixtures, *Electrophoresis,* **22**, 2844–2855.
25. Jensen, O.N., Podtelejnikov, A.V. & Mann, M. (1997) Identification of the components of simple protein mixtures by high-accuracy peptide mass mapping and database searching, *Anal. Chem.*, **69**, 4741–4750.
26. Jensen, O.N., Larsen, M. R. & Roepstorff, P. (1998) Mass spectrometric identification and microcharacterization of proteins from electrophoretic gels: strategies and applications, *Proteins Structure Function Genet. Suppl.*, **2**, 74–89.
27. Krause, E., Wenschuh, H. & Jungblut, P.R. (1999) The dominance of arginine-containing peptides in MALDI-derived tryptic mass fingerprints of proteins, *Anal. Chem.*, **71**, 4160–4165.
28. Brancia, F.L., Oliver, S.G. & Gaskell, S.J. (2000) Improved matrix-assisted laser desorption/ionization mass spectrometric analysis of tryptic hydrolysates of proteins following guanidination of lysine-containing peptides, *Rapid Commun. Mass Spectrom.*, **14**, 2070–2073.
29. Beardsley, R.L., Karty, J. A. & Reilly, J. P. (2000) Enhancing the intensities of lysine-terminated tryptic peptide ions in matrix-assisted laser desorption/ionization mass spectrometry, *Rapid Commun. Mass Spectrom.*, **14**, 2147–2153.
30. Hale, J.E., Butler, J.P., Knierman, M.D. & Becker, G.W. (2000) Increased sensitivity of tryptic peptide detection by MALDI-TOF mass spectrometry is achieved by conversion of lysine to homoarginine, *Anal. Biochem.*, **287**, 110–117.
31. Keough, T., Lacey, M.P. & Youngquist, R.S. (2000) Derivatization procedures to facilitate de novo sequencing of lysine-terminated tryptic peptides using postsource decay matrix-assisted laser desorption/ionization mass spectrometry, *Rapid Commun. Mass Spectrom.*, **14**, 2348–2356.

32. Karty, J.A., Ireland, M.M.E., Brun, Y.V. & Reilly, J.P. (2002) Defining absolute confidence limits in the identification of *Caulobacter* proteins by peptide mass mapping. *J. Proteome Res.*, **1**, 325–335.
33. Brun, Y.V. & Janakiraman, R. (2000) The dimorphic life cycle of *Caulobacter* and stalked bacteria, in *Prokaryotic Development* (eds. Y.V. Brun & L. J. Shimkets), pp. 297–317. ASM Press, Washington, DC.
34. Nierman, W.C., Feldblyum, T.V., Laub, M.T., Paulsen, I.T., Nelson, K.E., Eisen, J.A., Heidelberg, J.F., Alley, M.R., Ohta, N., Maddock, J.R. & 27 other authors. (2001) Complete genome sequence of *Caulobacter crescentus, Proc. Natl Acad. Sci. USA*, **98**, 4136–4141.
35. Grunenfelder, B., Rummel, G., Vohradsky, J., Roder, D., Langen, H. & Jenal, U. (2001) Proteomic analysis of the bacterial cell cycle, *Proc. Natl Acad. Sci. USA*, **98**, 4681–4686.
36. Fountoulakis, M. & Langen, H. (1997) Identification of proteins by matrix-assisted laser desorption ionization-mass spectrometry following in-gel digestion in low-salt, nonvolatile buffer and simplified peptide recovery, *Anal. Biochem.* , **250**, 153–156.
37. Beardsley, R.L. & Reilly, J.P. (2002) Optimization of guanidination procedures for MALDI mass mapping, *Anal. Chem.*, **74**, 1884–1890.
38. Kowalak, J.A. & Walsh, K.A. (1996) β-Methylthio-aspartic acid: Identification of a novel posttranslational modification in ribosomal protein S12 from *Escherichia coli, Protein Sci.*, **5**, 1625–1632.
39. Falick, A.M. & Maltby, D.A. (1989) Derivatization of hydrophilic peptides for liquid secondary ion mass spectrometry at the picomole level, *Anal. Biochem.*, **182**, 165–169.
40. Zhang, W. & Chait, B.T. (2002) ProFound: An expert system for protein identification using mass spectrometric peptide mapping information, *Anal. Chem.*, **72**, 2482–2489.
41. Sidhu, K.S., Sangvanich, P., Brancia, F.L., Sullivan, A.G., Gaskell, S.J., Wolkenhaue, O., Oliver, S.G. & Hubbard, S.J. (2001) Bioinformatic assessment of mass spectrometric chemical derivatization techniques for proteome database searching, *Proteomics*, **1**, 1368–1377.
42. Clauser, K.R., Baker, P.R. & Burlingame, A.L. (1994) Role of accurate mass measurement (±10 ppm) in protein identification strategies employing MS or MS/MS and database searching, *Anal. Chem.*, **71**, 2871–2882.
43. Wilkins, M.R. & Williams, K.L. (1997) Cross-species protein identification using amino acid composition, peptide mass fingerprinting, isoelectric point and molecular mass: a theoretical evaluation, *J. Theor. Biol.*, **186**, 7–15.
44. Young, M.A. & Desiderio, D.M. (1976) Detection of asparagine and glutamine in peptides sequenced by dipeptidyl aminopeptidase I via gas chromatography–mass spectrometry, *Anal. Biochem.*, **70**, 110–123.
45. Stewart, I.I., Thomson, T. & Figeys, D. (2001)$^{18}$O labeling: a tool for proteomics, *Rapid Commun. Mass Spectrom.*, **15**, 2456–2465.
46. Locke, S. & Figeys, D. (2000) Techniques for the optimization of proteomic strategies based on head column stacking capillary electrophoresis, *Anal. Chem.*, **72**, 2684–2689.
47. Karty, J.A., Ireland, M.M.E., Brun, Y.V. & Reilly, J.P. (2002) Artifacts and unassigned masses encountered in peptide mass mapping. *J. Chromatog. B*, **782**, 363–383.
48. Patel, K. & Borchardt, R.T. (1990) Chemical pathways of peptide degradation. II. Kinetics of deamidation of an asparaginyl residue in a model hexapeptide, *Pharm. Res.*, **7**, 703–711.
49. Patel, K. & Borchardt, R.T. (1990) Chemical pathways of peptide degradation. III. Effects of primary sequence of the pathways of deamidation of asparaginyl residues in hexapeptides, *Pharm. Res.*, **7**, 787–793.
50. Parker, K.C., Garrels, J.I., Hines, W., Butler E.M., McKee, A.H., Patterson, D. & Martin, S. (1998) Identification of yeast proteins from two-dimentional gels: working out spot cross-contamination, *Electrophoresis*, **19**, 1920–1932.
51. Andersen, J.S., Kuster, B., Podtelejnikoc, A., Morts, E. & Mann, M. (1999) Common peptide contaminants observed in low level sequencing of gel-separated proteins. In: *Proceedings of 47th ASMS Conference on Mass Spectrometry and Applied Topics,* Dallas, p. 405. American Society of Mass Spectometry, Dallas, Texas.
52. Vestling, M.M., Murphy, C.M. & Fenselau, C. (1990) Recognition of trypsin autolysis products by high-performance liquid chromatography and mass spectrometry, *Anal. Chem.*, **62**, 2391–2394.
53. Harris, W.A., Janecki, D.J. & Reilly, J.P. (2002) Use of matrix clusters and trypsin autolysis fragments as mass calibrants in matrix-assisted laser desorption/ionization time-of-flight mass spectrometry, *Rapid Commun. Mass Spectrom.*, **16**, 1714–1722.
54. Keller, B.O. & Li, L. (2000) Discerning matrix-cluster peaks in matrix-assisted laser desorption/ionization time-of-flight mass spectra of dilute peptide mixtures, *J. Am. Soc. Mass. Spectrom.*, **11**, 88–93.
55. Gibson, B.W. & Biemann, K. (1984) Strategy for the mass spectrometric verification and correction of the primary structures of proteins deduced from their DNA-sequences, *Proc. Natl Acad. Sci. USA*, **81**, 1956–1960.
56. Wassenberg, D., Wuhrer, M., Beaucamp, N., Schurig, H., Wozny, M., Reusch, D., Fabry, S. & Jaenicke, R. (2001) Local variability of the phosphoglycerate kinase–triosephosphate isomerase fusion protein from *Thermotoga maritima* MSB 8, *Biol. Chem.*, **382**, 693–697.

57. Belghazi, M., Bathany, K., Hountondji, C., Grandier-Vazeille, X., Manon, S. & Schmitter, J.M. (2001) Analysis of protein sequences and protein complexes by matrix-assisted laser desorption/ionization mass spectrometry, *Eur. J. Mass Spectrom.*, **7**, 101–109.
58. Smith, R.L. & Shaw, E. (1969) Pseudotrypsin: a modified bovine trypsin produced by limited autodigestion, *J. Biol. Chem.*, **244**, 4704–4712.
59. Keil-Dlouha, V., Zylber, N., Imhoff, J.M., Tong, N.T. & Keil, B. (1971) Proteolytic activity of pseudotrypsin, *FEBS Lett.*, **16**, 291–295.
60. Landon, M. (1977) Cleavage at aspartyl–prolyl bonds, In: *Methods in Enzymology*; Vol. 47, (ed. C.H.W. Hirs), pp. 145–149. Academic Press, NY.
61. Nielsen, H., Engelbrecht, J., Brunak, S. & von Heijne, G. (1997) Identification of prokaryotic and eukaryotic signal peptides and prediction of their cleavage sites, *Protein Eng.*, **10**, 1–6.
62. Pugsley, A.P. (1993) The complete general secretory pathway in Gram-negative bacteria, *Microbiol. Rev.*, **57**, 50–108.
63. Lundblad, R.L. (ed.) (1995) *Techiques in Protein Modification*, CRC Press, Boca Raton, FL.
64. Kannichts, C. (ed.) (2002) Posttranslational modification reactions: tools for functional proteomics, *Methods in Molecular Biology*, Vol. 194. Humana Press, Clifton, NJ.
65. Huddleston, M.J., Annan, R.S., Bean, M.F. & Carr, S.A. (1993) Selective detection of phosphopeptides in complex-mixtures by electrospray liquid-chromatography mass spectrometry, *J. Am. Soc. Mass Spectrom.*, **4**, 710–717.
66. Arnold, R.J., Polevoda, B., Reilly, J.P. & Sherman, F. (1999) The action of N-terminal acetyltranferases on yeast ribosomal proteins, *J. Biol. Chem.*, **274**, 37035–37040.
67. Carr, S.A., Huddleston, M.J. & Bean, M.F. (1993) Selective identification and differentiation of N-linked and O-linked oligosaccharides by liquid chromatography mass spectrometry, *Protein Sci.*, **2**, 183–196.
68. Lapko, V.N., Smith, D.L. & Smith, J.B. (2001) In vivo carbamylation and acetylation of water-soluble human lens alpha B-crystallin lysine 92, *Protein Sci.*, **10**, 1130–1136.
69. Gobom, J., Schuerenberg, M., Mueller, M., Theiss, D., Lehrach, H. & Nordhoff, E. (2001) alpha-cyano-4-hydroxycinnamic acid affinity sample preparation. A protocol for MALDI-MS peptide analysis in proteomics, *Anal. Chem.*, **73**, 434–438.
70. McCormick, J. P. & Thomason, T. (1978) Near-ultraviolet photooxidation of tryptophan – proof of formation of superoxide ion, *J. Am. Chem. Soc.*, **100**, 312–313.
71. Cohen, S. L. & Ward, G. (2002) Peptide and protein modification – there's something in the air, In: *Proceedings the 50th ASMS Conference on Mass Spectrometry and Applied Topics,* Orlando. American Society of Mass Spectrometry, Orlando, Florida.
72. Petersson, A.S., Steen, H., Kalume, D.E., Caidahl, K. & Roepstorff, P. (2001) Investigation of tyrosine nitration in proteins by mass spectrometry, *J. Mass Spectrom.*, **36**, 616–625.
73. Hall, S.C., Smith, D.M., Masiarz, F.R., Soo, V.W., Tran, H.M., Epstein, L.B. & Burlingame, A.L. (1993) Mass spectrometric and Edman sequencing of lipocortin I isolated by two-dimensional SDS/PAGE of human melanoma lysates. *Proc. Natl Acad. Sci. USA*, **90**, 1927–1931.
74. Hamdan, M., Bordini, E., Galvani, M. & Righetti, P.G. (2001) Protein alkylation by acrylamide and cross-linkers and its relevance to proteomics: a matrix assisted laser desorption/ionization-time of flight mass spectrometry study, *Electrophoresis*, **22**, 1633–1644.
75. Raftery, M.A. & Cole, R.D. (1966) On the aminoethylation of proteins, *J. Biol. Chem.*, **241**, 3457–3461.
76. Scigelova, M., Green, P.S., Giannakopulos, A.E., Rodger, A., Crout, D.H.G. & Derrick, P.J. (2001) A practical protocol for the reduction of disulfide bonds in proteins prior to analysis by mass spectrometry, *Eur. J. Mass Spectrom.*, **7**, 29–34.
77. Boja, E.S. & Fales, H.M. (2001) Overalkylation of a protein digest with iodoacetamide, *Anal. Chem.*, **73**, 3576–3582.
78. Wright, H.T. (1991) Nonenzymatic deamidation of asparaginyl and glutaminyl residues in proteins. *Crit. Rev. Biochem. Mol. Biol.*, **26**, 1–52.
79. Smith, D.L. & Zhou, Z.R. (1990) Strategies for locating disulfide bonds in proteins. In: *Methods in Enzymology*, Vol. 193, (ed. J.A. McCloskey) pp. 374–389. Academic Press, New York.
80. Gygi, S.P., Corthals, G.L., Zhang, Y., Rochon, Y. & Aebersold, R. (2000) Evaluation of two-dimensional gel electrophoresis-based proteome analysis technology, *Proc. Natl Acad. Sci. USA*, **97**, 9390–9395.
81. Zhang, Y., Figeys, D. & Aebersold, R. (1998) Purification of trypsin for mass spectrometric identification of proteins at high sensitivity, *Anal. Biochem.*, **261,** 124–127.
82. Kim, T. & Reilly, J.P., manuscript in preparation.
83. Keough, T., Youngquist, R.S. & Lacey, M.P. (1999) A method for high-sensitivity peptide sequencing using postsource decay matrix-assisted laser desorption ionization mass spectrometry, *Proc. Natl Acad. Sci. USA*, **96**, 7131–7136.
84. Keough, T., Youngquist, R.S. & Lacey, M.P. (2003) Sulfonic acid derivatives for peptide sequencing by MALDI MS, *Anal. Chem.*, **75**, 156A–165A.

# 5 Oligosaccharides

HÉLÈNE PERREAULT

## 5.1 Introduction

Carbohydrates are best and popularly known known as a food source for most organisms. Biologists, biochemists and chemists, however, recognize that they also serve as structural materials for plants and are involved in many other, frequently complex, biological functions. They take part in these functions when coupled to other classes of molecules with the combination being termed glycoconjugates. Protein glycosylation describes important relationships between proteins and oligosaccharides. It is one of the most important processes among all types of post-translational modifications [1] and produces a wide spectrum of biologically important molecules [2].

Investigation of the structure and function of glycoproteins is an active area of research, and the functional significance of the carbohydrate moieties has become increasingly apparent [3]. Glycoproteins encompass several important classes of macromolecules, including enzymes, hormones, immunoglobulins, transport proteins, cell adhesion molecules, and cytoplasmic proteins [2]. The roles of protein-bound carbohydrates in biological recognition, such as in host–pathogen, cell–cell and cell–molecule interactions generate considerable interest and investigation.

Clearly, elucidating the structure of various glycoproteins is an essential component of current biological study and constitutes a large part of the field of glycomics. In conjunction with genomics and proteomics, glycomics forms a triad of disciplines that investigate the structure and function of biological macromolecules. This triad holds great promise in advancing the understanding of biology and in the application of this knowledge to human welfare.

Although much remains to be discovered, there already exists knowledge about glycosylation of proteins. Generally, each glycosylation site on a protein bears different, although related, glycans, a phenomenon called microheterogeneity. Attachment sites may either be asparagine (Asn) or serine (Ser) and threonine (Thr) residues. The link to Asn occurs only in the consensus sequence Asn-AA-Ser/Thr, where AA is any amino acid except proline. In these cases oligosaccharides are termed *N*-glycans. If the link is through the oxygen of Ser or Thr, the glycoconjugates are the *O*-glycans [4]. Figure 5.1 shows, (A) the subgroups of *N*-linked sugar chains, (B) four types of structures found in *O*-linked sugar chains, and (C) the structures of monosaccharide units frequently encountered in glycoproteins from mammals [5]. A given glycoprotein may have *N*- and/or *O*-linked oligosaccharides on its chain.

Since the mid 1980s, fundamental studies in glycobiology have benefited from the availability of on-line high performance liquid chromatography/electrospray ionization–mass spectrometry (HPLC/ESI-MS) and direct-deposition matrix assisted laser desorption/ionization–mass spectrometry (MALDI MS). In addition, the requirements of therapeutic applications have pushed developments in numerous disciplines,

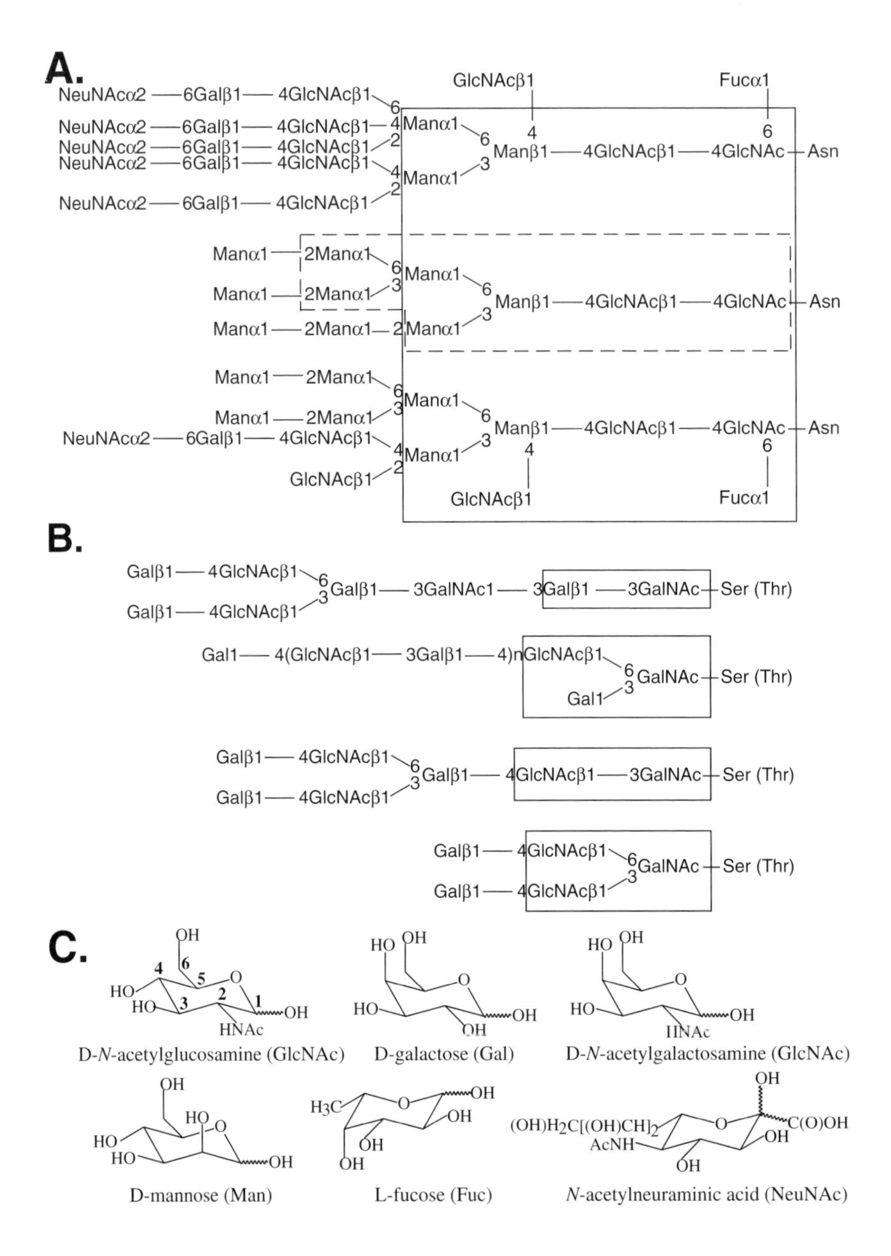


**Fig. 5.1** A. Subgroups of *N*-linked sugar chains, complex (top), high-mannose (middle), hybrid (bottom). In solid line: pentasaccharide structure common to all *N*-linked glycans. In dotted line: common heptasaccharide of high-mannose chains. Outside solid line: variable sugar chains. B. Four types of core structures in *O*-linked sugars. Cores are within the solid lines. C. Monosaccharides commonly found in mammalian glycoproteins [5].

including separation sciences and instrumentation. For instance, specific wild-type glycoproteins are difficult to isolate from cells in large quantities, so recombinant technologies have played a major role in the preparation of glycoproteins for therapeutic use. These products must be controlled very carefully before prophylactic use, and some of the control lies in the ability to verify oligosaccharide structures attached to the protein backbones. The combined needs of the basic and applied have, in recent years, resulted in newer technologies and high-throughput methods.

Despite the availability of very sophisticated instrumentation, quantitative or qualitative determination of oligosaccharides in diverse types of samples remains challenging. In the first instance, oligosaccharides often occur in mixtures and in complex biological matrices. They are generally thermolabile and thus not amenable to gas chromatography and to traditional mass spectrometric methods without derivatization, which is necessary to enhance their volatility. The lack of chromophores also makes it a difficult matter to analyse carbohydrates by HPLC with the standard detectors. Such analyses became feasible because of advances in the areas of liquid chromatography (LC) and mass spectrometry (MS), among other techniques, which have allowed the characterization of very small quantities of oligosaccharides without causing the sample thermal degradation encountered with traditional ionization techniques, i.e. electron impact ionization (EI) and chemical ionization (CI).

Regardless of the sophistication, sensitivity and specificity of the instruments, successful qualitative and quantitative determination of oligosaccharides is often predicated on high quality sample preparation. This is not surprising given the above description of the matrices and physicochemical characteristics of the analytes. The two techniques 'separation from matrix' and derivatization, become essential components of almost all sample preparation methods in glycomics. This chapter, therefore, describes some sample preparation methods aimed at analyzing free and conjugated oligosaccharides from several sources by electrospray ionization (ESI) and matrix assisted laser desorption/ionization (MALDI) MS aided by either on-line (for ESI) or off-line (for MALDI) high performance liquid chromatography (HPLC).

Yet these alone are not sufficient to deal with the full complexity of the analytical problems encountered in the study of glycomics. Chromatography and MS, for instance, either individually or combined, do not provide information on anomericity. For this reason, the present chapter also describes various chemical and enzymatic methods aimed at preparing oligosaccharide samples suitable for structural characterization by modern hyphenated MS methods.

## 5.2 Oligosaccharides from glycoproteins

### 5.2.1 *General aspects*

The isolation, purification, derivatization and analysis of protein-derived and other types of carbohydrates will be discussed. Generally, these will be preparatory to mass spectrometric measurements which can provide details on molecular mass. For general information on oligosaccharide sequence and branching pattern, tandem mass spectrometry (MS/MS) is very useful, although it will not provide details on anomericity, linkage positions and exact nature of each monosaccharide moiety, as there are always isomeric possibilities. When such details are needed, nuclear magnetic resonance (NMR) is a method of choice if the quantity of material available for analysis is in the milligram range or above. For smaller amounts of sample, assays have been developed using a combination of exoglycosidase reactions, gel electrophoresis and MALDI MS or HPLC. These assays necessitate elaborate sample preparation steps.

Glycomics has recently become a particularly active field as it has provided the tools for scientists to address post-translational modification of proteins. The advent of glycomics in relationship to proteomics and genomics has prompted the need for high throughput analytical methods, where fully-prepared kits are invaluable to keep procedures consistent. Not all methods discussed here have their reagents commercially available as kits, and because kits tend to be expensive, many researchers prefer to conduct reactions using materials purchased individually. The latter choice allows for optimization and adaptation of a method to a particular sample's needs. For this purpose, it is important to understand the chemical reactions taking place during sample transformation and preparation steps. Several types of all-inclusive sample transformation kits are available from companies, but unfortunately these kits do not contain detailed descriptions of the nature of the chemicals used and of the final products obtained.

### 5.2.2 *Glycoprotein purification and molecular mass measurement*

To obtain information on the nature of these glycans and their attachment sites, the original glycoprotein sample must be as pure as possible. Although protein purification is outside the scope of this chapter, it will be discussed briefly because buffers and other additives used in these methods can seriously hamper MS analysis. Glycoproteins have several biological origins and may be cleaned and purified according to the matrix. Most common purification methods include gel permeation columns, bio-affinity columns, electrophoretic gels, ion-exchange columns, dialysis, and combinations of these techniques. Denaturing agents such as sodium dodecyl sulfate or other detergents should be avoided or removed, because they can yield intense mass spectra themselves and quench glycoprotein-related signals at any subsequent stage of the analysis.

Figure 5.2 suggests some approaches to prepare glycoprotein samples for analysis by on-line HPLC/MS and off-line HPLC/MALDI MS. The text below provides more details pertaining to this diagram. Obtaining a molecular mass measurement on the purified glycoprotein is useful, although not always possible, especially if the amount of sample is limited and many sites are glycosylated and bear sialic acid residues. If the glycoprotein has been purified on a gel, it may be electroblotted onto a membrane and measured by MALDI, or electroeluted. It is suggested to skip extraction from the gel, and to proceed to the next step, i.e. in-gel digestion, without a molecular mass measurement. Yields for such extractions tend to be low. Linear MALDI-TOF instruments and most commercial ESI mass spectrometers will allow molecular mass measurements, although ESI is less tolerant than MALDI to the presence of residual salts, detergents or other additives. A successful molecular mass determination will sometimes clearly show the glycoforms of the protein, and in other cases produce a wide, unresolved peak, composed of all combined glycoforms. This analysis usually requires about femtomole to picomole amounts of glycoprotein.

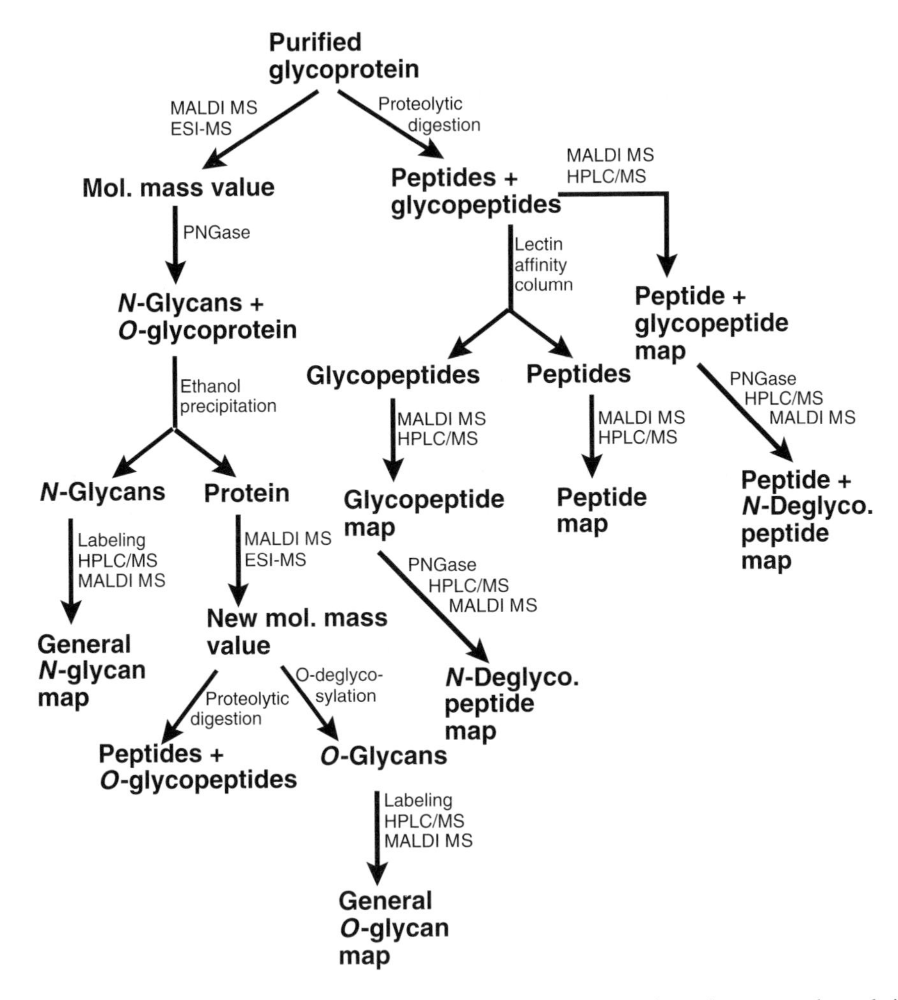


**Fig. 5.2** General procedure suggested to obtain information on protein identity and sequence, glycosylation sites and glycan molecular masses. For more specific information (anomericity, exact nature of residues, glycan sequence, branching patterns) more steps are required.

The rest of the sample may be split in two or more portions, for sugar removal and proteolytic digestion in parallel as suggested in Fig. 5.2. Proteolytic digestion is most commonly accomplished using trypsin, which cleaves proteins at the C-terminus of all arginine (Arg, R) and lysine (Lys, K) residues. Figure 5.3A illustrates this process. For detailed analysis of the total carbohydrate content, it is suggested the *N*- and *O*-glycans be detached from the protein rather than from tryptic glycopeptides, as indicated on the left hand side of Figure 5.2. Starting with the intact glycoprotein, detached glycans may be isolated at once by ethanol precipitation of the protein, whereas if tryptic glycopeptides are subjected to deglycosylation, ethanol does not precipitate all peptides and another type of separation method is needed to isolate the sugars. It is, however, useful to analyze glycopeptides and speculate on possible glycoforms and peptide sequences

**A.**

$NH_2$ – $(CH_2)_4$ – $(AA)_m$----HN-C-C(=O)----$(AA)_n$ —**Trypsin**→ $(AA)_m$----HN-C-C(=O)-OH + $H_2N(AA)_n$

**Lysine (K, Lys)**

$H_2N$–C(=NH) – NH – $(CH_2)_2$ – $(AA)_m$----HN-C-C(=O)----$(AA)_n$ —**Trypsin**→ $(AA)_m$----HN-C-C(=O)-OH + $H_2N(AA)_n$

**Arginine (R, Arg)**

**B.**

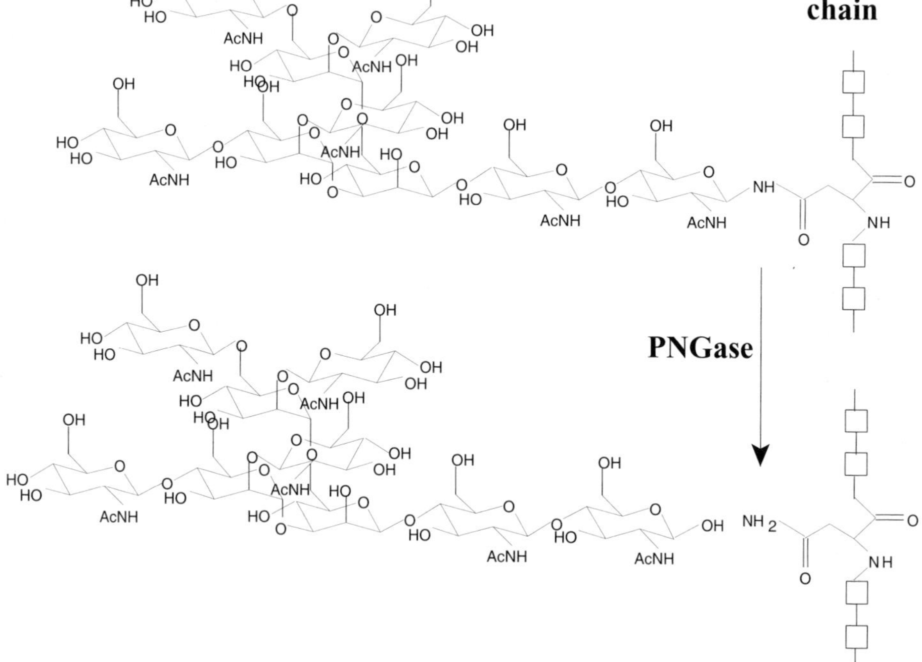


**Fig. 5.3** A. Schematic representation of tryptic cleavages at lysine and arginine residues. B. Cleavage of an *N*-linked glycan from a protein backbone by PNGase digestion.

from HPLC/MS and MALDI MS data (e.g.[6]). Figure 5.4A,B shows mass spectra of glycopeptides with high mannose (A) and complex-type oligosaccharides (B) [6].

### 5.2.3 *N-Glycosylation*

A popular and effective method for detaching *N*-linked sugars from the protein backbone is to use an enzyme called peptide *N*-glycosidase (PNGase) [7]. Two types are commonly used, PNGase-F and PNGase-A. PNGase-F releases most *N*-glycans, except those containing a fucose residue linked $\alpha$1–3 to the reducing GlcNAc terminal [8]. Such glycans are often found in plant glycoproteins and release necessitates use of PNGase-A [9]. PNGases, upon deglycosylation, transform the Asn attachment sites into aspartic acid (Asp) residues (see Fig. 5.3B). Often one or two repeated digestions are needed to detach all glycans from the protein chain. After each digestion, a small amount of the precipitated protein may be measured for molecular mass by ESI- or MALDI MS. This measurement indicates the level of protein homogeneity achieved and allows the original mass percentage of carbohydrate to be obtained, given that the molecular mass of the glycoprotein was measured.

At this stage, the deglycosylated protein may be digested with trypsin and the resulting peptides compared with intact glycoprotein tryptic digestion products. Although a preliminary comparison may be conducted by spotting both digestion mixtures onto MALDI targets and looking for spectral differences, the search for glycopeptides and the general peptide coverage are more successful when on-line (ESI) and direct-deposition (MALDI) reversed-phase HPLC/MS are used. It has been shown that glycopeptides have approximately the same reversed-phase retention times as their analogue deglycosylated peptides, although they tend to elute slightly earlier due to the polar character inherent to the sugar units (e.g. [6]). The advantage of using direct deposition on MALDI targets over on-line HPLC lies in the ability to conduct several MALDI analyses combined with on-target derivatization or digestion. On-line HPLC/ESI-MS experiments are dynamic and allow for one measurement, unless fractions are collected simultaneously.

Following proteolytic digestion of glycoproteins, lectin affinity columns have been used to separate peptides from glycopeptides and thus isolate the latter in a pool for further HPLC/MS or MALDI MS characterization (e.g. [10]). Glycopeptides were also isolated by covalent binding through hydrazide chemistry on a solid support [11] by first attaching the whole glycoprotein onto the support, and then digesting with trypsin. Free peptides were rinsed off the solid support, while the peptidic portion of bound glycopeptides could be released using PNGase digestion. This method does not yield information on the nature of sugars; it does, however, provide insights on the glycosylation sites.

Once the glycopeptide-containing fractions have been identified, MS/MS spectra may be acquired on the glycoforms of each peptide (see, for example, Fig. 5.4C [6]); for accurate identification of the glycosylaton sites within these peptides, PNGase may be used to transform glycosylated asparagines into aspartic acid residues (Fig. 5.2), which can then be located by MS/MS. Recently, two research groups have performed peptide deglycosylation in $^{18}$O-containing media followed by mass spectrometric identification of PNGase-deglycosylated peptides using the modified $^{18}$O/$^{16}$O isotope ratio as a marker [10,11].

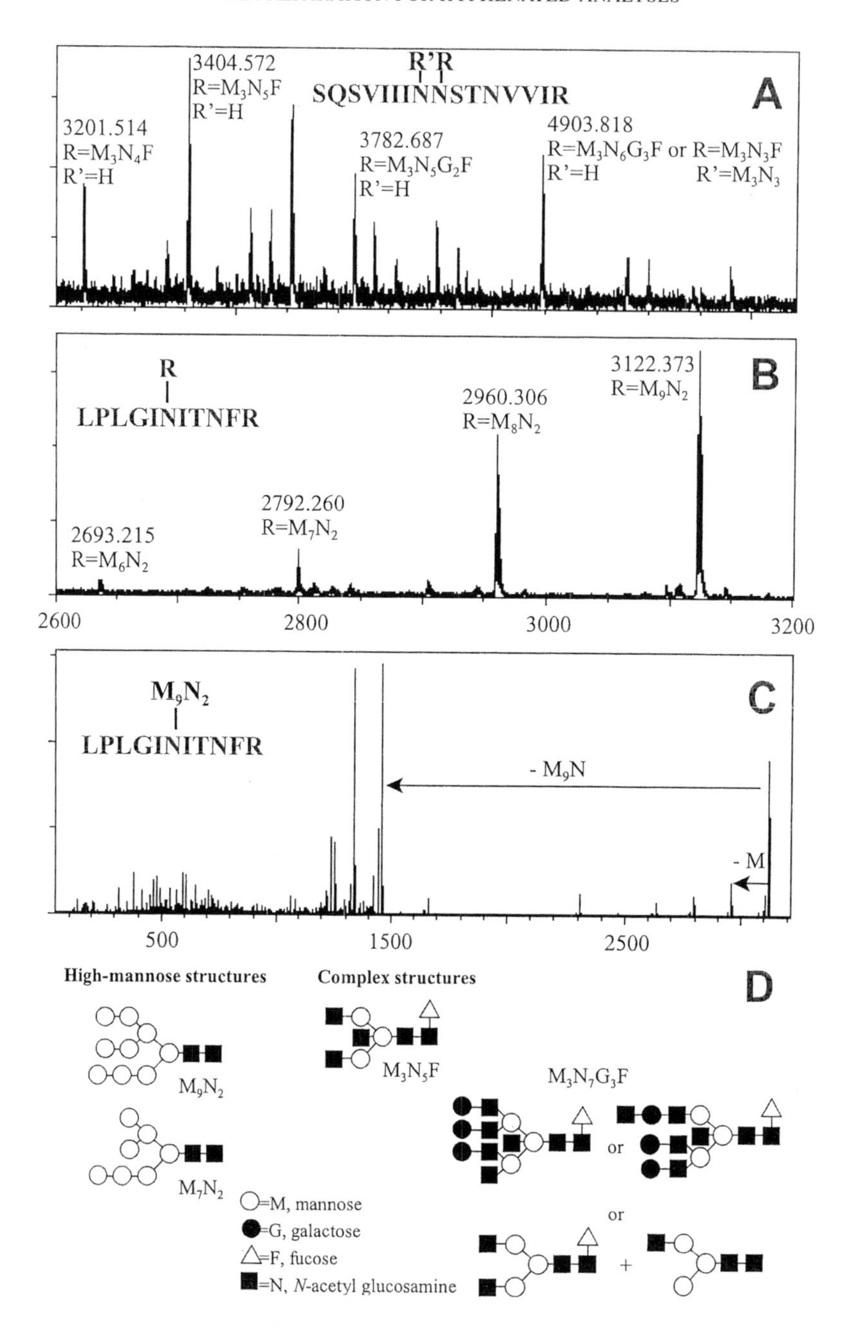


**Fig. 5.4** Direct MS (A and B) and MS/MS (C) detection of glycosylated tryptic peptides of the 139-kDa spike protein from the SARS-related coronavirus. (A) MS of HPLC fraction containing glycoforms of T111–126, showing complex glycans. (B) MS of HPLC fraction containing high-mannose glycoforms of T222–232. (C) MS/MS spectrum of ions from B to prove the presence of high mannose structures. (D) Suggested *N*-glycan structures. Adapted from [6].

If the glycoprotein is on a polyacrylamide gel, there are methods to deglycosylate in the gel and release the glycans [12,13], followed by in-gel tryptic digestion to release peptides from the gel [14]. The reverse order may also be followed, i.e. in-gel tryptic digestion first to recover peptides and glycopeptides. Some researchers prefer to conduct glycosylation analysis solely with glycopeptides, using a combination of exoglycosidase digestions and MS to determine the exact structures and compositions of glycans while still attached to the peptides (e.g. [15]). Software components exist to help determine the oligosaccharide composition on glycopeptides based on peptide sequences suitable for glycosylation [16].

Free oligosaccharides obtained by PNGase deglycosylation can be labeled at the reducing terminus using a choice of several reagents, depending on the types of analytical methods available in the laboratory. The label should be useful for UV-vis/fluorescence detection, and/or for ESI-MS and MALDI signal enhancement. Labeling can also improve chromatographic or electrophoretic separations, whether MS or another detection technique is used. Examples of such reagents include 2-aminobenzamide (2-AB) [17], 2-aminoacridone (2-AMAC) [18], 1-phenyl-3-methyl-5-pyrazolone (PMP) [19], phenylhydrazine (PHN) [20], 2-aminopyridine (2-AP) [21], to cite only a few examples. Other labels, precharged for the purpose of electrophoretic separation, include 8-aminonaphthalene-1,3,6-trisulfonate (ANTS) [22] and 1-aminopyrene-3,6,8-trisulfonate (APTS) [23]. These labels have been used for fluorophore-assisted carbohydrate electrophoresis (FACE) [22] and capillary electrophoresis (CE) separations rather than for on-line HPLC/ESI-MS. Figure 5.5 shows examples of reducing-end labeling; a more complete list may be found in ref. [24]. Use of a label is not always necessary, as it has been possible to conduct HPLC/ESI-MS experiments on underivatized *N*-glycans [25].

Labeled *N*-glycans can be analyzed for contents by HPLC-UV on normal phase columns [26] or reversed phase HPLC/ESI-MS, or on a MALDI target without further separation. Direct MS measurements for molecular mass allow estimation of the glycan compositions in the oligosaccharide pool in terms of the respective numbers of mannose (Man), *N*-acetyl glucosamine (GlcNAc), fucose (Fuc), galactose (Gal), neuraminic acid (NeuAc) and other possible residues contained in each glycan. Tandem MS measurements on parent ions of *N*-glycans yield a considerable amount of structural information on branching patterns or oligosaccharide sequences [27–32]. There are algorithms available for interpreting tandem spectra of *N*-glycans, mainly producing possible compositions and branching patterns based on biosynthetic rules around the trimannosyl core [16,33–35].

The use of arrays of exoglycosidases in combination with MALDI [36] or ESI-MS [15] yields more accurate results than MS and MS/MS alone for structural determination of *N*-glycans. These enzymes are really specific with respect to linkage position and anomericity, neither of which can be determined by MS/MS. As in the case of glycopeptides, the reducing termini are blocked, this time by the label, so exoglycosidases hydrolyze monosaccharide moieties only from the non-reducing end, thus simplifying the analysis.

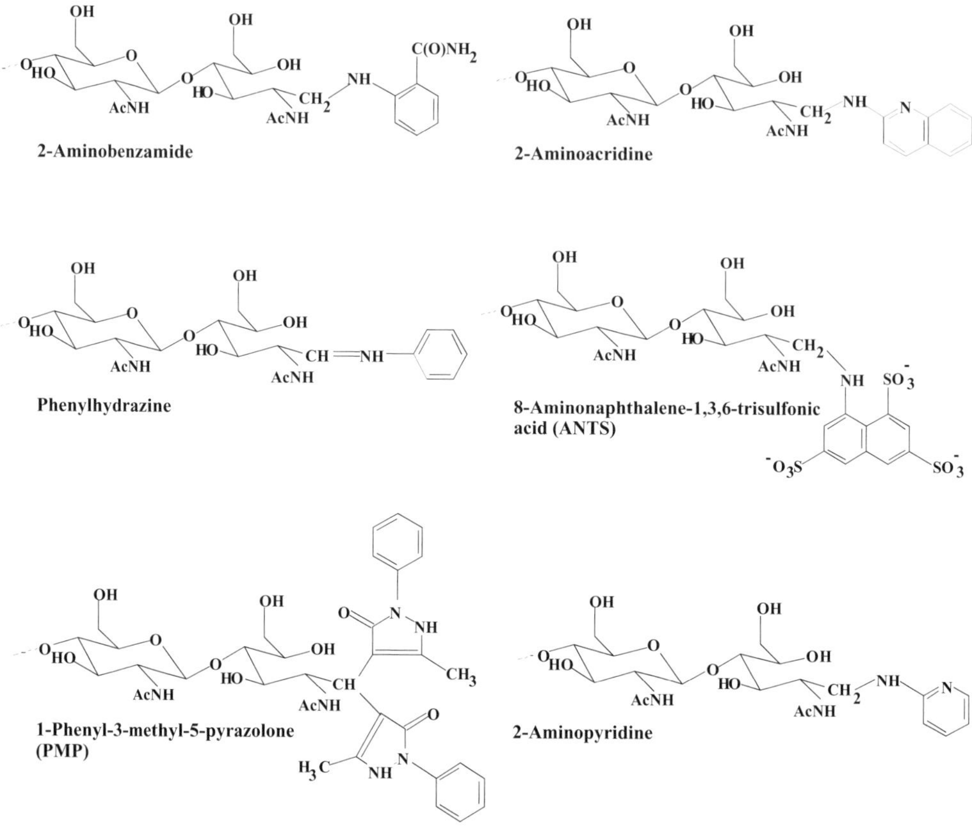

**Fig. 5.5** Some examples of reducing end labeling methods available for free oligosaccharides [17–22]. For a more complete list, see [24].

### 5.2.4 *O-Glycosylation*

After the required information has been obtained on glycosylation sites and *N*-glycan structures, the possibility of *O*-glycosylation may still need to be explored. There are published methods for *O*-glycosylation analysis of glycoproteins, some including MS (e.g. [37,38]). There is no established consensus sequence for *O*-linked glycosylation on Thr and Ser amino acids, but general trends have been observed; for example, increased numbers of Thr, Ser, Val, Ala, Gly and especially Pro tend to be found in the vicinity of occupied Ser or Thr residues [39,40]. The absence of a general *O*-glycosidase due to the existence of several core types in *O*-glycans (see Fig. 5.1) has made the analysis of these glycans and determination of *O*-glycosylation sites difficult compared to *N*-glycans.

Chemically, it is possible to release *O*-glycans from peptides or proteins using alkaline beta-elimination, and if this procedure is used under mild and controlled conditions the protein may remain intact [41]. Threonine sites are transformed into unsaturated

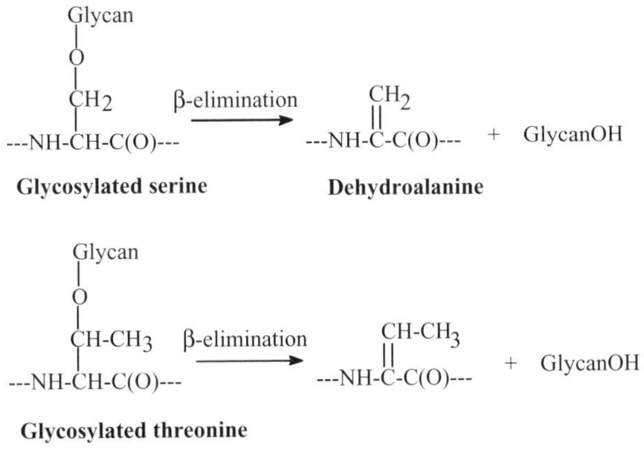

```
      Glycan
        |
        O
        |
       CH2       β-elimination        CH2
        |        ------------->       ||
---NH-CH-C(O)---                ---NH-C-C(O)---  +  GlycanOH
```

**Glycosylated serine** **Dehydroalanine**

```
      Glycan
        |
        O
        |
       CH-CH3    β-elimination        CH-CH3
        |        ------------->       ||
---NH-CH-C(O)---                ---NH-C-C(O)---  +  GlycanOH
```

**Glycosylated threonine**

**Fig. 5.6** Beta-elimination of *O*-linked glycans, giving rise to unsaturated residues easily characterized by mass spectrometry [42].

propyl chains, and serines are left as dehydroalanines (see Fig. 5.6) [42]. Often some peptides or proteins are degraded by this procedure, especially if a harsh base such as sodium hydroxide is used. When ammonium hydroxide is used as a base, ammonia reacts with dehydroalanine and aminates the residue, thus adding 17 Da to the mass [41]. A recent study by Hanisch *et al.* [43] describes *O*-glycan removal using methylamine and mild heating. Unsaturated sites left by beta-elimination on serine and threonine are immediately substituted with the alkylamino group, thus labeled for the identification of glycosylation sites. After deglycosylation has been performed, the protein or peptides which have either unsaturated sites [42], aminated sites [41] or alkylaminated sites [43] can be analyzed by MS, further digestion and MS/MS for glycosylation sites.

The *O*-glycans may be separated from proteins using ethanol precipitation, and from peptides by reversed-phase HPLC. For *O*-glycan analysis by MS and MS/MS, the same derivatization procedures as described for *N*-glycans may be applied. Because no universally common core structure belongs to all *O*-glycans, structural analysis necessitates a more careful approach than in the case of *N*-glycans with a common trimannosyl core. A combination of MS/MS and exoglycosidase arrays constitutes a valuable method, given that enough sample is available. Combining normal-phase and reversed-phase HPLC experiments is also a valuable method for sequencing *O*-glycans, as shown in Figure 5.7. *O*-Glycans were analyzed at each step of an exoglycosidase array digestion using HPLC with fluorescence detection. Retention times of the products were used to estimate the overall sugar sequence of specific glycans of rat CD48 glycoprotein expressed in Chinese hamster ovary (CHO) cells [12].

### 5.2.5 *Quantitative aspects*

Quantitative amounts of glycans removed from a glycoprotein are difficult to obtain without labeling the sugars, unless high pH anion exchange chromatography coupled with pulsed amperometric detection (HPAEC-PAD) is used [44,45]. Reducing-end labeling offers the advantage of incorporating one chromophore per molecule of glycan, such that each molecule absorbs or fluoresces the same intensity of light. Quantitative

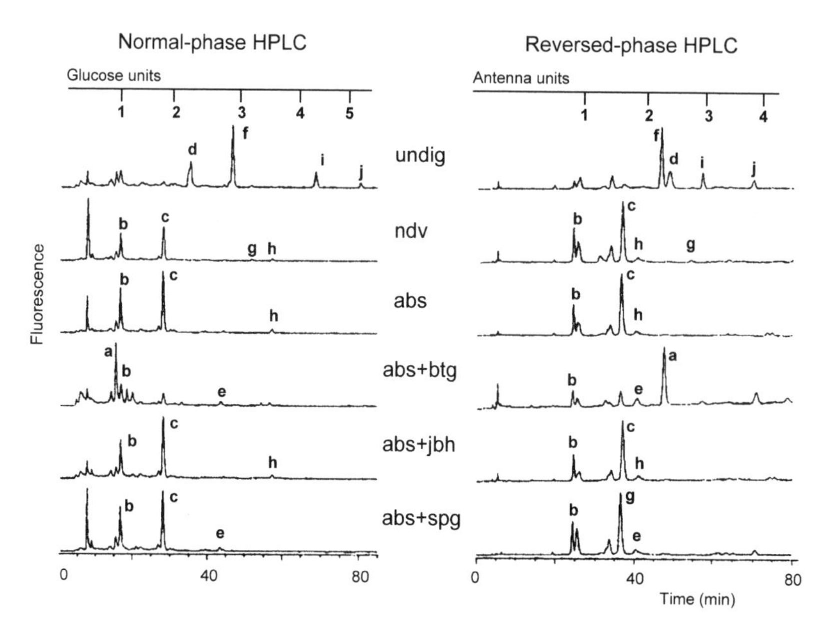


**Fig. 5.7** Separation of *O*-linked glycans using normal-phase (NP) and reversed-phase (RP) HPLC. Glucose unit (GU) and antennae unit (AU) values for digests of an *O*-linked tetrasaccharide in the small peak (i). Trace 2 from the top shows that the small peak (g) which results from digesting the glycan (i) with Newcastle disease virus sialidase (ndv) elutes at a smaller GU value than (i) whereas on a RP column both (i) and (g) elute in a similar position. The reverse situation occurs in the digests in trace 4 from the top using bovine testes beta-galactosidase (btg) where peak (h) is digested to peak (e). Thus the two methods are complementary, allowing two structures which co-elute on one system to be resolved on the other. abs, *Arthrobacter ureafaciens* sialidase; jbh, jack bean beta-*N*-acetylhexosaminidase; spg, *Streptococcus pneumoniae* beta-galactosidase; undig, undigested glycan pool. Adapted from [12] with permission.

experiments on fluorescent-labeled saccharides have been conducted mostly by FACE [46]. Most glycan profiling studies involving HPLC/MS in the literature have explored semiquantitative analytical aspects rather than quantitation [46,47].

Even with labeling, it is difficult to describe MALDI or ESI-MS as quantitative techniques for oligosaccharides because of $pK_a$ discrepancies between the different monosaccharide residues and variations in ionization efficiencies. Ideally the use of an internal standard for each substance studied would make these techniques quantitative; however, the price of most glycan standards would make this type of analysis financially prohibitive. In a case where one single oligosaccharide structure would be probed, quantitation would be appropriate with the use of the same molecule as internal standard. For complex mixtures of *N*-linked and *O*-linked oligosaccharides, MS can be exploited as a semiquantitative technique [29].

## 5.3 Oligosaccharides from other biological sources

### 5.3.1 *General aspects*

Carbohydrates of many varieties are ubiquitous in nature, either free or in the form of glycoconjugates. Milk is a well known source of free oligosaccharides, as are plants and fruits in general. Free oligosaccharides are also found biological fluids such as urine and plasma. It is often necessary to perform qualitative and/or quantitative analysis on the free oligosaccharide content of biological materials. Oligosaccharides are present at variable levels, mixed with complex biological matrices from which they must be isolated before analysis. Because sugars are devoid of good chromophores and relatively polar, their quantitation or differentiation by HPLC-UV is difficult. MS has proven to be a better detection method than UV for such compounds. On-line HPLC/MS has been a method of choice in the detection of oligosaccharides from human milk, and plant saccharides. In this section, sampling and analytical methods will be described for carbohydrates from milk, although reference will also be made to other types of oligosaccharide.

Other important sources of oligosaccharides are glycoconjugates. In these molecules, saccharides are covalently attached to other molecules, for example proteins (glycoproteins) that were discussed in the previous section, and lipids (glycolipids, including glycosphingolipids and related compounds).

Glycoconjugates have diverse biological functions, often made possible by the dual nature of these molecules. The roles of the carbohydrate portion may vary from anchor, signaling agent and structural hinge, to cationic capture agent, recognition template, etc., whilst, the non-carbohydrate portions of glycoconjugates may be trans-membrane fatty chains, enzymatic proteins, cell-surface anchors or receptors, among other possibilities. Sampling and analytical methods for these types of compounds have been developed and will also be reviewed in this chapter.

### 5.3.2 *Milk oligosaccharides*

So far, more than a hundred different oligosaccharide structures have been identified in human milk [48–50]. Carbohydrates up to 8000 Da molecular mass have been observed [51,52]. Most human milk oligosaccharides comprise lactose at the reducing end, linked to *N*-acetyllactosamine units of various sizes, branching patterns, linkage positions, and extents of sialylation and fucosylation. Figure 5.8 gives a few examples of human milk oligosaccharides. Because milk is a complex biological matrix and also a colloidal suspension, the isolation of oligosaccharides requires several steps that have been described in detail in the literature [52–55]. Briefly, pasteurization is followed by centrifugation, precipitation of proteins, preparative anion-exchange chromatography, HPLC, and gel permeation chromatography (GPC). Pasteurization consists of heating milk to a specific temperature for a specified period of time (e.g. 63°C for not less than 30 min, or 72°C for not less than 16 s) without allowing recontamination of that milk during the heat treatment process. It ensures destruction of pathogens and the enzyme phosphatase. Centrifugation is used to separate the fats, which can then be collected to facilitate the isolation of oligosaccharides. Because the solution still contains proteins (e.g. casein), precipitation is required and can be performed using cold ethanol, in which oligosaccharides remain soluble. Anion-exchange chromatography allows the preparation and separation of fractions according to p$K_a$ ranges characteristic to sugars, whereas HPLC can separate fractions further based on branching patterns or numbers of monosaccharide residues. As for GPC, it elutes sugars in reversed order of size or molecular mass.

Pfenninger *et al.* [56,57] have analyzed underivatized oligosaccharides from human milk by nano-ESI on a quadrupole-ion trap mass spectrometer using the positive and

**β-lactose**
Galβ1-4Glc

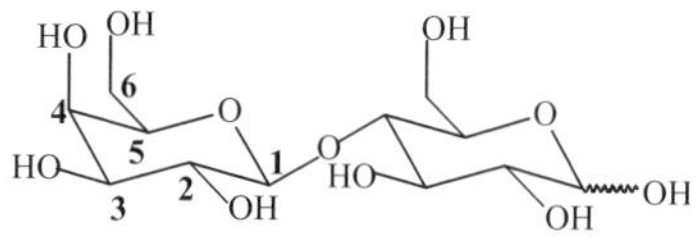

**2'-Fucosyllactose** Fucα1-2Galβ1-4Glc

**Lactodifucotretraose** Fucα1-2Galβ1-4Glc(3←Fucα1)

**Lacto-*N*-neotetraose**
Galβ1-4GlcNAcβ1-3Galβ1-4Glc

**Lacto-*N*-neohexaose**
Galβ1-4GlcNAcβ1-6 / Galβ1-4GlcNAcβ1-3 Galβ1-4Glc

**Fig. 5.8** Examples of free oligosaccharides found in human milk and urine.

negative ion modes. For positive mode, a solution of NaCl was added to help formation of $[M + Na]^+$ ions, whereas in the negative mode, the addition of ammonium acetate was necessary to help form $[M - H]^-$ species.

Chaturvedi *et al.* [58] used a combination of reduction to the alditol forms and perbenzoylation to quantitatively analyze oligosaccharides from human milk by reversed-phase HPLC-UV, after acidic and neutral species had been separated. They then performed *O*-debenzoylation followed by permethylation, and the samples had been investigated again for structure confirmation by ESI-MS. Figure 5.9 gives a chemical outline of this method with lactose as an example.

In a special issue on the analysis of various types of oligosaccharides and polysaccharides by capillary electrophoresis [59], Schmid *et al.* [60] reported on the use of nano-HPLC coupling with MS and of micellar electrokinetic chromatography (MEKC) for the analysis of human milk oligosaccharides. They used 4-amino benzoic acid butyl ester (4-ABBE), 4-amino benzoic acid methyl ester (4-ABME) and 4-amino benzoic acid ethyl ester (4-ABEE) to label the sugars at the reducing end to enhance the HPLC-UV, MS and MEKC detection. Figure 5.10 shows the corresponding generic lactose derivative.

Free oligosaccharides are also found in several other biological samples. For example, they are expressed in human seminal plasma, as shown by Chalabi *et al.* [61] in a study involving HPLC and MS, and found in urine, as observed by Klein *et al.* [62]. These authors first labeled the sugars with amino naphthalene trisulfonate (ANTS) (see Figure 5.5) and purified them using porous graphitized carbon.

### 5.3.3 *Glycolipids*

Glycosphingolipids such as gangliosides and other glycolipids can be analyzed as intact molecules by LC/ESI-MS or MALDI-TOF MS. Figure 5.11A shows a typical

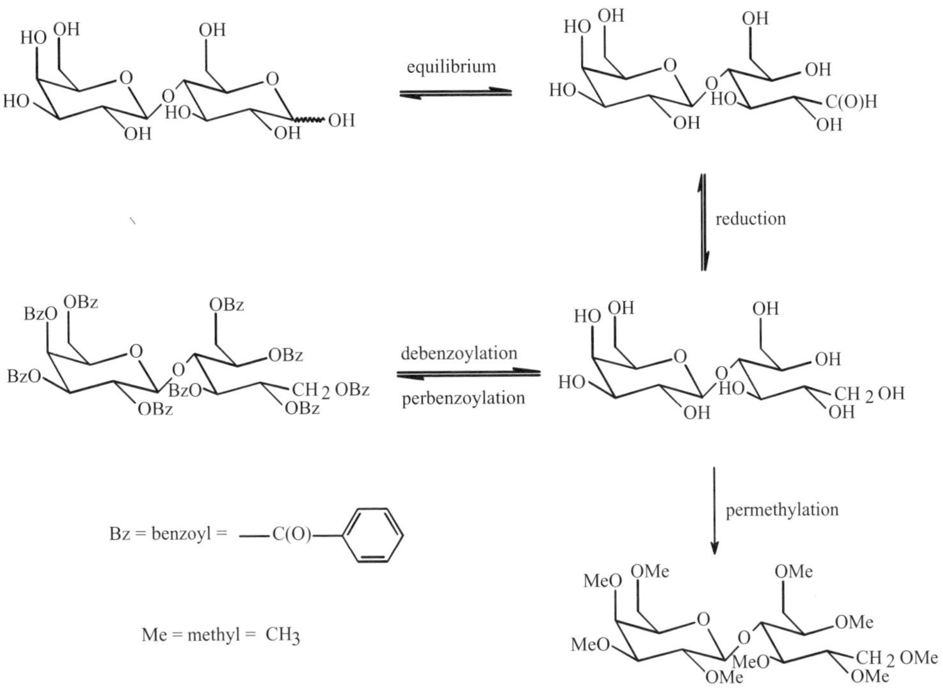

Fig. 5.9 Chemical transformations of milk oligosaccharides for quantitative analytical method using HPLC-UV and mass spectrometry [58].

ganglioside structure. Because in these molecules the reducing end is blocked by a lipid chain and because of the absence of a significant chromophore, HPLC-UV analysis requires precolumn derivatization with perbenzoylation [58] or incorporation of another UV absorbing group. Perbenzoyl derivatives are suitable for UV, and also for mass spectrometric detection [63]. Glycosphingolipids also yield enhanced ionization efficiencies when subjected to permethylation or peracetylation [64]. The isolation of these bipolar compounds for analysis from media such as cerebrospinal fluid (CSF),

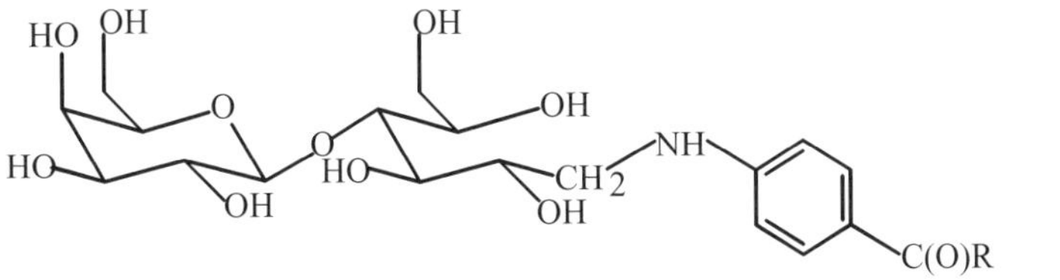

Fig. 5.10 Representative schematic of the labeling of beta-lactose with 4-aminobenzoic acid alkyl ester, where alkyl is methyl, or ethyl, or butyl [60].

**A**
GD1b ganglioside

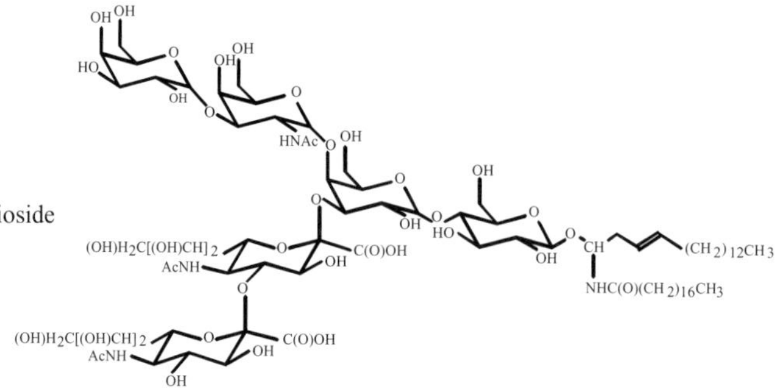

**B**
Heparin

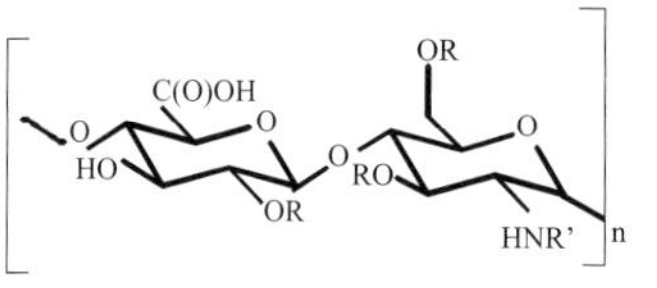

R = H or $SO_3^-$
R' = H or $SO_3^-$ or Ac

**C**
Cellulose

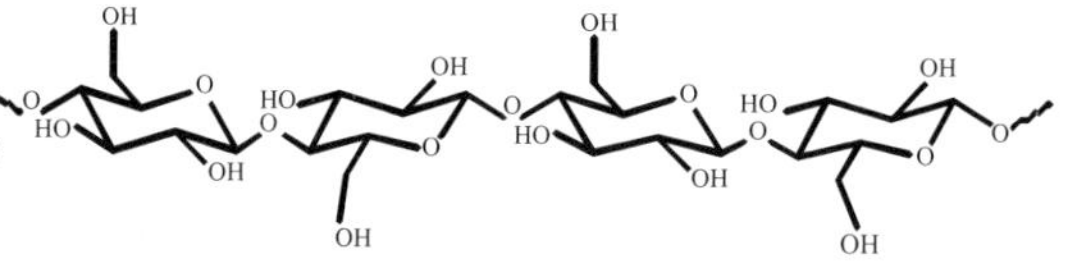

**D**
Amylose

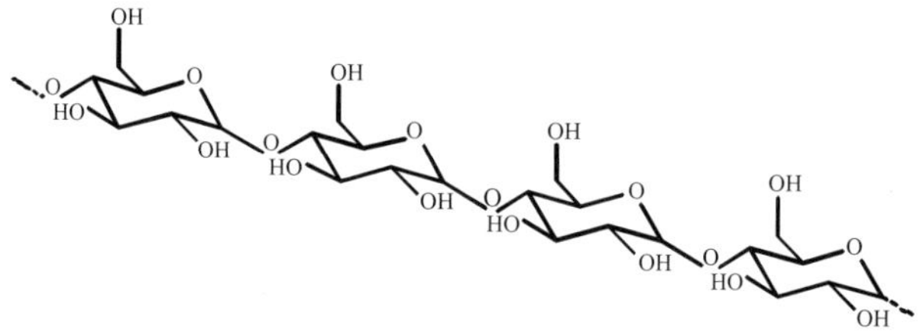

**E.**
Amylopectin, glycogen

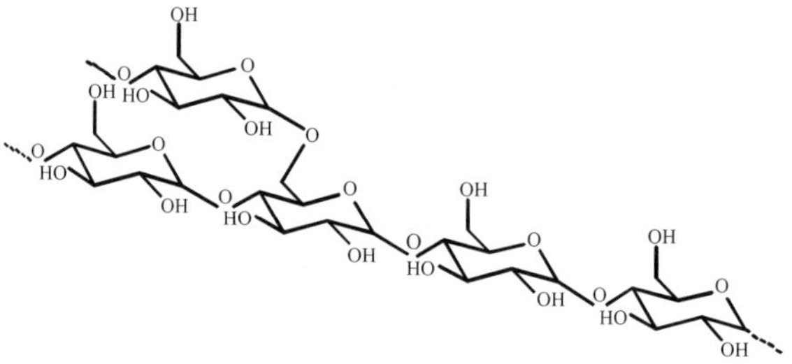

**F.**
Hyaluronic acid

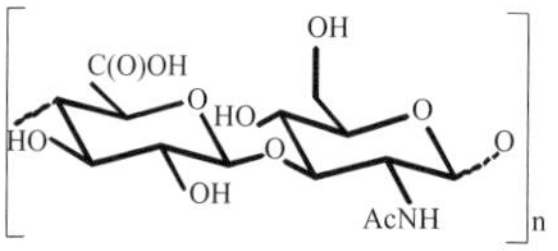

**Fig. 5.11** Examples of carbohydrate-containing compounds of diverse categories.

erythrocytes, kidney cells, and brain tissue has been reported and comprises several steps that will not be discussed here [65–69].

### 5.3.4 *Oligosaccharides as degradation products*

Oligosaccharides can result from degradation or hydrolysis of polysaccharides such as heparins/heparans, hyaluronic acid/hyaluronans, cellulose, starch, and glycogen (see Figure 5.11B–F). Each category calls for a different analytical pathway based on sample complexity, polarity and information needed.

Heparins are mucopolysaccharides with molecular masses ranging from 6000 to 40 000 Da. The polymeric chains are composed of repeating disaccharide units of D-glucosamine and uronic acid linked by 1••4 glycosidic bonds. The uronic acid residues can be either D-glucuronic acid or L-iduronic acid. Few hydroxyl groups on each of these monosaccharide residues may be sulfated, giving rise to negatively charged polymers. For a long time, highly charged heparin and heparan sulfate have been used therapeutically as commercial anticoagulants. It has been difficult to study the structure, mechanism, and structure–activity relationships of these molecules because they are hard to purify [70]. Kuberan *et al.* [70] used reversed-phase capillary HPLC with micro ESI-MS to characterize large heparan sulfate precursors (heparosans). A review has been published on the determination of structure and function of heparin and heparan sulfate [71]. Hyaluronic acid and hyaluronans are similar to heparins and heparans, but with a different linkage position in the dimer unit, i.e. the disaccharide unit is assembled by a 1••3 glycosidic bond.

Cellulose and starch are better known polymers. The former is made entirely of glucose arranged in a linear chain and is a major structural material in the plant kingdom. The term starch is used to describe a biopolymer system comprising predominantly two polysaccharides, amylose and amylopectin. These two polysaccarides are made of glucose monomers. The smaller one, amylose, is a linear molecule comprising 1••4 linked α-D-glucopyranosyl units. There is a small degree of branching through α1••6 linkages. The larger of the two components, amylopectin, has a similar composition but is more highly branched than amylose, with a much greater molecular mass. The primary structure of glycogen resembles that of amylopectin, but glycogen is even more highly branched, with branch points occurring every 8–12 glucose residues.

These materials will not be explored further in this section, as reviews have been written on their clean-up and analysis (e.g. [24]). The next section proposes some analytical methods for all types of oligosaccharides.

## 5.4 Methods

### 5.4.1 *General*

This section summarizes several methods and techniques used in carbohydrate analysis. Generally, these methods are also applicable to other classes of biological compounds, although some are specific to carbohydrates. Rather than being divided into subsections on different carbohydrate types, this portion of the chapter describes some analytical and/or clean-up methods based on the principal technique involved. Thus methods based on electrophoresis, liquid chromatography, derivatization and mass spectrometry, among other principles, will be listed and briefly discussed.

### 5.4.2 *Gel electrophoretic separations*

Polyacrylamide gel electrophoresis (PAGE) remains one of the most efficient methods for protein separation. One-dimension and 2D gels where the separation in each dimension is controlled by a different property, e.g. $pK_a$ or molecular mass, are still often preferred to chromatography or CE because of simplicity of operation and reproducibility.

Based on techniques developed for proteins, it is possible to separate carbohydrates based on size and charge, and thus mobility, on electrophoretic gels. The most popular method to achieve this type of separation in recent years has been FACE, first introduced by Jackson in 1990 [22]. This method was inspired by sodium dodecyl sulfate (SDS) PAGE, but instead of using SDS detergent bound to proteins by non-covalent interactions, sugars are covalently labeled with a trisulfated label, for example 8-aminonaphthalene-1,3,6-trisulfonate (ANTS) (see Fig. 5.5) This label confers a constant charge to the sugars, making migration mainly dependent on size and geometry. An elaborate system has been developed to link migration positions to oligosaccharide compositions [72,73]. This system is based on 'glucose units' (GU) and compares positions of migrating unknowns with those of dextran ladder chains with known numbers of glucose residues, corresponding to GU indices. Each monosaccharide residue present other than glucose in the unknown structure contributes differently to the GU system. For example, sialic acid residues and GlcNAc residues will behave in the opposite fashion when subjected to an electric field. Using an arithmetic sum of all pre-established contributions assigned to every monosaccharide in the carbohydrate under consideration, it is possible to predict very closely the migration distance from the gel top for a known oligosaccharide, and vice versa for an unknown oligosaccharide [72,73].

Typically, sugar samples subjected to separation by this method should be isolated as a group from other biological compounds before ANTS labeling and electrophoresis. After sugars have been separated on the gel and proposed structures have been assigned to each band, it is possible to cut the bands, elute the labeled sugars off the gel and thus use FACE as an isolation technique for particular targeted species. This has proven useful, especially in cases where isolated sugars were subjected to exoglycosidase digestion arrays and re-analyzed by FACE [47]. In such cases, FACE results can be compared with similar array studies performed using MALDI [36]. With FACE, detection of the bands is performed by fluorescence, produced by the ANTS group present on each separated sugar upon irradiation. Obtaining a picture of the final migration positions of the carbohydrates is possible by capturing the fluorescence signals on film. Algorithms have been designed to quantify the electrophoresed sugars from a scanned picture [46]. Figure 5.12 shows an example of FACE separation of oligosaccharides isolated from IgG glycoproteins [46].

### 5.4.3 *Capillary electrophoresis separations*

Most capillary electrophoretic (CE) methods rely on UV or fluorescence for the detection of analytes. The difficulty in analyzing native carbohydrates using CE-derived

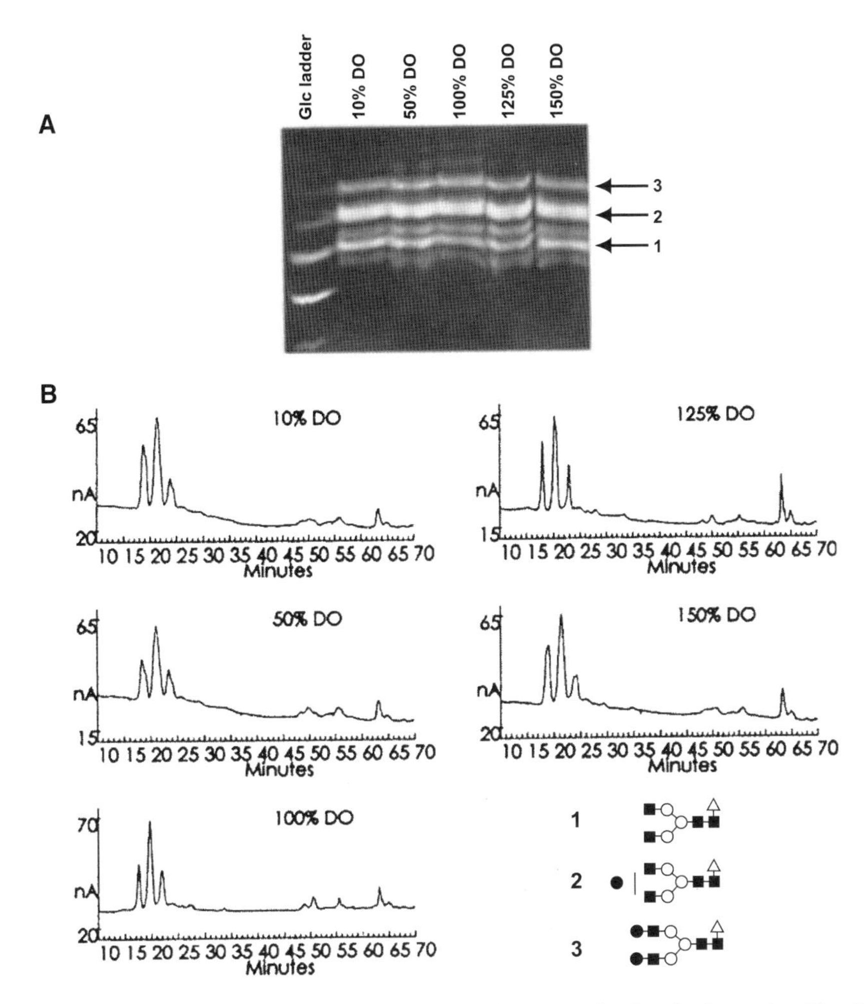


**Fig. 5.12** Glycan analysis of the *N*-linked oligosaccharides from monoclonal antibodies produced in a Bioreactor 2 in 10, 50, 100, 125, and 150% dissolved glycans. (A) FACE elecropherogram of 4 nmol of total released and ANTS-labeled glycans in each lane. (B) HPAEC-PAD chromatograms of 2 nmol of total released glycans in each chromatogram. Three major bands and peaks were observed and identified. See Fig. 5.4 caption for symbol legend. Adapted from [46] with permission.

techniques stems from the lack of suitable chromophores in these molecules. An entire issue of the journal *Biomedical Chromatography* (volume 16, 2002) [59] has been dedicated to CE of oligosaccharides and covers several aspects of CE applications and CE-related techniques; review and research articles describe the analysis of glycosaminoglycans in general [74,75], glycoprotein oligosaccharides [76], lipoglycans [77], heparin [78], neutral [79] and acidic [80] polysaccharides, sialic acid content [81], oligosaccharides from human milk [60], and hyaluronans [82]. Methods surveyed encompass detection techniques ranging from UV, fluorescence, laser-induced fluorescence (LIF), to electrochemistry and MS. Strousopolou *et al.* [81] emphasized the need for sample purification by ultrafiltration for the analysis of sialic acid content in blood

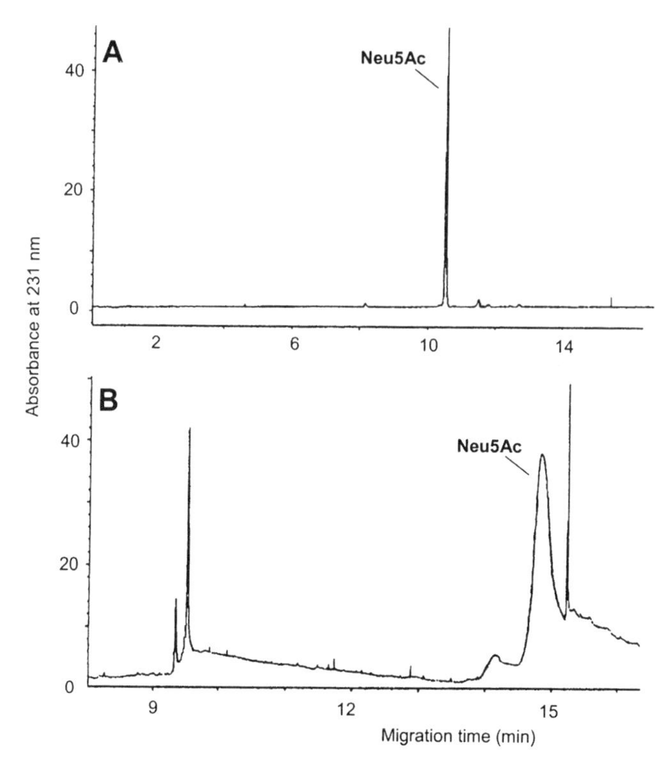


**Fig. 5.13** CZE analysis of per-*O*-benzoylated Neu5Ac in an $\alpha$1-acid glycoprotein solution that has been purified with ultrafiltration on Centricon-3 membrane before derivatization (A) and of the same sample analyzed without ultrafiltration (B). Adapted from [81] with permission.

sera. Figure 5.13 shows UV detection traces obtained for purified and non-purified perbenzoylated samples [81].

Che *et al.* [83] and Gennaro *et al.* [84] have been successful at separating ANTS-labeled oligosaccharides in dextran ladder mixtures [83] and *N*-linked glycans from glycoproteins [84] with ESI-MS detection. Other groups have also used CE/MS for the separation and detection of other classes of oligosaccharides, e.g. glycosaminoglycans [85,86]. On-line CE/MS restricts the choice of electrophoretic buffer to volatile compounds such as ammonium formate or acetate, in order to avoid interferences in the ionization process and excessive background signals. For better results, oligosaccharide samples injected in CE systems should be free from other types of biological molecules. The presence of low molecular mass salt components in samples may be acceptable, if not in excessive concentration, because of their high CE mobility relative to analytes. Problems arising from the presence of analytes other than sugars may be circumvented by the use of specific ion monitoring (e.g. monitoring of label-related ions for sugars) or of a specific UV wavelength to which other analytes do not respond. Also, samples should be as concentrated as possible to minimize diffusion effects during injection. Preconcentration techniques have been developed to reduce the width of the sample plug and thus improve the efficiency of CE injection and separation [87].

### 5.4.4 *Anion exchange chromatography*

The ionic nature of some types of oligosaccharides makes them good candidates for separation using methods based on anion exchange chromatography. Native neutral and acidic oligosaccharides have often been analyzed by HPAEC-PAD, as originally described by Townsend *et al.* [88]. This method takes advantage of the presence of reducing termini of sugars and their ability to become oxidized easily. This transformation causes a current change at the detection electrodes and is recorded as a quantitative signal, reflecting the amount of material being oxidized. The high pH values are used to help deprotonation of the oligosaccharides according to their respective p$K_a$ values and thus help to control separation. This method has been applied to the separation of oligosaccharides of different origins into subgroups of neutral, acidic, sulfated, phosphorylated sugars. For example, acidic sugars from glycoproteins have been separated into monosialylated, tri- and disialylated using HPAEC-PAD [44–47], as shown in Fig. 5.12B. With this method, labeling cannot be used, because the reducing end of carbohydrates must not be blocked. This is in most cases an advantage, and if needed for further analyses the fractions can be dialyzed and labeled after separation.

The direct on-line operation of HPAEC and ESI-MS has been attempted and successfully utilized by Conboy and Henion [89]. The main difficulty was to eliminate the potassium or sodium hydroxide used to give the high pH values. Many researchers have turned to weak anion-exchange chromatography instead of HPAEC, because the buffers used are milder and more amenable to UV and MS detection, which implies that sugars should be labeled before analysis. This method also allows sugars to be classified in groups of neutral and acidic compounds.

Ion exchange resin beads are commonly used for sample clean-up in MS. Without loading them into a column, the beads can be used as an additive to the solution to capture salts as a sample purification step. Beads can then be centrifuged or filtered out of the solution before ESI or MALDI analysis. In other cases, beads may be used to capture the analyte itself, among other classes of molecules.

### 5.4.5 *Affinity chromatography*

Some compounds have natural affinities for carbohydrates and can be useful in the development of separation methods based on affinity chromatography. A very simple example is graphitized carbon which Packer *et al.* [90] showed is useful for desalting glycans samples *before* chromatography and/or MS. More on the biological side, lectin-affinity columns such as concanavalin A, lentil lectin, *Phaseolus vulgaris* erythroagglutinin, *Ricinus communis* agglutinin I, *Triticum vulgaris* agglutinin, *Glycine max* agglutinin and *Ulex europaeus* agglutinin have been used in a sequential scheme by Hoja-Lukowicz *et al.* [91] to study branched oligosaccharides in rat liver beta-glucuronidase. Affinity chromatography allowed first proteolytic glycopeptides to be isolated from nonglycosylated materials, and then separation of glycopeptides into subclasses [91].

Hirabayashi *et al.* [92] studied the oligosaccharide specificity of galectins by frontal affinity chromatography [92]. Galectins have been described as widely distributed sugar-binding proteins whose basic specificity for beta-galactosides is conserved by evolutionarily preserved carbohydrate-recognition domains (CRDs). Lectin affinity chromatography has also played an important role in determining the carbohydrate structures of the human immunodeficiency virus (HIV) recombinant envelope glycoprotein gp120 produced in Chinese hamster ovary (CHO) cells [93]. Recently, Kaji *et al.* [10] published a study describing the lectin affinity capture, isotope-coded tagging and MS to identify *N*-linked glycoproteins. Concavalin A was used as the lectin medium to first isolate glycoproteins from other materials in a complex biological matrix, and a second time to isolate glycopeptides from other peptides following proteolysis [10].

### 5.4.6 *HPLC and HPLC/ESI-MS*

Several studies have reported the use of direct interfacing of HPLC with MS for carbohydrate analysis. Fortuitously, most derivatives developed for sugar separation by HPLC-UV absorption or fluorescence measurements also serve to enhance MS signals in ESI and MALDI modes. Labeling methods developed for HPLC of carbohydrates are numerous, and most rely on Schiff-base chemistry followed by reduction. It is thus important to select a reaction which is specific to the reducing termini of sugars, quantitative, simple to carry out, and which has a simple post-reaction clean-up procedure. Desialylation or other types of defunctionalization i.e. reactions using excessively high or low pH values, should be avoided. One of the most popular methods for HPLC is the use of 2-AB labeling [17]. Other labels used for HPLC include trimethyl(*p*-aminoethyl)amine (TMAPA) [21], ABEE [94] and its aminoethyl version (ABDEAE) [95], 2-AP [21], 2-AMAC [18], PMP [17], and PHN [20,30]. Figure 5.5 shows the structures of some of the aforementioned derivatives. These derivatization methods are applicable to *N*-linked glycans from glycoproteins, maltopentose moieties, dextran ladders, oligosaccharides from milk, and any oligosaccharide with a reducing end available for labeling. Most reducing-end labeled oligosaccharides yield limited extents of separation on reversed phase columns ($C_{18}$ or $C_8$); however, separation can be enhanced if a low polarity normal phase packing is used, for example an amino- or cyano-column (e.g. [26,47,96]).

Labeling with phenylhydrazine as a reducing end tag has recently been applied to the determination of ovalbumin glycans by $C_{18}$ reversed-phase HPLC/ESI-MS [30], as shown in Figure 5.14 by several selected ion chromatograms.

HPLC/MS of glycoconjugates, e.g. glycosphingolipids, also necessitates derivatization. If MS only is used as the detection method, permethylation and peracetylation are often employed to make the compounds more volatile and enhance MS sensitivity, but if UV detection is considered, perbenzoylation yields better results on compounds with blocked reducing termini [58].

Protocols have been developed for HPLC determination of glycosaminoglycans, e.g. with conductivity detection [97] and UV detection using PMP labeling [98]. Other studies combined CE and HPLC [99], or else HPLC and gel electrophoresis [100,101].

A recent HPLC/ESI-MS study published by Kühn *et al.* [86] showed quantitation of three different hyaluronic acid fragment preparations (HAF) derived from hyaluronic acid (HA) present in pharmaceutical formulations. These authors succeeded in carrying out HAF analysis in less than 8 min; no labeling was involved in their procedure [86].

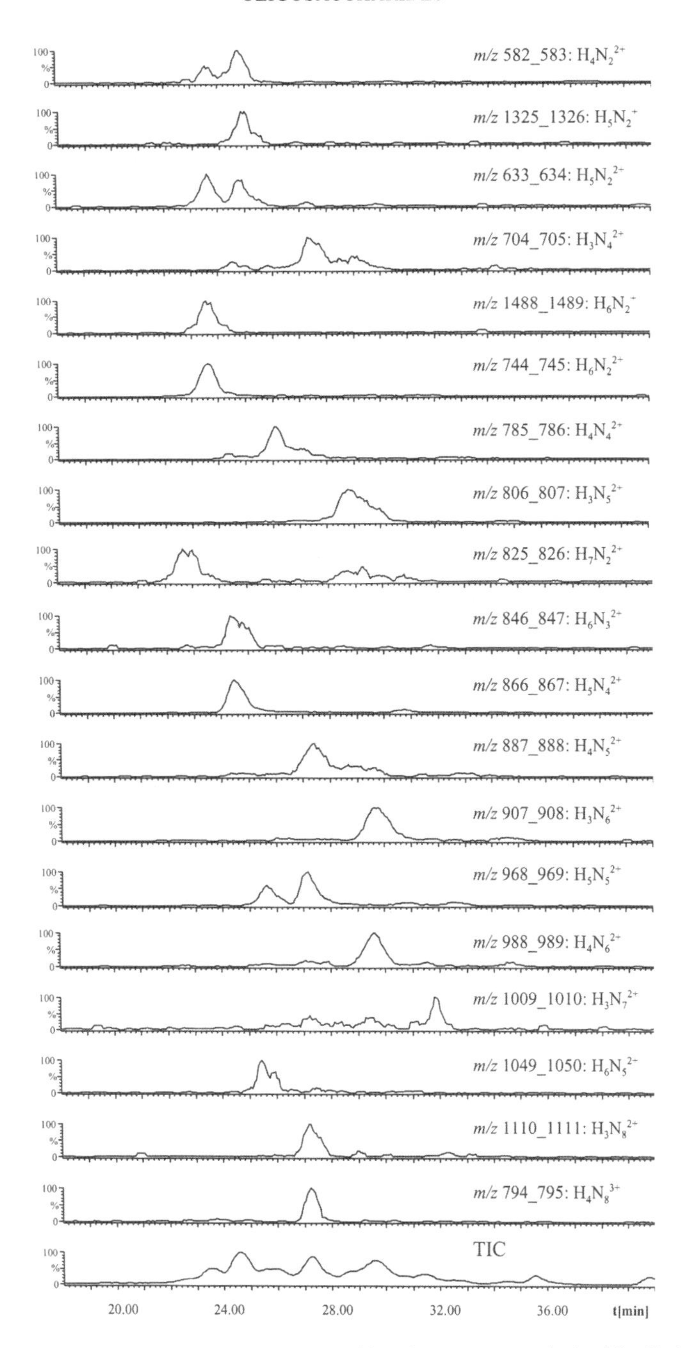


**Fig. 5.14** Reversed-phase HPLC/MS selected and total ion chromatograms obtained for *N*-glycans detached from Sigma grade VII hen ovalbumin and labeled with phenylhydrazine. H = hexose (mannose, galactose), N = *N*-acetylglucosamine. Adapted from [30].

### 5.4.7 *Off-line HPLC/MALDI-MS*

Since its development by Karas, Hillenkamp and coworkers in the mid to late 1980s [102–104], matrix-assisted laser desorption/ionization mass spectrometry (MALDI MS) has become increasingly popular as an off-line detection method in several separation techniques. Hillenkamp and coworkers [105] were also among the first researchers to prove that proteins separated on electrophoretic gels and electroblotted onto membranes could be analyzed by MALDI and identified from individual spots. These studies and others led to in-gel digestion techniques for proteomics [106] and to well accepted approaches that are now routinely applied in protein analyses, e.g. [107]. In-gel digestion procedures have been extended to deglycosylation, which allows sugars to be released and analyzed off-line by MALDI or another technique [12,13,108].

Besides gel electrophoresis as a main sample separation technique, other groups have worked on developing off-line CE sample fraction collection devices for MALDI targets [109,110], and off-line HPLC-type collection devices [111–113]. In these devices, MALDI matrix is either predeposited on the target or mixed with the analyte as a solution during deposition. There are advantages to on-line HPLC/MS analysis, e.g. no sample is lost to obtain maximal signal, and a direct relationship is available between retention time and molecular mass or nature of the eluting substances. However, the main advantage of using off-line collection for MALDI is the ability to re-analyze the samples if desired. MALDI is also more tolerant of buffers and other impurities than ESI-MS, and therefore some additives may be present, in low concentration, in the HPLC or CE eluents, without significantly interfering with the MALDI analyte signals. Provided that buffers and additives are volatile and remain at relatively low concentration, any CE or HPLC separation scheme may be used for deposition, because the solvent will not be involved in the ionization process as in the case of ESI-MS. So far, most off-line MALDI studies have reported peptide separations, where HPLC or CE have been used following 1D or 2D gel separations and in-gel proteolytic digestions. ZipTips have been used as an alternative to HPLC with $C_{18}$ packing or affinity packing with solvent gradient to provide elution of different classes of peptides or proteins [112–114].

Successful separations of labeled carbohydrates by HPLC, e.g. [12,26,30,47,96] suggest that off-line sample deposition using packed pipet tips would be valuable for MALDI, because the fractions collected would be simpler to analyze than entire oligosaccharide pools, for any type of oligosaccharide sample.

Another different yet simple off-line sample clean-up method consists of removing detergent and salt additives from samples before MALDI analysis. The medium used is a hydrophobic resin, which works based on micro-adsorption [115]. This method can precede ZipTip or HPLC fractionation and can generate samples suitable for MALDI MS.

## 5.5 Summary and general considerations

The preparation of oligosaccharide samples for either HPLC/ESI-MS or off-line HPLC/MALDI-MS often includes several purification steps. As for any other analytes, the number of steps tends to increase with the complexity of the biological matrix under investigation for target substances. Because HPLC is a purification technique in itself, and as ESI-MS is very specific detector, the hyphenation of these two techniques allows researchers to combine steps and thus shorten analysis time. Carbohydrates, being polar and in some cases ionic, are in principle well suited to ESI-MS and MALDI MS conditions. These thermolabile molecules, however, tend to break down easily and generally yield low ionization efficiencies. If other compounds are present in a sample, carbohydrate signals are often suppressed, especially if the other components are peptides. Preparation of sugar derivatives is thus commonly used, and the products are either less polar or more volatile than their native counterparts. These properties enhance both MALDI and ESI signals.

It is important to minimize the use of detergents for purification procedures, because they interfere with ionization processes. During HPLC separations, detergents tend to coat column packings and elute constantly and together with the analytes. They should be avoided or removed before separation. Nonvolatile buffers, in contrast, can be removed by HPLC if their concentration is not excessive. Volatile buffers are compatible with ESI and MALDI, especially with the latter, which is more impurity-tolerant than ESI.

Bipolar compounds, including glycolipids, do not contain a sugar residue with a reducing terminus. They may be difficult to isolate due to their ambivalent properties. A combination of permethylation (for MS), perbenzoylation (for HPLC and MS) can be useful in their determination.

The study of *O*-linked and *N*-linked glycans from glycoproteins is constantly gaining in popularity and importance. The completion of several genomic maps has contributed to raise the level of interest in understanding how the building blocks of organisms, proteins, function together. Researchers working in the area of proteomics have already established the partial or complete proteomes of several types of cells. As a result, glycomics, which is the study of protein glycosylation in cells under specific expression conditions, is also gaining popularity as the roles of protein carbohydrates are becoming better understood and thus somewhat controllable. At the cellular level, glycoprotein concentrations are relatively low, and this has brought cell engineers to use recombinant technologies to produce larger amounts of specific glycoproteins by overexpression in cells cultured under controlled conditions. Several recombinant glycoproteins are of great interest to pharmaceutical companies as agents for therapeutic use [116]. Glycobiologists, analytical chemists and mass spectrometrists can take advantage of the larger amounts of recombinant glycoproteins available to better study the structures of *N*- and *O*-glycans, and understand their roles and functions within the cell.

Overall, oligosaccharides are ubiquitous in most biological environments. They form a major portion of the food we ingest, constitute a major form of energy in most living organisms, and are important indicators of disease in certain pathological states. Whether they are found in the free or conjugated form, the roles, structures and functions of oligosaccharides continue to interest numerous scientists. Mass spectrometry has certainly played an important role in raising this level of interest. The on-line and off-line hyphenation of separation techniques such as HPLC, PAGE, CE, AEC and AC to ESI-MS or MALDI MS offers an invaluable tool to complement other analytical techniques commonly used in glycobiology. Sample preparation is and will remain a subject of primary importance, and the development of protocols to enable oligosaccharide analysis by HPLC-MS for different kinds of sample matrices is still under investigation in several research groups around the world.

## Acknowledgements

The author thanks the National Science and Engineering Council of Canada (NSERC), the Canadian Foundation for Innovation (CFI) and the Canada Research Chair Program for funding.

## References

1. Schäffer, C., Graninger, M. & Messner, P. (2001) Prokaryotic glycosylation, *Proteomics*, **1**, 248–261.
2. Bahl, O.P. (1992) Introduction to glycoproteins, in *Glycoconjugates: Composition, Structure, and Function*, (eds H.J. Allen & E.C. Kisailius), Marcel Dekker, New York. pp. 1–12.
3. Varki, A. (1993) Biological roles of oligosaccharides: all of the theories are correct, *Glycobiology*, **3**(2), 97–130.
4. Haynes, P.A. (1998) Phosphoglycosylation: a new structural class of glycosylation? *Glycobiology*, **8**, 1–5.
5. Kobata, A. (1992) Structures and functions of the sugar chains of glycoproteins, *Eur. J. Biochem.*, **209**, 483–501.
6. Krokhin, O., Li, Y., Andonov, A., Feldman, H., Flick, R., Jones, S., Stroeher, U., Bastien N., Dasuri, K.V.N., Cheng, K., Simonsen, J.N., Perreault, H., Wilkins, J., Ens, W., Plummer, F., & Standing, K.G. (2003) Mass spectrometric characterization of proteins from the SARS virus: a preliminary report, *Mol. Cell. Proteomics*, **2**, 346–356.
7. Mort, A.J., & Pierce, M.L. (2002) Preparation of carbohydrates for analysis by modern chromatography and electrophoresis, J. Chromatogr. Library Vol 66 on *Carbohydrate Analysis by Modern Chromatography and Electrophoresis*, (ed., Z. El Rassi), Elsevier, Amsterdam., pp. 3–38.
8. Tretter, V., Altmann, F. & Marz, L. (1991) Peptide-N4-(*N*-acetyl-beta-glucosaminyl) asparagine amidase F cannot release glycans with fucose attached alpha 1–3 to the asparagine-linked *N*-acetylglucosamine residue, *Eur. J. Biochem.*, **199**, 647–652.
9. Wilson, I.B.H., Zeleny, R., Kolarich, D., Staudacher, E., Stroop, C.J.M., Kamerling, J.P. & Altmann F. (2001) Analysis of Asn-linked glycans from vegetable foodstuffs: widespread occurrence of Lewis A, core 1,3-linked fucose and xylose substitutions", *Glycobiology.*, **11**, 261–274.
10. Kaji, H., Saito, H., Yamauchi, Y., Shinkawa, T., Taoka, M., Hirabayashi, J., Kasai, K., Takahashi, N. & Isobe, T. (2003) Lectin affinity capture, isotope-coded tagging and mass spectrometry to identify *N*-linked glycoproteins, *Nature Biotechnol.*, **21**, 667–672.
11. Zhang, H., Li X., Martin, D.B. & Aebersold, R. (2003) Identification and quantification of *N*-linked glycoproteins using hydrazide chemistry, stable isotope labeling and mass spectrometry, *Nature Biotechnol.*, **21**, 660–666.
12. Rudd, P.M., Colominas, C., Royle, L., Murphy, N., Hart, E., Merry, A.H., Hebestreit H.F. & Dwek R.A. (2001) A high-performance liquid chromatography based strategy for rapid, sensitive sequencing of *N*-linked oligosaccharides: modifications to proteins in sodium dodecyl sulphate polyacrylamide electrophoresis gel bands, *Proteomics*, **1**, 285–294.
13. Kuster, B., Krogh, T.N., Mortz, E. & Harvey, D.J. (2001) Glycosylation analysis of gel-separated proteins, *Proteomics.*, **1**, 350–361.
14. van Montfort, B.A., Canas, B., Duurkens, R., Godovac-Zimmermann, J. & Robillard, G.T. (2002) Improved in-gel approaches to generate peptide maps of integral membrane proteins with matrix-assisted laser desorption/ionization time-of-flight mass spectrometry, *J. Mass Spectrom.*, **37**, 322–330.
15. Stimson, E., Hope, J., Chong, A. & Burlingame, A.M. (1999) Site specific charactrerization of the *N*-linked glycans of murine prion protein by high-performance liquid chromatography/electrospray mass spectrometry and exoglycosidases digestions, *Biochemistry.*, **38**, 4885–4895.
16. Cooper, C.A., Gasteiger, E. & Packer, N.H. (2001) GlycoMod - a software tool for determining glycosylation compositions from mass spectrometric data, *Proteomics* **1**, 340–349.
17. Guile, G.R., Rudd, P.M., Wing, D.R., Prime, S.B. & Dwek R.A. (1996) A rapid high-resolution high-performance liquid chromatographic method for separating glycan mixtures and analyzing oligosaccharide profile. *Anal. Biochem.*, **240**, 210–226.

18. Okafo, G.N., Burrow, L.N., Carr, S.A., Roberts, G.D., Johnson, W. & Camilleri, P. (1996) A coordinated high performance liquid chromatographic, capillary electrophoretic and mass spectrometric approach for the analysis of oligosaccharide mixtures derivatised with 2-aminoacridone, *Anal. Chem.*, **68**, 4424–4430.
19. Honda, S., Akao, E., Suzuki, S., Okuda, M., Kakehi, K. & Nakamura, J. (1989) High-performance liquid chromatography of reducing carbohydrates as strongly ultraviolet-absorbing and electrochemically sensitive 1-phenyl-3-methyl-5-pyrazolone derivatives, *Anal. Biochem.*, **180**, 351–357.
20. Lattova, E. & Perreault, H. (2003) Labelling saccharides with phenylhydrazine for electrospray and matrix-assisted laser desorption/ionization mass spectrometry, *J. Chromatogr. B*, **793**, 167–179.
21. Kondo, A., Suzuki, J., Kuraka, N., Hase, S., Kato, I. & Ikenaka, J. (1990) Improved method for fluorescence labeling of sugar chain with sialic acid residues, *Agric. Biol. Chem.*, **54**, 2169–2170.
22. Jackson, P. (1990) The use of polyacrylamide-gel electrophoresis for the high-resolution separation of reducing saccharides labelled with the fluorophore 8-aminonaphthalene-1,3,6-trisulphonic acid, *Biochem. J.*, **270**, 705–713.
23. Suzuki, H., Muller, O., Guttman A. & Karger B.L. (1997) Analysis of 1-aminopyrene-3,6,8-trisulfonate-derivatized oligosaccharides by capillary electrophoresis with matrix-assisted laser desorption/ionization time-of-flight mass spectrometry, *Anal. Chem.*, **69**, 4554–4559.
24. Harvey, D.J. (2003) Matrix-assisted laser desorption/ionization mass spectrometry of carbohydrates and glycoconjugates, *Int. J. Mass Spectrom.*, **226**, 1–35.
25. Huang, L. & Riggin, M.R. (2000) Analysis of nonderivatized neutral and sialylated oligosaccharides by electrospray mass spectrometry, *Anal. Chem.*, **72**, 3539–3546.
26. Charlwood, J., Birrell, H., Bouvier, E.S.P., Langridge, J. & Camilleri P. (2000) Analysis of oligosaccharides by microbore high-performance liquid chromatography, *Anal. Chem.*, **72**, 1469–1474.
27. Creaser, C.S., Reynolds, J.C. & Harvey D.J. (2002) Structural analysis of oligosaccharides by atmospheric pressure matrix-assisted laser desorption/ionization quadrupole ion trap mass spectrometry, *Rapid Commun. Mass Spectrom.*, **16**, 176–184.
28. Sagi, D., Peter-Katalinic, J., Conradt, H.S. & Nimtz M. (2002) Sequencing of tri- and tetreantennary *N*-glycans containing sialic acid by negative mode ESI QTOF tandem MS, *J. Am. Soc. Mass Spectrom.*, **13**, 1138–1148.
29. Saba, J.A., Kunkel, J.P., Chan, D.C.H., Ens, W., Standing, K.G., Butler, M., Jamieson, J. C. & Perreault, H. (2002) A study of IgG glycosylation in monoclonal and polyclonal species by electrospray and matrix-assisted laser desorption/ionization mass spectrometry, *Anal. Biochem.*, **305**, 16–31.
30. Lattova, E. & Perreault H. (2003) Profiling of *N*-linked oligosaccharides using phenylhydrazine derivatization and mass spectrometry, *J. Chromatogr. B*, **793**(1), 167–179.
31. Muehlecker, W., Gulati, S., McQuillen, D.P., Ram, S., Rice, P.A. & Reinhold, V.N. (1999) An essential saccharide binding domain for the mAb 2C7 established for *Neisseria gonorrhoeae* LOS by ES-MS and MS$^n$, *Glycobiology*, **9**, 157–171.
32. Cipollo, J.F., Costello, C.E. & Hirschberg, C.B. (2002) The fine structure of *Caenorhabditis elegans N*-glycans, *J. Biol. Chem.*, **277**, 49143–49157.
33. Kornfeld, K. & Kornfeld, S. (1985) Assembly of asparagine-linked oligosaccharides, *Annu. Rev. Biochem.*, **54**, 631–664.
34. Ethier, M., Saba, J.A., Ens, W., Standing, K.G. & Perreault, H. (2002) Automated structure assignment of derivatized complex *N*-linked oligosaccharides from tandem mass spectra, *Rapid Commun. Mass Spectrom*, **16**, 1743–1754.
35. Mizuno, Y., Sasagawa, T., Dohmae, N. & Takio, K. (1999) An automated interpretation of MALDI/TOF postsource decay spectra of oligosaccharides. 1. Automated peak assignment, *Anal. Chem.*, **71**, 4746–4771.
36. Mechref, Y. & Novotny, M. (1998) Mass spectrometric mapping and sequencing of *N*-linked oligosaccharides derived from submicrogram amounts of glucoproteins, *Anal. Chem.*, **70**, 455–463.
37. Roule, L., Mattu, T.S., Hart, E., Langridge, J.I., Merry, A.H., Murphy, N., Harvey, D.J., Dwek, R.A. & Rudd, P.M. (2002) An analytical and structural database probides a strategy for sequencing *O*-glycans from microgram quantities of glycoproteins, *Anal. Biochem.*, **304**, 70–90.
38. Novotny, M., Huang, Y. & Mechref, Y. (2001) Microsale nonreductive release of *O*-linked glycans for subsequent analysis through MALDI/TOF mass spectrometry and capillary electrophoresis, *Anal. Chem.*, **73**, 6063–6069.
39. Wilson, I.B.H., Gavel Y. & von Heije G. (1991) Amino acid distributions around *O*-linked glycosylation sites, *Biochem. J.*, **275**, 529–534.
40. Hansen, J.E., Lund, O., Tolstrup, N., Gooley, A.A., Williams, K.L. & Brunak, S. (1998) NetOglyc: prediction of mucin-type *O*-glycosylation sites based on sequence context and surface accessibility, *Glycoconj. J.*, *15*, 115–130.
41. Rademaker, G.J., Havekamp, J. & Thomas-Oates, J. (1993) Determination of glycosylation sites in *O*-linked glycopeptides: a sensitive mass spectrometric protocol, *J. Org. Mass Spectrom.*, **28**, 1536–1541.

42. Greis, K.D., Hayes, B.K., Comer, F.I., Kirk, M., Barnes, S., Lowary, T.L. & Hart, G.W. (1996) Selective detection of glycopeptides and site mapping by electrospray mass spectrometry, *Anal. Biochem.*, **234**, 38–49.
43. Hanisch, F.G., Jovanovic, M. & Peter-Katalinic J. (2001) Glycoprotein identification and localization of *O*-glycosylation sites by mass spectrometric analysis of deglycosylated/alkylaminylated peptide fragments, *Anal. Biochem.*, **290**, 47–59.
44. Kunkel, J.P., Jan, D.C.H., Jamieson, J.C. & Butler, M. (1998) Dissolved oxygen concentration in serum-free continuous culture affects *N*-linked glycosylation of a monoclonal antibody, *J. Biotechnol.*, **62**, 55–71.
45. Kunkel, J.P., Jan, D.C.H., Butler, M. & Jamieson, J.C. (2000) Comparisons of the glycosylation of a monoclonal antibody produced under nominally identical cell culture conditions in two different bioreactors, *Biotechnol. Progr.*, **16**, 462–470.
46. Kunkel, J.P. & Jamieson, J.C. (2003) Quantitation of fluorophore-assisted carbohydrate electrophoresis gels: comparison with high-pH anion-exchange chromatography and pulsed amperometric detection, *Anal. Biochem.*, **316**, 142–146.
47. Yang, M. & Butler, M. (2002) Effects of ammonia and glucosamine on the heterogeneity of erythropoietin glycoforms, *Biotechnol. Prog.* **18**, 129–138.
48. Yamashita, K.,Tachibana, Y. & Kobata, A. (1977) Oligosaccharides of human milk. Structural studies of two new octasaccharides, difucosyl derivatives of para-lacto-*N*-hexaose and para-lacto-*N*-neohexaose, *J. Biol. Chem.*, **252**, 5408–5411.
49. Bruntz, R., Dabrowski, U., Dabrowski, J., Ebersold, A., Peter-Katalinic, J. & Egge H. (1988) Fucose-containing oligosaccharides from human milk from a donor of blood group O Le(a) nonsecretor, *Biol. Chem. Hoppe Seyler.*, **369**, 257–273.
50. Strecker, G., Wieruszeski, J.M., Michalski J.C. & Montreuil J. (1989) Primary structure of human milk nona- and decasaccharides determined by a combination of fast atom bombardment mass spectrometry and $^{1}$H-/$^{13}$C-nuclear magnetic resonance spectroscopy. Evidence for a new core structure, iso-iacto-*N*-octaose *Glycoconj. J.*, **6**, 169–182.
51. Stahl, B., Thurl, S., Zeng J., Karas M., Hillenkemp F., Steup M. & Sawatzki G. (1994) Oligosaccharides from human milk as revealed by matrix-assisted laser desorption/ionization mass spectrometry, *Anal. Biochem.*, **223**, 218–226.
52. Finke, B., Stahl, B., Pfenninger, A., Karas, M., Daniel, H. & Sawatsky, G. (1999) Analysis of high-molecular-weight oligosaccharides from human milk by liquid chromatography and MALDI-MS, *Anal. Chem.*, **71**, 3755–3762.
53. Stahl, B., Setup, M., Karas, M. & Hillenkamp, F. (1991) Analysis of neutral oligosaccharides by matrix-assisted laser desorption ionization mass spectrometry, *Anal. Chem.*, **63**, 1463–1466.
54. Thurl, S., Offermanns, J., Mueller-Werner, B. & Sawatzki, G. (1991) Determination of neutral oligosaccharide fractions from human milk by gel permeation chromatography, *J. Chromatogr.*, **568**, 291–300.
55. Thurl, S., Mueller-Werner, B. & Sawatzki, G. (1996) Quantification of individual oligosaccharide compounds from human milk using high-pH anion-exchange chromatography, *Anal. Biochem.*, **235**, 202–206.
56. Pfenninger, A., Karas, M., Finke, B. & Stahl, B. (2002) Structural analysis of underivatized neutral human milk oligosaccharides in the negative ion mode by nano-electrospray $MS^n$. Part 1: Methodology, *J. Am. Soc. Mass Spectrom.*, **13**, 1331–1340.
57. Pfenninger, A., Karas, M., Finke, B., & Stahl, B. (2002) Structural analysis of underivatized neutral human milk oligosaccharides in the negative ion mode by nano-electrospray $MS^n$. Part 2: Application to isomeric mixtures, *J. Am. Soc. Mass Spectrom.*, **13**, 1341–1348.
58. Chaturvedi, P., Warren, C.D., Altaye, M., Morrow, A.L., Ruiz-Palacios, G., Pickering, L.K. & Newburg, D.S. (2001) Fucosylated human milk oligosaccharides vary between individuals and over the course of lactation, *Glycobiol.*, **11**, 365–372.
59. Karamanos, N. K., (ed.) (2002) Special issue on analysis of carbohydrate macromolecules by capillary electrophoresis (glycoproteins, proteoglycans/glycosaminoglycans and glycolipids). *Biomed. Chromatogr.*, **16**, 75–161 (10 articles).
60. Schmid, D., Behnte, B., Metzger, J. & Kuhn, R. (2002) Nano-HPLC–mass spectrometry and MEKC for the analysis of oligosaccharides from human milk, *Biomed. Chromatogr.*, **16**, 151–156.
61. Chalabi, S., Easton, R.L., Patankar, M.S., Lattanzio, F.A., Morrison, J.C., Panico, M., Morris, H.R., Dell A. & Clark G.F., (2002) The expression of free oligosaccharides in human seminal plasma, *J. Biol. Chem.*, **277**, 32562–32570.
62. Klein, A., Lebreton, J., Lemoine, J., Perini, J.M., Roussel, P. & Michalski, J.C. (1998) Identification of urinary oligosaccharides by matrix-assisted laser desorption ionization time-of-flight mass spectrometry, *Clin. Chem.*, **44**, 2422–2428.
63. Perreault, H., & Costello, C.E. (1994) "LSIMS, MS/MS and MALDI-TOF MS characterization of glycosphingolipid derivatives", *Org. Mass Spectrom.*, **29**, 720–735.

64. Perreault, H., Hronowski X.L., Koul O., Street J., McCluer R.H. & Costello, C.E. (1997) High sensitivity mass spectral characterization of glycosphingolipids from bovine erythrocytes, mouse kidney and fetal calf brain, *Int. J. Mass Spectrom. Ion Processes* **169/170**, 351–370.
65. Chien, J.L., Li S.C., Laine, R.A. & Li Y.T. (1978) Characterization of gangliosides from bovine erythrocyte membranes, *J. Biol. Chem.*, **253**, 4031–4035.
66. Chien, J.L., Li, S.C. & Li, Y.T. (1979) Isolation and characterization of a heptaglycosylceramide from bovine erythrocyte membranes, *Lipid Res.*, **20**, 669–673.
67. Sekine, M., Suzuki, M., Inagaki, F., Suzuki, A. & Yamakawa, T. (1987) A new extended globoglycolipid carrying the stage specific embryonic antigen-1 (SSEA-1) determinant in mouse kidney, *J. Biochem.*, **101**, 553–562.
68. Williams, M.A., Gross, S.K., Evans, J.E. & McCluer, R.H. (1988) Glycolipid stage-specific embryonic antigens (SSEA-1) in kidneys of male and female C57BL/6J and beige adult mice, *J. Lipid Res.*, **29**, 1613–1619.
69. Nudelman, E., Levery, S.B. & Hakomori, S. (1986) Novel fucolipids of human adenocarcinoma: characterization of the major ley antigen of human adenocarcinoma as trifucosylnonaosyl ley glycolipid (III3FucV3FucVI2FucnLc6), *J. Biol. Chem.*, **261**, 11247–11253.
70. Kuberan, B., Lech, M., Zhang, L., Wu Z.L., Beeler, D.L. & Rosenberg, R.D. (2002) Analysis of heparan sulfate oligosaccharides with ion pair-reverse phase capillary high performance liquid chromatography-microelectrospray ionization time-of-flight mass spectrometry, *J. Am. Chem. Soc.* **124**, 8707–8718.
71. Rabenstein, D.L. (2002) Heparin and heparan sulfate: structure and function. *Nat. Prod. Rep.*, **19**, 312–331.
72. Jackson, P. (1996) The analysis of fluorophore-labeled carbohydrates by polyacrylamide gel electrophoresis, *Mol. Biotechnol.*, **5**, 101–123.
73. Rudd, P.M., Guile, G.R., Kuster, B., Harvey, D.J., Opdenakker, G. & Dwek R.A. (1997) Oligosaccharide sequencing technology, *Nature* **388**, 205–207.
74. Mao, W., Thanawiroon, C. & Linhardt, R.J. (2002) Capillary electrophoresis for the analysis of glycosaminoglycans and glycosaminoglycan-derived oligosaccharides, *Biomed. Chromatogr.*, **16**, 77–94.
75. Lamari, F.N., Militsopoulou, M., Mitropoulou, T.N., Hjerpe, A. & Karamanos, N.K. (2002) Analysis of glycosaminoglycan-derived disaccharides in biological samples by capillary electrophoresis and protocol for sequencing glycosamingolycans, *Biomed. Chromatogr.*, **16**, 95–102.
76. Kakehi, K., Kinoshita, M. & Nakano, M. (2002) Analysis of glycoproteins and the oligosaccharides thereof by high-performance capillary electrophoresis – significance in regulatory studies on biopharmaceutical products, *Biomed. Chromatogr.*, **16**, 103–115.
77. Lamari, F.N., Gioldassi, X.M., Mitropoulou, T.N. & Karamanos, N.K. (2002) Structure analysis of lipoglycans and lipoglycan-derived carbohydrates by capillary electrophoresis and mass spectrometry, *Biomed. Chromatogr.*, **16**, 116–126.
78. Duchemin, V., Le Potier, I., Troubat, C., Ferrier, D. & Taverna M. (2002) Analysis of intact heparin by capillary electrophoresis using short end injection configuration, *Biomed. Chromatogr.*, **16**, 127–133.
79. Tokarz, M. Gustavasson P. & Stefansson, M. (2002) Employment of Detergent-tag/solute interactions in capillary electrophoresis of neutral polysaccharides, *Biomed. Chromatogr.*, **16**, 134–140.
80. Kinoshita M., Shiraishi, H., Muranushi, C., Mitsumori N., Ando, T., Oda, Y. & Kakehi, K. (2002) Determination of molecular mass of acidic polysaccharides by capillary electrophoresis, *Biomed. Chromatogr.*, **16**, 141–145.
81. Strousopoulou, K. Militsopoulou, M., Stagiannis, K., Lamari, F.N. & Karamanos, N. K. (2002) A. capillary zone electrophoresis method for determining *N*-acetylneuraminic acid in glycoproteins and blood sera, *Biomed. Chromatogr.*, **16**, 146–150.
82. Theocharis A.D. & Theocharis D.A. (2002) High-performance capillary electrophoretic analysis of hyaluronan and galactosaminoglycan-disaccharides in gastrointestinal carcinomas. Differential disaccharide composition as a possible tool-indicator for malignancies, *Biomed. Chromatogr.*, **16**, 157–161.
83. Che, F.Y., Song, J.F., Zeng, R., Wang, K.Y. & Xia Q.C. (1999) Analysis of 8-aminonapthalene-1,3,6-trisulfonate-derivatized oligosaccharides by capillary electrophoresis–electrospray ionization mass spectrometry, *J. Chromatogr. A*, **858**, 229–238.
84. Gennaro, L.A., Delaney, J., Vouros, P., Harvey, D.J. & Domon B. (2002) Capillary electrophoresis/electrospray ion trap mass spectrometry for the analysis of negatively charged derivatized and underivatized glycans, *Rapid Commun. Mass Spectrom.*, **16**, 192–200.
85. Zamfir, A., Seidler, D.G., Kresse, H. & Peter-Katalini, J. (2002) Structural characterization of chondroitin/dermatan sulfate oligosaccharides from bovine aorta by capillary electrophoresis and electrospray ionization quadrupole time-of-flight tandem mass spectrometry, *Rapid Commun. Mass. Spectrom*, **16**, 2015–2024.
86. Kühn, A.V., Rüttinger, H.H., Neubert, R.H.H. & Raith, K. (2003) Identification of hyaluronic acid oligosaccharides by direct coupling of capillary electrophoresis with electrospray ion trap mass spectrometry, *Rapid Commun. Mass Spectrom.*, **17**, 576–582.

87. Burgi, D.S. & Chien, R.L. (1996) Application and limits of sample stacking in capillary electrophoresis, *Methods Mol. Biol.*, **52**, 211–226.
88. Townsend, R.R., Hardy, M.R., Hindsgaul, O. & Lee Y.C. (1988) High performance anion-exchange chromatography of oligosaccharides using pellicular resins and pulsed amperometric detection, *Anal. Biochem.*, **174**, 459–470.
89. Conboy, J.J. & Henion, J. (1992) High-performance anion-exchange chromatography coupled with mass spectrometry for the determination of carbohydrates, *Biol. Mass Spectrom.*, **21**, 397–407.
90. Packer, N.H., Lawson, M.A., Jardine, D.R. & Redmond J.W. (1998) A general approach to desalting oligosaccharides reseased from glycoproteins, *Glycoconj. J.*, **15**, 737–747.
91. Hoja-Lukowicz, D., Litynska A. & Wojczyk, B.S. (2001) Affinity chromatography of branched oligosaccharides in rat liver beta-glucuronidase, *J. Chromatogr. B*, **755**, 173–183.
92. Hirabayashi, J., Hashidate, T., Arata, Y., Nishi, N., Nakamura, T., Hirashima, M., Urashima, T., Oka, T., Futai, M., Muller, W.E., Yagi, F. & Kasai, K. (2002) Oligosaccharide specificity of galectins: a search by frontal affinity chromatography, *Biochim. Biophys. Acta*, **1572**, 232–254.
93. Mizuochi, T., Spellman, M.W., Larkin, M., Solomon, J., Basa, L.J. & Feizi T. (1988) Carbohydrate structures of the human-immunodeficiency-virus (HIV) recombinant envelope glycoprotein gp120 produced in Chinese-hamster ovary cells, *Biochem. J.*, **254**, 599–603.
94. Wang, W.T., LeDonne, N.C., Jr. & Ackerman, B. (1984) Structural characterization of oligosaccharides by high-performance liquid chromatography, fast-atom bombardment-mass spectrometry, and exoglycosidase digestion, *Anal. Biochem.*, **141**, 366–381.
95. Yoshino, K.I., Takako, T., Muruta, H. & Shimonishi, Y. (1995) Use of the derivatizing agent 4-aminobenzoic acid 2-(diethylamino)ethyl ester for high-sensitivity detection of oligosaccharides by electrospray ionization mass spectrometry, *Anal. Chem.*, **65**, 4028–4031.
96. Saba, J.A., Shen, X., Jamieson, J.C. & Perreault, H. (2001) Investigation of different combinations of derivatization, separation methods and electrospray ionization mass spectrometry for standard oligosaccharides and glycans from ovalbumin, *J. Mass Spectrom.*, **36**, 563–574.
97. Chaidedgumjorn, A., Suzuki, A., Toyoda, H., Toida T., Imanari T. & Linhardt R.J. (2002) Conductivity detection for molecular mass estimation of per-O-sulfonated glycosaminoglycans separated by high-performance size-exclusion chromatography, *J. Chromatogr. A.*, **959**, 95–102.
98. Negishi, M., Sakamoto, H., Sakamaki, T., Ishikawa, O., Kanda, T., Tamura, J., Kurabayashi, M. & Nagai, R. (2003) Disaccharide analysis of glycosaminoglycans synthesized by cardiac myxoma cells in tumor tissues and in cell culture, *Life Sci.*, **73**, 849–856.
99. Alkrad, J.A., Merstani, Y. & Neubert, R.H. (2002) New approaches for quantifying hyaluronic acid in pharmaceutical semisolid formulations using HPLC and CZE, *J. Pharm. Biomed. Anal.*, **30**, 913–919.
100. Passos, C.O., Onofre, G.R., Martins, R.C., Graff, D.L., Pagani, E.A., Sodre, C.T. & Silva, L.C. (2002) Composition of urinary glycosaminoglycans in a patient with relapsing polychondritis, *Clin. Biochem.*, **35**, 377–381.
101. Maccari F. & Volpi, N. (2003) Direct and specific recognition of glycosaminoglycans by antibodies after their separation by agarose gel electrophoresis and blotting on cetylpyridinium chloride-treated nitrocellulose membranes, *Electrophoresis*, **24**, 1347–1352.
102. Karas, M., Bachmann, D. & Hillenkamp, F. (1985) Influence of the wavelength in high-irradiance ultraviolet laser desorption mass spectrometry of organic molecules, *Anal. Chem.*, **57**, 2935–2939.
103. Karas, M., Bachmann, D., Bahr, U. & Hillenkamp, F. (1987) Matrix-assisted ultraviolet laser desorption on non-volatile compounds, *Int. J. Mass Spectrom. Ion Processes*, **78**, 53–68.
104. Karas, M. & Hillenkamp, F. (1988) Laser desorption ionization of proteins with molecular masses exceeding 10,000 daltons, *Anal. Chem.*, **60**, 2299–2301.
105. Eckerson, C., Strupat, K., Karas, M., Hillenkamp, F. & Lottspeich, F. (1992) Mass spectrometric analysis of blotted proteins after gel electrophoretic separation by matrix-assisted laser desorption/ionization, *Electrophoresis*, **13**, 664–665.
106. Schevchenko, A., Wilm, M., Vorm, O. & Mann, M. (1996) Mass spectrometric sequencing of proteins from polyacrylamide gels, *Anal. Chem.*, **68**, 850–858.
107. Zhou, H., Ranish, J.A., Watts, J.D. & Aebersold, R. (2002) Quantitative proteome analysis by solid-phase isotope tagging and mass spectrometry, *Nature Biotechnol.*, **20**, 512–515.
108. Charlwood, J., Skehel, J.M. & Camilleri, P. (2001) Immobilisation of antibodies in gels allows the improved release and identification of glycans, *Proteomics*, **1**, 275–284.
109. Preisler, J., Hu, P., Rejtar, T., Moskovets, E., & Karger, B.L. (2002) Capillary array electrophoresis–MALDI mass spectrometry using a vacuum deposition interface, *Anal. Chem.*, **74**, 17–25.
110. Minarik, M., Klepárnik, K., Gilar, M., Foret, F., Miller, A.W., Sosic, Z. & Karger, B.L. (2002) Design of a fraction collector for capillary array electrophoresis, *Electrophoresis*, **23**, 35–42.

111. Krokhin, O., Qian, Y., McNabb, J.R., Spicer, V., Ens, W. & Standing, K.G. (2002) An off-line interface between HPLC and orthogonal MALDI-TOF, In: *Proceedings of 50th ASMS Conference on Mass Spectrometry and Allied Topics*, Orlando. American Society of Mass Spectrometry, Orlando, Fla.
112. Chen, V.C., Keding, K., Krohkin, O., Lee C.H., Ens W., Standing K.G., Nagy J.I., and Perreault, H. (2003) Off-line on-target micro-LC vacuum deposition device for the characterization of connexin and β-casein phoshorylated peptides by MALDI-TOF-MS and MALDI-QqTOF-CID-MS, In: *Proceedings. 51st ASMS Conference on Mass Spectrometry and Allied Topics*, Montreal. American Society of Mass Spectrometry, Montreal, Canada.
113. Millipore Corp. (2000) Fractionation of complex peptide or protein mixtures prior to MALDI-TOF MS using ZipTipμ-$C_{18}$ and ZipTip$C_4$ piptette tips, Technical Note TN226. Millipore Corporation, Milford, Mass.
114. Gygi, S.P. & Aebersold, R. (2000) Mass spectrometry and proteomics, *Curr. Opin. Chem. Biol.*, **4**, 489–94.
115. Huang, Y., Mechref, Y., Tian, J., Gong, H., Lennarz, W.J. & Novotny, M.V. (2000) A simple sample preparation for enhancing the sensitivity of mass spectrometric oligosaccharide determinations through the use of an adsorptive hydrophobic resin, *Rapid Commun. Mass Spectrom.*, **14**, 1233–1237.
116. Al-Rubeai, M. (ed.) (2002) *Cell Engineering, Vol. 3: Glycosylation*, Kluwer Academic, Dordrecht, The Netherlands.

# 6 Hyphenated techniques in drug discovery: purity assessment, purification, quantitative analysis and metabolite identification

JENNY KINGSTON, DESMOND O'CONNOR, TIM SPAREY
and STEVEN THOMAS

## 6.1 Challenges associated with sample analysis and purification in support of early drug discovery

### 6.1.1 *Introduction*

At every stage of the drug discovery process, analytical chemistry provides a vital support function. Synthesized samples must be separated and characterized, components quantified and pure target analytes isolated. Recent developments, particularly in the field of high-throughput library synthesis, have led to a vast increase in sample production. This has necessitated a change in analytical approach from the tailor-made solutions used in the past to the widespread adoption of more generic analytical methodology.

This section describes the analytical techniques utilized within the pharmaceutical industry to support high-throughput library synthesis. Analytical methods applicable to compounds with wide-ranging physicochemical characteristics are explained. Emphasis is placed on techniques used to maximize sample throughput, and efforts undertaken to optimize generic methodologies. In particular, the use of hyphenated techniques for characterization and purification in the context of library synthesis is detailed. Consideration will be given to techniques which have proven successful in analyzing, quantifying and purifying multiple samples in a high-throughput environment. Analytical, semipreparative and preparative chromatographic techniques are covered, with emphasis placed upon highlighting the benefits and limitations of each combination of separation system and detector. In particular, hyphenated techniques such as liquid chromatography–ultraviolet detection/mass spectrometry (LC-UV/MS) and supercritical fluid chromatography–ultraviolet detection/mass spectrometry (SFC-UV/MS) are covered, with alternative detection techniques such as evaporative light scattering detection (ELSD) and chemiluminescent nitrogen detection (CLND) also discussed. The applicability of these techniques to an open-access environment is also covered.

### 6.1.2 *Analysis in support of discovery chemistry*

6.1.2.1 *High performance liquid chromatography (HPLC)* Reverse-phase HPLC is the principal technique used within the pharmaceutical industry for separation of reaction mixtures before characterization or purification of active components. HPLC is a powerful separation technique and benefits from having a proven track record within all areas of the pharmaceutical industry. Both software and hardware applications are, therefore, well established, and method optimization techniques are familiar to the majority of end-users.

*Choice of stationary phase* There are a large range of separation columns commercially available, which, in conjunction with the adaptability of mobile phase constituents, have allowed the separation of many diverse compounds. The majority of pharmaceutical compound separations make use of hydrophobic, silica-based stationary phases. These have been gradually refined and adapted by column manufacturers by end-capping of free silanol groups and incorporation of polar-embedded functionality to produce the range of highly efficient columns currently available to the analytical chemist. The recommended pH range for using silica-based columns is 2–8. At more basic pH silica is prone to dissolve with gradual loss of column efficiency and eventual voiding of the stationary phase. This can present difficulties for the analysis of basic drug-like compounds, where adaptation of mobile-phase to high pH conditions enhances both the retention and peak shape of analytes by suppressing ionization. One answer to this problem is the development of hydrophobic reverse-phase columns based on a polymeric stationary phase. Styrene divinyl benzene (SDB) analytical and semipreparative columns have been commercially available for some time; they have yet to find widespread applicability for generic HPLC methodology because their separation efficiencies are lower than those of their silica-based counterparts.

*Rapid HPLC analysis in support of combinatorial chemistry* Within the pharmaceutical industry, one of the principal challenges facing the analytical chemist in recent years has been the need to develop rapid techniques for quality assessment of the samples produced by parallel synthesis techniques. With ever increasing numbers of diverse compounds requiring characterization, it is clearly not possible to develop separation methods tailored to the physicochemical properties of each individual sample.

The principal approach taken within the pharmaceutical industry to address this issue has been that of rapid, reverse-phase HPLC coupled with hyphenated detection systems suitable for analysis of the majority of expected analytes. These assays have several primary functions: identification of target compounds within the sample, assessment of sample purity, and quantification of desired product. To maximize sample throughput, fast analysis cycle-times have been adopted for chromatographic separations. As a general trend, reverse-phase gradient HPLC analysis cycles have been reduced by incorporating high-flow rates [1], elevated temperatures [2] and shorter narrow-bore columns [3,4]. Using these techniques, adequate resolution can be achieved with a typical analysis time of 5–10 min per sample. Increased flow rates have the effect of minimizing the loss of resolution due to reduced column length. In addition, the use of smaller stationary-phase particle size has the effect of increasing the resolving power of the HPLC method. The speed and efficiency of the developed methods have allowed variable impurity and degradation profiles to be recorded during early drug development, thus providing rapid feedback to the synthetic chemist. Recent developments in silica manufacture using sol–gel technology have led to the development of silica-rod monolith reverse-phase columns which combine the efficient separation power of a modern $C_{18}$ stationary phase with extremely low back-pressures [5]. These have been particularly useful for the development of high flow-rate, rapid gradient analysis methods.

The routine use of a mass spectrometer as the detector of choice for generic HPLC analysis, particularly of crude reaction mixtures, provides specificity to each analytical run. Products, starting materials, and side reaction products can be clearly identified during a short analytical run. A further advantage to using a generic approach with MS detection is that it lends itself to an open-access environment. Open-access LC-UV-MS systems are now widely available in all drug discovery environments, providing the nonexpert user with a tool for gathering sample component information.

A recent development is the use of high-resolution MS as an HPLC detection technique; this allows valuable structural information to be obtained by providing accurate molecular mass information for separated product and impurity components.

6.1.2.2 *Supercritical fluid chromatography* Supercritical fluid chromatography (SFC) is a complementary separation technique to both HPLC and gas chromatography (GC). SFC uses a mobile phase comprising mainly a supercritical fluid, e.g. $CO_2$ held at high pressure. Supercritical fluids have density and solvating power approaching that of a liquid, but viscosity similar to that of a gas. The advantages of using such a mobile phase system over HPLC is that the low viscosity of the solvent results in low column back-pressures and high solute diffusion coefficients relative to HPLC. Higher efficiency chromatographic separations than those of HPLC are therefore theoretically possible.

The suitability of this technique for separation of pharmaceutical compounds is well established and it offers clear advantages over HPLC in terms of speed and efficiency. Both capillary and packed column SFC have been used to separate pharmaceutical compounds. Indeed, there has been a growing trend over recent years for the substitution of SFC in place of HPLC for MS applications in combinatorial library analysis [6].

6.1.2.3 *High performance capillary electrophoresis (CE)* Capillary electrophoresis is an orthogonal technique to HPLC and is thus capable of providing complementary data. CE separations are achieved by means of the differential migration rates of charged analytes in an electrolyte solution under the influence of an applied voltage. This technique offers the advantage of highly efficient separations in a short time. Generic CE methods have been implemented for analysis of pharmaceutical compounds. However, CE-MS is not widely used for the analysis of combinatorial libraries. Cheng and Hochlowski [6] report the potential of CE-MS for miniaturized biochip separations.

6.1.2.4 *Quantifying parallel synthesis products – comparison of different detection techniques* With large numbers of parallel synthesis products being generated there is a need both to analyse purified samples and also to determine the amount of material available for biological screening. At present, this is usually carried out using automated weighing of dried samples following chromatographic analysis. Several detection techniques have been shown to provide a response for a wide range of analytes. However, to use this detection to quantify sample concentrations accurately the detector response must be proportional to sample mass and uniform across diverse compound classes. McCrossen *et al.* [7] compared the ability of different chromatographic detectors to determine organic impurities. To compare different detection techniques identical chromatographic conditions are necessary.

*Liquid chromatography with ultraviolet detection (LC-UV)* UV absorbance is the most common technique for detecting and quantifying pharmaceutical compounds following chromatographic separation. UV absorbance detectors have been coupled to CE, HPLC and SFC systems. The continued popularity of this detection technique arises from its ease of use, dynamic linear range, high sensitivity, and precision. Detection and quantification of target compounds takes place by the simple means of measuring the amount of UV radiation absorbed at a particular wavelength or range of wavelengths.

However, UV detection has an important drawback when included as the sole detection mechanism during rapid, high throughput pharmaceutical analyses. Some small pharmaceutical molecules do not contain UV absorbing chromophores and will therefore have little or no response in a UV detector; conversely, other molecules absorb UV radiation very strongly. Response is therefore not uniform across different compound classes and this technique is thus not suitable for determining either the concentration or purity of a pharmaceutical sample based on a single chromatographic analysis.

*MS* HPLC analysis is compatible with modern mass spectrometers. The development of the Z-spray design interface allows for analysis of even crude reaction mixture with little need for sample preparation or intervention. The choice of volatile buffers for the HPLC mobile phase is essential, and care must be taken to choose an appropriate mobile phase that does not suppress ionization of target compounds in the mass spectrometer. Use of trifluoroacetic acid (TFA) is therefore not generally recommended for LC-MS applications, and basic mobile-phase constituents are usually not a suitable choice when basic analytes are to be characterized. The majority of generic LC-MS analytical procedures make use of a simple gradient analysis on a silica-based reverse-phase column using a mobile phase comprising acetonitrile as the organic component and 0.1% formic acid as the aqueous component.

The advantage of using MS is that this detection technique is specific. Compound identity can be confirmed rapidly together with the identity of any sample impurities, and preliminary purity assessments can be made. However, a mass spectrometer does not provide a uniform response to all classes of compound and, as such, is not suitable for providing a measure of sample amount from a single analytical run.

*Evaporative light scattering detection (ELSD)* ELSD measures the scatter produced by vaporized sample particles passed through a light beam. Analyte particles eluting from a chromatographic source are nebulized and then vaporized in a heated drift tube. The resultant small particles are passed through a beam of light and the scatter produced is measured at a photodiode.

ELSD has been investigated as a possible universal HPLC detection system. Indeed, under controlled mobile-phase conditions, the response to a set of structurally similar non-volatile compounds has been reported to be uniform. However, detector response is dependent on the quantity and nature of particles produced during desolvation and, although near uniform response to compounds within a class is seen, it is affected by factors such as volatility and charge. More importantly, mobile-phase composition will affect the response of a compound to the detector, rendering this technique unsuitable for quantification of library products using gradient HPLC.

Despite these drawbacks, ELSD has been employed to estimate sample concentrations using flow-injection analysis with uniform solvent composition.

*Chemiluminescent nitrogen detection (CLND)* The use of CLND as a 'universal' detection technique for HPLC has been the subject of recent interest. In this detector, HPLC effluent is combusted in an oxygen rich furnace to convert eluting compounds into gaseous oxides, including nitric oxide. This nitric oxide is reacted with ozone to produce nitrogen dioxide in an excited state which emits photons upon return to the ground state [8].

The response of a compound using a CLND detector has been found to be proportional to the number of moles of nitrogen in each eluting peak. As the majority of small drug-like molecules do contain nitrogen, this detector has been investigated as a possible HPLC 'universal' detector for on-line quantification of compounds and impurities. Nussbaum *et al.* [9] used gradient HPLC coupled with CLND detection to measure the UV response factors of five structurally diverse compounds. In a direct comparison with ELSD detection, they found that the accuracy of response using this detector was between 100.6% and 107.6%. In comparison, the accuracy of response of the ELSD detector under the same gradient elution conditions was found to vary from 134% to 1740%. Lewis *et al.* [10] investigated the accuracy of this detection technique for quantifying parallel synthesis samples following purification; NMR and CLND quantifications of sample concentration were carried out using external standard calibration and both techniques produced results within 5% of the weighed value. Where a larger error occurred, both techniques reported less material than expected, indicating an error in solution concentration as a consequence of adsorbed water in the solid sample or lower than reported purity. Taylor *et al.* [8] evaluated the CLND detector as a universal detector for nitrogen-containing molecules, reporting a linear response over the range 25–6400 pmol of nitrogen and a 10% error in the response rate amongst the compounds tested. Unlike ELSD, the response of the CLND detector was largely independent of mobile-phase composition and it therefore lends itself more readily to gradient HPLC analysis. Thus a single hyphenated LC-UV-CLND-MS analysis of a crude or purified sample will provide information on purity, structural confirmation and quantification of the final compound. The exception to this universal response is the response for compounds which can decompose thermally to form molecular nitrogen. Compounds such as azides or tetrazoles which contain N—N bonds and decompose in this pattern will give a lower than expected response in the CLND detector.

In addition to this limitation, use of any nitrogen-containing solvents or additives (e.g. MeCN, $NH_3$, alkylamines) is precluded. However, despite these limitations, this detector has proved useful for quantification of library-synthesized samples where sample molecular formulae are known. CLND has also been coupled with SFC and SFC-MS systems. Using SFC rather than HPLC removes the disadvantage of limitations in mobile-phase composition apparent for HPLC analysis where acetonitrile is precluded from use with the CLND detector.

### 6.1.3 *Purification of reaction mixtures*

The increase in the number of compounds being synthesized by pharmaceutical and outsourcing companies has necessitated the development of high-throughput techniques for purification to provide distinct entities for primary *in vitro* evaluation. Combinatorial chemistry has developed from the production of vast numbers of compounds of often ill-defined purity to the isolation of fewer well characterized, single compounds. To cope, traditional purification techniques such as HPLC, column chromatography and recrystallization have been replaced by hyphenated techniques centered upon mixed eluents coupled to a detector. Other technologies have also had to develop alongside purification, namely liquid handling devices, tube/plate manipulation and evaporation facilities, to cope with the increased number of crude reaction samples and clean fractions generated from this process. The most commonly used purification systems are preparative LC-UV and LC-MS. Recently, HPLC has been replaced by SFC, and, in addition, a range of alternative detectors are being used, either in-line or off-line. The main driving force for change has been the challenge of widening the purification bottleneck associated with serial-based techniques, caused by parallel synthesis. One industry-wide solution has been the establishment of groups dedicated to library synthesis and/or purification, providing services ranging from library work-ups and purification to post-purification sample handling. Many of these groups developed from combinatorial chemistry teams, but focus on the synthesis of pure, fully characterized, single compounds within small, targeted libraries (10–500) as opposed to large libraries (1000–100 000) of often mixed pools of compounds. The purification can be performed by the chemist using open-access facilities or solely by purification groups. The advantages of the former are that the chemist is in control of the process and timelines, and will better understand the characteristics of the samples, whereas for the latter, in-depth chromatography expertise can be applied to problem samples. Generally open-access facilities rely on generic options incorporating a low number of gradient and/or isocratic methods, limited eluent types (low and high pH) and column choices. However, a limited number of combinations of the above can cover the majority of purification needs.

#### 6.1.3.1 *LC-UV/MS*

*UV or MS detection* The utility of HPLC, either in normal or reverse-phase mode, is well established. It is a highly efficient method for purifying compounds, and has been developed by being coupled to a detector, such as a UV spectrophotometer or mass spectrometer. The choice of detector is important, as inherently with UV detection, within a library of compounds, the chromophoric and physiochemical properties of samples are often unknown, although predicted values can be predetermined [11]. Consequently, not all samples will behave similarly. Thus the application of generic approaches to setting threshold levels above which collection is triggered, may be unsuccessful. Alternative methods involve prescreening some, or all, samples from the library, using analytical LC-MS to determine if the product exists, and if so, to what extent. This can be done using short run times (1–5 min) providing rapid sample interrogation. However, variable ionizability, even within a library, can make the application of generic methods difficult. Analytical LC-UV can also be used for a scouting run, although no confirmation can be given of the product existence. The analytical information can then be extrapolated to preparative scale giving options such as a fixed time-window for collection, a fixed time-window for collection coupled to a threshold level, or simply a threshold level. These options help reduce the number of fractions that UV detection produces, this being one of the main problems associated with this method. Simple systems that collect all fractions in a run above preset threshold levels result in multiple fractions needing considerable care in postpurification processing. However, software developments now mean that sample–fraction relationships can be mapped accurately.

The introduction of LC-MS to the purification of libraries has resulted in this technique replacing LC-UV as the most popular within the pharmaceutical industry. The inherent benefit of collecting only peaks that have the correct mass reduces fraction numbers and the need for post-purification analysis. The miniaturization of high throughput screening (HTS) assays has resulted in libraries aimed at delivering smaller quantities of compound and when purified delivered directly into 96- or 384-well plates for biological evaluation. Target quantities for libraries aimed at identifying hits have gone down (<3 mg); similarly, hit-to-lead libraries have also targeted smaller quantities (1–10 mg). Such small amounts, coupled to library size produced at this stage of the drug discovery process, favor the utilization of LC-MS producing low fraction numbers.

*Column considerations* There is now a wide choice of columns available on the market. Columns can be selected depending on whether the library compounds are polar or lipophilic, acidic or basic, or on relative purity. The choice of stationary phase will greatly affect the retention time of compounds, depending on eluent pH, eluent temperature and system pressure. Often, prescreening in parallel with several column types will rapidly identify the preferred option.

One factor contributing to run times for samples is the time needed for equilibration of the column at the end of the run. A solution to this has been to use two identical columns in parallel, and during a run on one column, the spare column can be equilibrated using a third pump. This reduces run times and also improves column lifetime.

*Eluent pH* Most compounds produced in the pharmaceutical industry are either basic or neutral; hence the choice of eluent pH can be important. Early on, eluents modified with TFA, formic acid or other nonmineral acids to attain low pH, were the mobile phase of choice. A wider range of pH is used now, notably high pH being achieved by using ammonia or ammonium bicarbonate buffered with ammonia. The retention times of basic compounds at high pH will usually be longer than at low pH (when the compound is protonated), often translating into better resolution. Additionally, the combination of pHs during analytical prescreening can provide powerful options for tackling impure samples. Furthermore, for basic compounds, using high pH eluent returns compounds as free bases, an important property because some assays require certain salt forms. Additionally, there are reports of compounds stored as TFA salts undergoing degradation during freeze–thaw cycles [12].

*Throughput* The first published preparative LC-MS system for the purification of libraries was developed by Kassel and coworkers [13,14]; it was capable of running in both analytical and preparative mode, allowing ‘walk-up’ assessment and purification. Later modifications allowed around 200 analytical runs per day followed by 200 preparative overnight runs, producing impressive throughput. Since then, numerous publications have emerged describing differing approaches to preparative LC-MS [15–17]. A more recent development has been the use of multiplexing (MUX), a way of further increasing throughput by using two, four or eight columns and UV detectors in parallel coupled to one mass spectrometer [18,19]. Using such systems, sample throughputs of up to 350 000 per year are possible.

*Loadability and solubility* At-column dilution (ACD) is a technique for increasing the loadability of a system by premixing the sample solution with the organic eluent to minimize the effect of the carrier solvent (e.g. DMSO) on chromatography. Initially this required an additional pump to feed in the organic eluent; however, a two-pump set-up has been made possible [20]. The key issue of insolubility of samples can also be reduced with this method. Clogging of injector ports, tubing, fraction collection heads, and so on, are the main problems found with preparative LC-UV/MS systems. Prepurification filtration reduces the amount of insolubles in samples, but often with overnight runs, on standing in the autosampler, fresh precipitate can appear. HPLC systems are generally intolerant of all but the smallest amounts of particulate material. Furthermore, the incidence of sample 'breakthrough,' i.e. sample running through the column at the solvent front, can be minimized by dilution of the DMSO solution; hence, sample losses are kept to a minimum. Alternative sample vehicles such as methanol, acetonitrile and mixtures with dimethyl sulfoxide (DMSO) can be used when sample solubility is less of an issue.

One problem with ACD is over-loading the detector, in particular the MS. A solution to this has been the use of an active splitter [21]. A 1:1000 split is the most commonly used ratio post UV, allowing one part to the MS detector, and the remainder to waste or the fraction collector. The passive splitting does not take into account the variation in loading within a library and the trend towards increasing loading through a variety of methods. Therefore, active splitting allows the amount of sample that is directed to the detector to be varied, significantly reducing overloading; this reduces tailing of sample on the detector, giving cleaner samples and fewer fractions, and also reduces time devoted to cleaning the MS source.

An alternative to ACD is the 'DMSO slug' approach, whereby the sample is buffered between eluent, air gaps and DMSO [22]. This maintains sample solubility at the injector and precolumn tubing by not allowing the eluent to mix with the DMSO solution. This has been reported to result in a significant reduction in system downtime.

#### 6.1.3.2 *SFC-UV/MS*

*Technical problems* SFC has been used for many years in the food and other chemical industries, from the isolation of natural products [23] to the removal of caffeine in the coffee industry [24]. The principle is simple; replace the mobile phases with liquid $CO_2$. In practice, the application to the purification of library compounds within the pharmaceutical industry has been relatively slow. Several technical issues have proved troublesome, including maintaining consistent pressures across a system, mainly because of the high incidence of leaks caused by the high pressure of liquid $CO_2$, and fraction collection. The latter consisted of two problems, both of which have proved difficult to overcome. The first problem is that of containing rapidly expanding gas in a controlled way due to high pressure liquefied $CO_2$ exiting the system. Termed aerosoling, this can result in significant sample losses, together with jamming of valves due to icing. A number of solutions have been attempted, including off-line depressurization and cyclone technology [25]; however, these are unsuited to high-throughput purification of libraries. Success has been found with simple solutions such as wrapping foil over each collection tube, allowing the collection needle to pierce the foil and deposit sample without loss through aerosoling and pressurization of the collection vessel. Good recoveries and minimal cross contamination of samples have been reported [26]. Second, most commercial systems have limited fraction collection facilities, but modifications such as coupling to a Gilson 215 fraction collector [26], a robotic workstation [25] or other custom solutions [27] have increased fraction collection capacity.

*SFC or HPLC?* There are two main advantages of SFC over reverse-phase HPLC; the small amount of solvent needing dry-down after fraction collection [28], and the enhanced chromatographic speed and resolution. The latter is a result of the increased diffusivity and reduced viscosity of supercritical $CO_2$. A modifier, usually methanol, is needed to elute most compounds, because $CO_2$ has a lower polarity than the stationary phase. In addition, additives such as TFA or diethylamine are also included to further enhance the polarity of the eluent, and also for similar reasons as discussed above for LC-UV/MS. After fraction collection, depending on the ratio of methanol to $CO_2$ (most commonly 10–30%), only relatively small quantities of methanol (and any additive) remain. Dry-down is therefore quick and cheap compared to removing the aqueous acetonitrile mixtures found with HPLC. Further, disposal costs are much less, as methanol can be incinerated, whereas reverse-phase eluent waste requires specialist removal.

*Column considerations* Column choice in SFC should be limited to specialist SFC columns capable of withstanding the higher pressures found with this technique, but some users have found normal HPLC columns adequate; others prefer to pack their own columns. The main use of SFC has been in chiral separations, using the enhanced resolution power to isolate single enantiomers [29]. In addition, SFC is increasingly being used to purify libraries of compounds, a trend that is set to continue.

*UV triggered collection* Fraction collection triggered by UV is currently the main method of sample isolation. Some reports describe using MS triggering, but this method is still relatively uncommon [26,29]. Prescreening library samples using analytical SFC/UV or SFC/MS allows the use of a selection of collection methods, including set time-windows, threshold or gradient-based triggering, or indeed combinations of these methods. When the target compounds have been identified within the crude mixtures, usually a snapshot of the library (10%), method selection is often generic. For some libraries, prescreening of all samples may be necessary; for this the increased speed of SFC is particularly beneficial.

*SFC/MS* Using SFC/MS has required several system design modifications. In addition to a third make-up flow pump after the UV detector to ensure good recoveries in fraction collection, a fourth make-up pump has been incorporated post splitting to the MS to prevent tailing and provide an even baseline. The choice of additive is also important because some basic additives reduce signal-to-noise ratios. One additive used has been ethyldimethylamine, which has been shown to have minimal influence on baseline consistency compared to isopropylamine under ESI [26].

*Solubility* One of the main problems for sample loading and collection arises from poorly soluble samples blocking injector ports and tubing. Because of the low dielectric constant of liquid $CO_2$, many pharmaceutical compounds are poorly soluble under the conditions employed. The use of modifiers and additives helps improve this, such as for sample loading where injection into the methanol stream is normal. Samples can be injected as DMSO solutions or DMSO/methanol mixtures; however, as this is a normal phase set-up, DMSO can be a complicating factor in fraction collection. Also, chromatographic efficiency can be reduced. Regular preventative maintenance is essential with SFC systems to help reduce downtime caused by blockages.

*Future prospects for SFC* SFC is poised to become the dominant technique for purification of library samples. Further developments are required, including improved hardware capable of operating robustly over extended time-periods. Furthermore, SFC/MS is still relatively underused. Coupling the speed and efficiency and reduced fraction volumes associated with SFC to MS detection, inherently producing few fractions, will put SFC/MS at the forefront of library purification. Open-access usage of preparative SFC systems is as yet unreported; however, technical developments and improved reliability will ensure that this is soon possible.

### 6.1.4 *Conclusion*

The challenge of high-throughput library sample analysis and purification has been largely met using a combination of rapid, generic chromatography, coupled with multiple detection techniques and increased use of automation both pre- and post-purification. Key contributors to the speed and efficiency of this process have been the development of multiplexing for parallel purification, and the introduction of SFC to replace HPLC, bringing in the advantages of reduced solvent handling and increased resolution.

## 6.2 Quantitative bioanalysis

### 6.2.1 *Introduction*

Within the pharmaceutical industry, both quantitative concentration data and information about metabolic routes are required at all stages of the drug discovery process. The precise nature of these requirements changes as projects progress from early discovery through safety testing to clinical development. At the early stages, many individual compounds are profiled, and the ability to generate data rapidly in order to influence medicinal chemistry programs in real time is important. Thus, at this stage, a certain degree of accuracy may be justifiably sacrificed in favour of throughput. In the later stages of the drug discovery process, the emphasis is on generating extensive and definitive data for a small number of compounds. Here, accuracy is paramount and must not be compromised. This work must be performed at increasingly faster rates, both to accommodate the increased output of medicinal chemistry departments and to decrease the time taken for products to reach the market, while maintaining sufficient accuracy for the intended purpose [30–32]. Of the available analytical techniques, reverse-phase high performance liquid chromatography (RP-HPLC) coupled with triple quadrupole mass spectrometry (MS/MS), usually referred to as LC-MS/MS, has emerged as the most versatile means of achieving these goals. The quality of a bioanalytical method can be judged by its robustness, throughput, sensitivity, selectivity and accuracy, and the aim of the analyst is to achieve sufficient degrees of these criteria to meet the demands of each particular application.

The process of acquiring bioanalytical data with LC-MS/MS detection can be divided into four stages: sample generation, sample preparation, chromatography and mass spectrometry.

In this section of the chapter the emphasis will be on sample preparation, but the other aspects will also be discussed in order to put sample preparation into the correct context. This section will begin with short introductions to some of the issues that will later be referred to in the discussion on sample preparation. This includes descriptions of the types of biological samples typically analyzed by quantitative LC-MS/MS, the working principles of LC-MS/MS, factors affecting MS and HPLC robustness, and the phenomenon of ion suppression. Following this background material, the methodologies for sample preparation for LC-MS/MS analysis will be discussed. Included will be both traditional techniques and novel technologies which allow sensitivity and accuracy to be attained while maintaining throughput.

### 6.2.2 In vitro *samples*

An ever increasing range of *in vitro* studies are being performed to support drug discovery programs, many of which require quantitative measurement of drug levels. Drug concentrations are typically in the micromolar region, and the matrix is relatively clean, consisting of a physiological buffer with the addition of relatively pure biological material on which the assay is based. Metabolites of the test compound may be present, as may be other drugs which form part of the assay. Typical examples include the following:

*Microsomal stability* The metabolic stability of test compounds is measured in incubations in the presence of a liver extract. Typically compounds are tested at 1 μM, and the incubation contains 0.5 mg/ml protein, and is performed in a physiological buffer. The most common use of this technique is to study metabolism by P450 enzymes, where metabolism is initiated by the addition of the cofactor NADPH, and is terminated by addition of a protein denaturing agent (usually organic solvent or acid). Test compound concentrations in these samples are compared with concentrations in control samples which have not been treated with cofactor to reveal the metabolic stability.

*P450 inhibition* This is a means of predicting whether test compounds would be likely to cause drug interactions via inhibition of the enzymes important for metabolism of co-administered drugs. The methodology is similar to that for metabolic stability. In this case a probe substrate, i.e. a drug known to be metabolized by a certain enzyme, is incubated in the presence of increasing concentrations of the test compound. The concentration of the metabolite of the probe substrate is the target of the analysis.

*Cell permeability* Cell lines such as Caco-2 or MDCK (derived from human colon carcinoma or dog kidney cells, respectively) are grown to form confluent monolayers in a medium of a physiologically realistic buffer on transwell plates (multiwell plates where each well contains two chambers separated by a semipermeable membrane, on which the cell monolayer is grown.) The test compound is added to one side of the monolayer and sampled from the other side. Analysis of the receiver compartment gives an indication of the permeability of the test compound.

### 6.2.3 In vivo *samples*

Measuring drug concentrations in plasma is a major activity for most bioanalytical groups, and this is the biological fluid which will form the basis for many of the examples in this chapter. Clinical trials and safety studies generate large numbers of samples for individual test compounds, which must be analyzed using well validated methodologies. In contrast, in the early drug discovery phase, small numbers of samples of many different analytes are analyzed to support optimization of pharmacokinetic and pharmacokinetic-pharmacodynamic (PK-PD) properties of test compounds. In these analyses, analytes must be determined at nanomole levels in an extremely complex matrix. Plasma proteins, at concentrations of 60–80 mg/ml, constitute the most concentrated endogenous species. The range of other endogenous molecules includes cholesterol at 1–2 mg/ml and glucose at 7–11 mg/ml. A variety of inorganic ions are also present, sodium and chloride at 140 mM and 100 mM, respectively, being the most abundant [33]. In addition, metabolites of the test compound and components of the dosing vehicle would be expected to be present.

### 6.2.4 *LC-MS/MS*

Reverse-phase high performance liquid chromatography (RP-HPLC) coupled with triple quadrupole mass spectrometry (MS/MS) [34], usually referred to as LC-MS/MS, has emerged as the key technique for bioanalysis. Initial efforts to interface the high volume liquid output of an HPLC instrument with traditional high vacuum gas phase mass spectrometer ionization sources were crude. An example of an effort to produce a sample compatible with MS is the moving belt interface, where the HPLC eluent was dried before entering the source as a solid sample. Spectral data were obtained after thermal desorption from the belt and electron impact ionization. Eventually, focus shifted from designing the LC output to fit the mass spectrometer to designing MS sources that could take a flowing liquid sample. Original thermospray sources gave way to the robust atmospheric pressure ionization sources of today, using chemical ionization (CI) and electrospray ionization (ESI) [35]. The availability of these interfaces means the dilemma of how to combine HPLC having flow rates of up to 1 ml/min, with mass spectrometers working under high vacuum is no longer an issue.

A triple quadrupole mass spectrometer contains two mass analyzers in series separated by a collision cell, usually containing argon. Parent ions are transmitted through the first mass analyzer, usually as mono-protonated or deprotonated ions, but sometimes as adducts with ammonia or metal ions which are present in the HPLC mobile phase. In the collision cell these parent ions fragment into one or more product ions, and one of these is selected to pass through the second mass analyzer to the ion detector. The complete process is known as multiple reaction monitoring (MRM). The task of developing the MRM method involves optimizing conditions for transmission of parent ions to the collision cell, then determining the optimum product ion for sensitivity, and the mass spectrometer conditions needed to form this ion. This optimization process is commonly known as 'tuning up'. Mass spectrometer manufacturers supply software which allows this process to be performed in an automated manner, via introduction of pure standard solutions of the analyte into the mass spectrometer, either by injection or an infusion pump. Alternatively the process can be performed by the analyst. In either case it takes a few minutes to optimize conditions for each analyte, though it is possible to optimize several analytes in parallel.

The HPLC mobile phase must be volatile; hence, buffers such as ammonium formate are commonly used. Either isocratic or gradient elution may be used. Gradient elution is common in early drug discovery where many different test compounds are encountered, because it is possible to devise generic conditions under which the majority of analytes will elute with a good peak shape. Additionally, the narrow peaks obtained with gradient elution lead to increased signal-to-noise ratios, and lower limits of detection. However, isocratic elution is an inherently more rapid technique, as the chromatographic run does not require an equilibration phase, so may be preferable when large numbers of samples of the same analyte are to be analyzed. Parallel HPLC techniques, which allow faster analyses to be performed with relatively long run-times, are described below.

### 6.2.5 *Mass spectrometer robustness*

Involatile components in the HPLC mobile phase have an adverse effect on mass spectrometer robustness because they are physically deposited in the ion source, particularly around the sampling cone, the inlet to the high vacuum region of the mass spectrometer. These involatiles fall into two main categories: first, inorganic salts, which are present in both the buffers used to prepare *in vitro* samples and *in vivo* biological fluids; second, endogenous biological macromolecules, which are also constituents of both *in vitro* and *in vivo* samples. Improvements in MS sensitivity mean smaller quantities of biological samples need to be introduced into the instrument, and mass spectrometer source designs have evolved to minimize contamination of the sampling cone. However, it remains a key goal of the analyst to reduce the introduction of these components into the mass spectrometer. This can be achieved both in the sample preparation stage, and in the chromatography stage.

The problem of involatile buffer salts can be overcome in two distinct ways. In reversed-phase chromatography, these salts elute at the solvent front; using a switching valve, the first portion of the chromatogram can be diverted to waste, thus protecting the mass spectrometer. This is effective with all sample preparation methods. Alternatively, sample preparation by either liquid–liquid extraction (LLE) or solid phase extraction (SPE) (see below) generally produces samples free from buffer salts.

A major source of involatile macromolecules in biological samples is proteins. A large proportion of proteins in a sample can be removed by precipitation with organic solvent, acid, base or strong salt solution, followed by centrifugation. For many *in vitro* applications where the matrix is relatively clean, this is usually all the sample preparation that is required, particularly if the LC-MS/MS is operated with a switching valve as described above. This procedure may also be used to analyze *in vivo* samples while maintaining mass spectrometer performance for an acceptable period; however, a more extensive sample preparation, such as LLE or SPE, or an on-line technique, should allow instrument performance to be maintained for a longer period. Protein precipitation and other sample preparation techniques are discussed in detail below.

6.2.5.1 *HPLC column robustness* Traditional HPLC columns are generally robust to buffer salts, but not biological macromolecules. This is because the channels in a column with the typical particle size of 3–5 μm are too small to allow free passage of large protein molecules, leading to contamination of the stationary phase and deterioration of chromatographic performance. Additionally, although these macromolecules may be soluble under aqueous conditions, they may not be when the proportion of organic solvent in the mobile phase increases. Precipitation of these macromolecules will cause backpressure increases (which can eventually lead to failure of analytical runs if column fittings fail) and again reduce chromatographic performance. A variety of columns are now available that do allow direct injection of plasma. These columns contain large through channels, which allow free passage of macromolecules under aqueous mobile phase conditions, while analytes are retained by the stationary phase material. Analytes are subsequently eluted in a mobile phase with increased organic content. Variations of this technique are described in section 6.2.7.4 On-line sample preparation.

6.2.5.2 *Ion suppression* The selectivity of MRM detection means that typically LC-MS/MS chromatograms only contain one chromatographic peak, that of the analyte. This is in contrast to previous bioanalytical techniques, such as HPLC with UV detection, where considerable method development was required to devise chromatographic separations of the analyte and various UV-active endogenous components which also appeared in the chromatograms. This selectivity, combined with the inherent sensitivity of MRM detection, makes the initial stages of development of an LC-MS/MS analytical method quite straightforward in the majority of cases. However, to achieve a high degree of sensitivity and analytical precision can be more challenging. The primary reason for this is the phenomenon of ion suppression.

While the chromatogram may show only the components of interest, a variety of other undetected species will be co-eluting. Competition for ionization means that in samples with higher concentrations of these co-eluting components, the MS signal for the analyte will be reduced, leading to reduced limits of detection. Additionally, if the ion suppression effect varies between samples, there will be a loss in analytical accuracy.

As would be expected, ion suppression is greatest when minimal sample preparation is employed. Hence plasma samples prepared by protein precipitation (PP) showed a greater degree of ion suppression than those prepared by the more rigorous method of LLE [36]. This same paper demonstrated that the effect of ion suppression was compound dependent, and also introduced a method for monitoring ion suppression effects in a chromatogram. The MS/MS signal of the analyte is monitored while the analyte is infused post-column at a constant rate into the LC stream produced following injection of a prepared plasma blank. With this procedure, dips in the baseline indicate chromatographic regions of increased ion suppression. This methodology has been used to minimize the ion suppression effect by manipulating chromatography such that the analytes do not elute in regions of high ion suppression [37].

It has been shown that ion suppression can originate from several sources in addition to endogenous plasma components. Residual polymeric dosing vehicle, such as polyethylene glycol (PEG) in the plasma samples can lead to drug concentrations being underestimated by up to threefold [38]. The anticoagulant Li-heparin can be a cause of ion suppression. The effects of the alternative anticoagulant ethylenediaminetetraacetic acid (EDTA) are reported to be smaller [39]. Mei *et al.* [39] also showed that simply storing sample plasma and the control plasma used to prepare calibration curves in different kinds of plastic tubes can lead to different ion suppression effects between samples and standards, and hence errors in quantification.

A detailed example of how ion suppression can be investigated and minimized has been described for the drug finasteride [40]. The extent of ion suppression was measured by comparing LC-MS/MS peak areas of pure standards to standards spiked into extracted control plasma. Despite a thorough sample preparation method (LLE), variations in response were observed between different batches of plasma. Thus, it is clear that the concentrations of components of plasma that cause ion suppression can vary between batches of plasma. The problem was overcome by either increasing the chromatographic run-time (thus separating the unseen ion suppressing species from the analyte) or employing a more selective extraction procedure (basification before LLE). Additionally, in this case, the ion suppression effect was not observed when the MS ionization technique was changed from ESI to the alternative ionization technique of atmospheric pressure chemical ionization (APCI).

For studies in early drug discovery, such extensive validation is not feasible. In this case, certain procedures can be observed to minimize detrimental effects of ion suppression on quantitative accuracy. Control and sample plasma should be obtained and stored as identically as possible. An internal standard should be chosen that is a close structural analogue of the analyte, and the chromatography system should not separate the analyte and internal standard. This should ensure that the analyte and internal standard are subject to the same ion suppression effects.

### 6.2.6 *Traditional sample preparation techniques*

The traditional approaches to conversion of plasma samples to analytical samples suitable for LC-MS/MS are SPE, LLE and PP. The principles of these techniques are described below. Additionally the 'dilute and shoot' method, which involves essentially no sample preparation, will be discussed.

6.2.6.1 *Dilute and shoot* Biological fluids that are essentially aqueous systems, such as urine and saliva, are amenable to LC-MS/MS analysis with practically no sample preparation. In some cases drug levels may be so high in such samples that dilution is required to bring analyte concentration within the linear range of the MS detector, hence the name 'dilute and shoot.' This technique is usually not employed with plasma samples, since neat plasma is compatible with neither reverse-phase chromatography nor MS. However, if sample drug concentrations are sufficiently high that the samples can be significantly diluted before analysis, then acceptable results can be obtained with minimal sample preparation. An example is an analysis of dextromethorphan, where sample preparation simply consisted of dilution of a 15 μl plasma aliquot with 485 μl of a diluent comprising water, methanol and formic acid (70:30:0.1, v/v/v) [41]. One thousand 5 μl injections onto a reverse-phase HPLC column with 3.5 μm particle size were made with no deterioration in analytical performance. This methodology was far less sensitive than alternative sample preparation by LLE, but the simplicity of sample preparation stage resulted in a 50-fold increase in sample throughput.

6.2.6.2 *Protein precipitation (PP)* PP produces samples devoid of most of the proteins in the sample, but lacks any additional selective isolation of analytes over other sample components. The two most commonly used methods of PP are addition of an organic solvent and acidification [42]. Addition of an organic solvent lowers the dielectric constant of the solution, causing protein aggregation by increasing intermolecular protein–protein interactions. Acidification of proteins in solution protonates basic residues leading to precipitation of proteins as insoluble salts. A less widely used method of PP is addition of solutions containing highly hydrated ions, such as ammonium sulfate; these cause a reduction in availability of water molecules at the protein surface, leading to protein aggregation by increasing intermolecular protein–protein interactions. Finally, addition of solutions of heavy metal cations, such as zinc salts, leads to reduced solubility and precipitation of proteins as the metal ions coordinate with acidic protein residues. In all cases, the samples generated are centrifuged to yield a supernatant suitable for analysis.

In practice, PP with acetonitrile is most commonly used. This is despite the fact that in order to precipitate more than 99% of plasma proteins, a volume ratio of 3:1 for acetonitrile:plasma must be employed. Whilst the high concentrations of acetonitrile in the samples may be deleterious to chromatography, the solvent can be either evaporated or diluted with aqueous mobile phase. Acids, such as a 10% aqueous solution of trichloroacetic acid (TCA), are more effective precipitation agents than organic solvents. A volume ratio of 0.6:1 for 10% TCA:plasma leads to precipitation of more than 99% of plasma proteins. However, the extremes of pH created may have consequences for analyte stability. Addition of salts and metal ions is rarely reported, most probably because the samples produced would contain high concentrations of involatile inorganic ions, which would adversely affect MS robustness.

Of the traditional sample methods, PP gives the least amount of sample clean-up, and also, if final evaporation and reconstitution steps are omitted, produces diluted rather than concentrated final samples. As mentioned above, the poor sample clean-up of PP can lead to errors in quantification by MS detection, due to the presence of co-eluting compounds in the analytical samples which can suppress ionization of the analyte, possibly in a manner which varies between samples. Despite this caveat, PP is a widely used method, and is particularly applicable in the drug discovery environment, where absolute accuracy is less important than in later stages of drug development. Additionally, PP is ideal for most *in vitro* applications. *In vitro* experiments may be readily designed such that inter-sample variation in ion suppression should not occur, and the drug concentrations are generally higher than for *in vivo* samples. For experiments involving a metabolizing system, a protein precipitation step has the additional benefit that it is an ideal way of terminating reactions at the end of the incubation.

6.2.6.3 *Liquid–liquid extraction (LLE)* With LLE, ionizable analytes are first converted into their neutral form. Thus samples containing basic analytes are basified, and samples containing acidic analytes are acidified. Then an immiscible organic solvent is added and the samples are mixed and centrifuged. The organic phase is then transferred to a new container and the solvent removed by evaporation, leaving the analyte to be reconstituted in mobile phase for subsequent analysis. The choice of organic solvent can affect the extraction recovery. For instance, analytes with polar functionality tend to extract into more polar solvents such as ethyl acetate, while analytes without polar functionality tend to extract into apolar solvents such as hexane. This methodology gives very clean samples, because inorganic ions, polar organics and proteins do not partition into organic solvents, and also offers a concentration step. Thus 1 ml plasma samples can be converted into 100 μl analytical samples with many of the analytical interferences removed. The sample concentration and clean-up associated with LLE meant that it was commonly employed when analysis by the inherently insensitive and unselective method of LC-UV was the norm. The prominence of LLE diminished with the advent of the more sensitive and selective technique of LC-MS/MS. However, as noted above, LLE still has a role to play as the clean samples produced are far less prone to ion suppression than those produced by PP [36].

6.2.6.4 *Solid phase extraction (SPE)* SPE is a technique in which an analyte is selectively retained on a solid adsorbent while other sample components are discarded to waste. The analyte is then eluted in a purified, and concentrated, form. Thus, like LLE, SPE offers good clean-up coupled with a concentration step. The technique has been traditionally performed with stationary phases designed for either reverse phase (non-polar) or ion exchange retention mechanisms, using hardware consisting of cartridges packed with sorbent consisting of a derivatized silica. Recently, organic polymers offering stability over a wider range of pH have become available, as have a wider range of hardware options, including 96-well plate and extraction disc formats [43].

The simplest example is the analysis of a lipophilic analyte using a reverse-phase sorbent. The steps are as follows:

(i) prime cartridge with organic solvent
(ii) condition cartridge with an aqueous wash
(iii) load sample (usually neat biological fluid)
(iv) aqueous wash to remove polar contaminants
(v) elution step using organic solvent
(vi) dry down sample and reconstitute in HPLC mobile phase for analysis.

Simple variations exist for ionizable analytes. For example, with a basic analyte the conditioning and wash steps can be with a high pH buffer, and the sample itself basified, to ensure the analyte is in a neutral form and thus successfully retained. The elution step can be performed under acidic conditions to ensure complete recovery. Greater selectivity can be imparted to the technique if method development is performed to identify wash solvent with the maximal eluting strength which does not elute the analyte, and an elution solvent with the minimal elution strength required to elute the analyte.

A range of ion exchange stationary phases allow clean-up of samples containing ionizable analytes. Strong cation exchange (e.g. sulfonic acids) and anion exchange (e.g. quaternary amines) phases are permanently negatively and positively charged, respectively. The mode of operation is to manipulate the pH of the mobile phase at the load, wash and elution steps such that the analyte is charged for loading and washing, and neutral for elution. For instance, for analysis of a basic analyte using strong cation exchange the loading should be performed at low pH and elution at high pH. Thus these phases are suitable for weak acids and bases, where the ionization state of the analyte can be readily manipulated.

For strong acids and bases, which are permanently ionized in aqueous conditions, weak cation and anion exchange phases are required. These consist of silica derivatized with weak acids and bases, respectively, allowing the ionization state of the stationary phase to be modified by pH manipulation. Examples of these phases are carboxylic acid derivatives (p$K_a$ 5) for weak cation exchange, and primary or secondary amines (p$K_a$ 10) for weak anion exchange. For a strong acid, weak anion exchange phase operated at low pH would retain the analyte, while at high pH the stationary phase would be neutral and elution of the analyte would occur.

Most prospective drug candidates are both lipophilic and ionizable; thus, in principle, both reverse-phase and ion exchange sample preparation could be effective. However, ion exchange purifications are rarely used in bioanalysis. The biological samples to be analyzed contain many ionizable species together with the analyte; thus, ion exchange will not give a great deal of purification. It is generally the higher lipophilicity of the analyte that distinguishes it from the majority of endogenous components in the sample; thus, reverse-phase is the most effective purification method. However, mixed mode phases are now commercially available. These are reverse-phase materials with an additional ion exchange component. For instance, a mixed mode reverse phase/strong cation exchange phase can give enhanced selectivity for purification of amines. The analyte is loaded and retained under acidic conditions. A wash step with acidified organic solvent will remove acidic and neutral contaminants, but not the analyte. The analyte is finally eluted with basified organic solvent. Using simple reverse-phase extraction the analyte would not be separated from acidic and neutral contaminants of similar lipophilicity.

### 6.2.7 *Automated preparation of* in vitro *and* in vivo *samples*

Recent years have seen widespread adoption of the 96-well plate format, often in conjunction with programmable liquid handling workstations. These units can perform multi and single channel pipeting operations, and can be programmed to transfer samples and reagents between storage reservoirs and assay plates, allowing the automation and unattended operation of bioanalysis. The incorporation of 37°C incubators into these systems means that entire *in vitro* experiments may be performed on the workstation. For a comprehensive review of approaches to high throughput sample preparation for bioanalysis, containing numerous examples and advice on automation strategies, see the recent book by Wells [44].

In the area of bioanalysis, the use of programmable workstations means that large numbers of samples can be processed with minimal user intervention, and that validated procedures can easily be transferred between users. The traditional approaches to plasma sample preparation mentioned above (PP, LLE and SPE) have all been at least partially automated. The complete range of mixing, centrifugation and evaporation steps necessary with these techniques is not accommodated by the robotic systems in common use; therefore, there are generally some off-line steps, necessitating manual intervention, in published examples of automated bioanalytical procedures. However, the liquid handling steps are readily amenable to parallel processing, and in-house designed applications allowing varying degrees of automated sample processing are now being reported regularly.

One important consideration in automated bioanalysis is the physical nature of plasma samples. Plasma is prepared by centrifugation of blood samples which have been collected into tubes containing anticoagulant. Plasma will be frozen until the time of analysis. This apparently simple process introduces a problem when fully automated analytical procedures, to include pipeting of plasma, are being designed. This is because when frozen plasma is thawed, fibrinogen clots may be produced. When dealing with such samples manually, the analyst will manoeuvre the clot away from the pipet tip to ensure accurate pipeting. With an automated system, this is generally not possible, so pipeting errors can occur unless precautions are taken to remove clots. Centrifugation of plasma samples after thawing will remove the majority of, but probably not all, the fibrinogen clots.

In order to facilitate automated handling of plasma samples, several workers have investigated ways to reduce clot formation. The use of EDTA as anticoagulant produced fewer clots than the more widely used anticoagulant Li-heparin [45]. Plasma snap frozen to, and stored at, −80°C appears to suffer from fewer clots on thawing than plasma samples stored frozen at higher temperatures [46]. However, even following optimization of anticoagulant and sample storage conditions and extensive centrifugation, some clots tend to remain. 96-Well filter plates, using 20 μm polypropylene filters, have been used to produce clot-free plasma ideal for automated analysis [47]. Plasma from the subject is aliquoted directly onto the plate, with a seal in place beneath the filters. The entire assembly is frozen, and when the samples are to be analyzed the seal is removed. As the plasma thaws it is filtered into the collecting plate placed underneath.

6.2.7.1 *Automated PP* Automation of PP methods is generally easier than automation of SPE and LLE, because less specialized equipment and fewer off-line steps are required. As mentioned above, PP is an ideal method to prepare samples from *in vitro* experiments. An example is the use of PP to terminate reactions in a highly automated P450 inhibition screening experiment, which produces samples directly amenable to LC-MS/MS analysis [48].

Typical examples of PP applied to preparation of plasma samples for LC-MS/MS analysis are the methods used in our laboratory for analysis of either large sample sets of a single analyte, or small sample sets from several analytes [49,50]. In each case, PP is performed on a liquid handling workstation by addition of acetonitrile, and there is an off-line component to the method where samples are mixed and centrifuged. To improve chromatographic performance, a portion of the supernatant is added to mobile phase buffer, again by the liquid handling system, to give final samples suitable for LC-MS/MS analysis. An additional feature of these methods is that dilutions of standard solutions and spiking of these into control plasma are also performed automatically by the workstation.

In an alternative approach to PP [51], plasma and acetonitrile are added to 96-well filter plates, vacuum is applied, and the supernatants captured on a collection plate, the precipitated plasma proteins being retained by the filters. A 3:1 acetonitrile:plasma ratio was found to be necessary to effect full precipitation of proteins. In this case, the supernatants were dried and reconstituted in water before LC-MS/MS analysis.

6.2.7.2 *Automated LLE* Of the three main sample preparation techniques, LLE is the most technically demanding to automate because of the requirements for vigorous mixing, centrifugation and evaporation steps. Currently these are not performed on commercially available liquid handling workstations, so several manual intervention steps are required in automated LLE protocols.

Examples are methodologies for the analysis of estrogen receptor modulators [52] and fluconazole [53] in plasma. In summary, each method mixes plasma with extraction solvent on 96-deep-well plates. The plate are sealed, mixed and centrifuged, and a portion of the organic layer transferred to a new 96-well plate. The organic solvent is evaporated by heating under a stream of nitrogen, and the sample reconstituted for LC-MS/MS analysis. In each case, all liquid transfer steps are performed by a liquid handling workstation, but transferring plates for mixing, centrifugation and drying-down is performed by the analyst. With the estrogen receptor modulators method, 2000 samples per day can be prepared and analyzed, using a 30 s analytical run-time, with no degradation in analytical performance. This is made feasible by the excellent clean-up afforded by LLE.

The need for manual intervention has been decreased by the solid-supported LLE technique. This involves performing LLE in a 96-well filter plate packed with inert diatomaceous earth particles of average diameter 1 mm. Examples of this technique include the analysis from plasma of carboxylic acid based protease inhibitors [54] and an indolocarbazole [55]. In each case, plasma samples are loaded directly onto the diatomaceous earth particles (with prior acidification of the carboxylic acids to convert them to the neutral form) and left for a few minutes. Coating the support particles in this manner leads to a high surface area for the aqueous phase. Extraction solvent is added, and either allowed to flow through to a collection plate under gravity, or pulled through the aqueous phase under gentle vacuum, with the large area of the aqueous–organic interface leading to efficient extraction. The organic phase is evaporated and the samples reconstituted for LC-MS/MS analysis. This methodology avoids the need for the vigorous mixing and centrifugation steps required with traditional LLE.

6.2.7.3 *Automated SPE* Complete automation of SPE is achievable using commercially available 96-well extraction plates and programmable liquid handling equipment. The only additional apparatus to a standard liquid handling workstation is a vacuum manifold or centrifuge to draw liquids through the extraction plate. This can either be incorporated into and be controlled by the workstation, or can be used in an off-line manner. For complete automation, the liquid handling workstation must have the ability to physically move plates around its workspace. If this is not the case, manual intervention will be required to replace the collection plate for wash steps with a collection plate for the elution step.

A typical application is the development of a validated assay to detect a protease inhibitor in plasma [56]. Extraction was performed using commercially available extraction plates with SPE conditions (i.e. volume and composition of wash and elution solvents) manually optimized for recovery and selectivity using portions of a plate. A vacuum manifold was used to pull liquids through the extraction plate, and an off-line evaporation step using a heating block was required to dry-down the eluted samples before reconstitution in a low volume of solvent suitable for LC-MS/MS analysis.

As mentioned above, mixed mode SPE extraction sorbents allow enhanced purification for analytes that are both lipophilic and ionizable. Such a phase has been used in the 96-well format for the analysis in plasma of the basic drug fentanyl [57]. The dual retention mechanism, which combines reverse phase and strong cation exchange functionality, means that both acid and methanol washes are possible, leading to particularly highly purified extracts, which are eluted under basic conditions. A manually controlled vacuum manifold is used to control the SPE process, and there is again an evaporation step to remove the elution solvent before reconstitution for LC-MS/MS analysis.

The requirement for an evaporation step in the above methods is a consequence of the relatively large elution volumes required with the 96-well SPE plates employed. For instance, the fentanyl method used plates with 25 mg packing material per well, and required a total elution volume of 0.75 ml. The samples produced would need to be further diluted with HPLC buffer if evaporation and reconstitution steps were to be avoided. However, the large dilutions involved would give unacceptably low limits of detection. Recently plates with 2 mg of stationary phase per well have become available [58]. Up to 0.75 ml plasma can be loaded, but the elution volume can be as low as 25 μl, thus negating the need for the evaporation and reconstitution steps. In practice, the samples produced with these plates would then be diluted with HPLC buffer to improve chromatographic performance.

A different approach to avoiding the evaporation and reconstitution steps is the use of normal phase chromatography. Using either silica or cyano columns, methods for a variety of drugs have been developed using acidified mobile phases with a high organic:aqueous ratio [59]. While not giving the same degree of sample concentration as the low elution volume plates, samples in organic solvent can be injected directly with no deleterious effect on chromatographic performance.

6.2.7.4 *On-line sample preparation* Volumes of neat plasma sufficient to afford ng/ml sensitivity for most analytes cannot be injected onto normal LC-MS/MS systems without quickly blocking the column, particularly because the high percentages of organic solvent employed to elute the analyte would denature plasma proteins while they are in the column. However, the plasma sample clean-up and concentration steps achieved off line by the SPE and LLE methodologies described above can also be performed in an on-line fashion. The most common method is turbulent flow chromatography, which has been commercialized for several years. This technique utilizes short, narrow columns (5 cm long, 1 mm internal diameter) packed with large particle size stationary phase (typically 50 μm), through which macromolecules such as plasma proteins pass without chromatographic retention, while drug analytes can either be retained or eluted depending on mobile phase composition. In general, several hundred injections of neat, or diluted, plasma can be made onto these systems before column failure.

In the simplest mode of operation, plasma spiked with internal standard is introduced to the extraction column [60]. The initial mobile phase is aqueous buffer flowing at 4 ml/min, that washes out plasma proteins and salts, and is directed via a switching valve to waste. At 1 min the mobile phase is changed to 95:5 acetonitrile:buffer, flowing at 1.5 ml/min, to elute retained analytes, which are directed, following a 5:1 split, to

metabolic routes (e.g. oxidized M+16 species). Full scan data can also be obtained, to allow detection of unexpected metabolites if they are at sufficiently high concentration. With access to advanced instrumentation, more elaborate strategies, taking advantage of the full range of capabilities of modern MS, are possible [77,78]. As will be discussed below, a variety of selective detection techniques are available to enhance the ability to detect particular metabolites in complex matrices. The spectral data available with these techniques may be sufficient to characterize fully the metabolism of the test compound. However, often the more definitive technique of LC-NMR will be required, which will necessitate sample concentration and purification of the metabolites partially characterized by LC-MS. In this case, targeted extractions can be performed and monitored by LC-MS to ensure successful isolation of the desired metabolites.

There will be occasions when the initial MS analysis does not yield useful data. This is particularly true when laboratories do not have access to the latest MS technologies. In these cases extractions must be performed 'blind' in an attempt to increase the signal-to-noise ratio for as yet unidentified metabolites in the sample. Because the analytes themselves are unknowns, the sample preparation and analytical steps must be carefully designed so as not to leave any metabolites undetected. The overall preparation and analysis phase should be capable of identifying compounds related to the parent compound, but with a wide variety of physicochemical properties, against a complex biological background. There is no universal solution to this problem, and in practice a thorough analysis will require several approaches to be performed in parallel. Thus this is in reality a fractionation process, where from a single complex biological sample several simpler samples are produced, with the aim of making detection and characterization of metabolites more tractable.

The availability of appropriately radiolabeled test compounds certainly facilitates metabolite identification, as a radio-HPLC trace of an unadulterated biological sample directs the analyst to all the metabolites present in the sample (assuming that metabolism has not resulted in loss of the radiolabel), and allows easy tracking of the progress of metabolite isolations. However, radiolabels are generally not available in the early drug discovery phase, and the analytical scientist must be aware of the potential metabolic processes the parent compound may undergo, and the physicochemical properties of the metabolites.

The process of metabolite identification can be divided into sample generation, sample manipulation, i.e. purification and concentration, and sample analysis. These will be discussed in turn.

### 6.3.2 *Sample generation*

Generation of metabolites can be simply an *in vivo* collection process of animal plasma or excreta (bile, urine or faeces). The absolute amounts of metabolites generated *in vivo* can be quite high. For instance, dosing a compound at 3 mg/kg to a 300 g rat introduces approximately 1 mg of material into the animal. If the entire dose was converted into one metabolite and excreted over several hours into bile, a concentration of up to 0.5 mg/ml would be expected. Such samples have the most *in vivo* relevance, and are ultimately required for complete characterization of a drug candidate. However, samples derived from *in vivo* dosing can be complex: first, the background of endogenous material will be high; second, a complex array of metabolites encompassing multiple phase I reactions and their associated phase II conjugates may be present.

Particularly in early drug discovery, the key aim is often to determine the primary metabolic event, and an *in vitro* approach may be a simpler solution. *In vitro* systems are much cleaner, and the range of reactions can be controlled by manipulation of experimental conditions. *In vitro* experiments are also preferable from an ethical perspective, but care must be taken that the *in vitro* system has *in vivo* relevance for the compound under investigation. For instance, decisions must be taken as to whether it is necessary to study both phase I and phase II metabolism in whole cells such as hepatocytes, or if a relevant metabolic profile could be achieved by studying phase I metabolism only in a subcellular liver fraction such as microsomes.

In the drug discovery phase, plate-based *in vitro* metabolic stability screens, typically at 1–10 μmol/l drug concentration, will yield metabolites in sufficient concentration for preliminary analysis by LC-MS. More detailed studies, particularly NMR studies, may require higher concentrations, and scaled up *in vitro* reactions are often required. A typical incubation protocol for such a scale-up is as follows:

- 50 μl 5 mmol/l substrate (final concentration 50 μmol/l)
- 1 ml liver microsomes (final concentration 0.5 mg/ml protein)
- 0.5 ml 10 mmol/l NADPH (final concentration 1 mmol/l)
- 3.45 ml phosphate buffered saline pH 7.4.

Incubations are usually performed at 37°C, for 30 min to several hours depending on the metabolic stability of the test compound, and can be terminated by the addition of an equivalent volume of cold acetonitrile. Metabolite yields from *in vitro* incubations can be maximized by following certain guidelines. First, concentrations of organic solvent should be restricted to the minimum required to solubilize the test compound. This is because many solvents, and in particular DMSO, are inhibitors of metabolizing enzymes at concentrations greater than 1% v/v. Second, many metabolic reactions have $K_m < 50$ μmol/l. Thus, absolute quantities of metabolites may be increased by performing large volume, low concentration incubation, as opposed to smaller scale incubations at high drug concentration. Yields can be improved by performing multiple small volume incubations instead of a single large experiment. For instance, multi-well plate technology can be used to perform 96 identical incubations, which can then be combined for further processing and analysis. Alternatively, large-scale bioreactors, using highly concentrated, artificially expressed enzymes, can be used to scale-up reactions in a highly specific manner [79].

### 6.3.3 *Sample purification and concentration*

In general, the first examination of a metabolite sample will be by LC-MS. Analysis is usually carried out on a sample prepared by PP with organic solvent. PP is a simple technique that does not introduce any selectivity (apart from removal of protein molecules), with the benefit that no metabolites will be lost. Also, avoidance of extremes of pH or elevated temperature means labile metabolites should not be degraded. Further sample processing may be required to increase the signal-to-noise ratio available to subsequent analytical techniques. Several methods are available, all of which have the aim of both purifying and concentrating the metabolites.

Traditionally, purification and concentration have been performed by either LLE or SPE. In LLE the original metabolite-containing solution is mixed with an immiscible organic solvent. Sufficiently lipophilic metabolites partition into the organic solvent. It is possible to improve extraction efficiencies and specificities by manipulating the pH and changing the extraction solvent. With SPE the metabolite-containing solution is applied to the top of a cartridge typically containing a lipophilic alkyl-chain-based phase on a silica support. By washing through with an aqueous medium of low eluting power the metabolites can partition between the stationary and mobile phases. A wide choice of sorbent means that more flexibility is achieved than is possible with LLE. Reverse phase, normal phase and ion exchange phases are available. Further control over the separation process is achieved through the choice of eluent. In reverse-phase SPE, the lipophilic components adhere to the sorbent, while more polar components are eluted with the aqueous wash. Slightly lipophilic molecules can be encouraged to elute by the presence of a small percentage of methanol in the wash solvent. This can be incrementally increased until neat methanol is used to obtain the most lipophilic moieties. One further advantage over LLE is the small eluting volumes required, typically less than 1 ml. This facilitates and speeds up the drying process and reduces the environmental impact of large quantities of organic waste. This technique can be taken further by using automated systems. Parallel incubations can be applied robotically to an array of cartridges and elution aided by a vacuum manifold.

Further enhancements in SPE have seen the advent of solid phase microextraction (SPME). Instead of introducing the sample to be separated to the sorbent, the sorbent is introduced to the sample. A polymer or adsorbent coated fused silica fiber is introduced into the liquid, usually as an insert within a septum-piercing needle. Organic compounds then adsorb and become enriched on the fiber, ready to be removed for analysis. SPME was originally designed for headspace sampling of volatile samples, with the analytes being subsequently released via thermal desorption in the heated injector of a gas chromatograph. Since the necessary diffusion is many times faster in the gas phase than liquid, SPME from aqueous media requires vigorous stirring. Also since thermal desorption would be detrimental for thermally labile metabolites, analytes are removed with a high organic wash [80].

When the aim is to isolate a metabolite which has already been detected, these extractions can be readily monitored by the original analytical method (typically LC-MS or radio-HPLC), ensuring a satisfactory conclusion. When extractions are being performed to simplify complex samples, the analyst must be aware of the likely selectivities of each technique in relation to the physicochemical nature of the metabolites likely to be in the sample. For instance, performing LLE under alkaline conditions would be expected to satisfactorily extract phase I metabolites of a basic or neutral test compound, provided they have some degree of lipophilicity. However, if the test compound has also undergone phase II metabolism to highly polar water-soluble species then these may not be extracted under these conditions or, in fact, may even be pH labile.

To generate metabolites in the purest form, preparative HPLC remains the best technique. In particular, preparative HPLC with contrasting mobile and stationary phase conditions to those to be employed in the subsequent analytical step can remove components that would otherwise co-elute with the metabolites of interest, thus leading to introduction of purer material into the MS or NMR analyser. Of note in the field of metabolite isolation is the ability to perform mass directed fractionation, thereby automatically generating relatively pure samples for further characterization at the same time as MS profiling is performed [81].

After the compounds of interest have been separated from the bulk of the biological matrix a concentration step is usually required. This is typically evaporation to dryness *in situ*, since unnecessary transfers of metabolites from one vessel to another invariably results in the loss of precious material. This can occur either through inefficient pipeting, or an affinity of the molecules for the laboratory-ware employed. Metabolites can adsorb onto the walls of glass vials, pipet tips and peek tubing. Care should be taken to avoid thermal decomposition of labile metabolites, and only heat sufficient to aid evaporation under a steady stream of nitrogen should be applied.

In circumstances of exceptionally simple phase 1 metabolism in an *in vitro* matrix, analysis by 'extraction NMR' is possible [82]. Following LLE or SPE, the incubate is reconstituted and analyzed as a standard NMR sample. Using conventional $^1$H NMR, the technique is particularly successful for identifying and assigning the positions of aromatic hydroxylations. Using two-dimensional methods, even aliphatic modifications, usually swamped by microsomal matrix, can be identified if correlations to an aromatic system can be observed.

Traditionally, NMR was a stand-alone technique to which samples were delivered following purification by preparative HPLC. However, for several years HPLC has been coupled to NMR [83]. Using the original stop-flow methodology the success of this technique for metabolite identification was limited by the amount of metabolite that could be delivered to the NMR probe, with one limitation being that the elution volumes of the chromatographic peaks were generally larger than the probe's flowcell. Peak broadening can be minimized, and sensitivity thus increased, by using high efficiency monolithic silica columns for delivery of sample to the flowcell [84].

Recent developments in LC-NMR technology have seen chromatographic peaks stored off-line in loops for later NMR analysis, a technique known as 'peak parking.' A further enhancement is the replacement of the loops with SPE cartridges. The peak of interest is diverted to the cartridge, with the eluent being diluted with aqueous buffer to ensure retention, from where the analyte can later be eluted into the NMR probe with deuterated organic solvent [85]. Several successive injections can be trapped onto the same SPE cartridge before analysis, allowing sensitive analyses in cases when larger injection volumes would lead to poor chromatography, or where turnover to the metabolite of interest is low.

An advantage of performing the chromatography in the deuterated solvents necessary to simplify the NMR analysis is the production of deuterated species. With an integrated LC-MS-NMR system, the metabolites detected by MS for LC-NMR will have gained one mass unit for each exchangeable proton they possess. This is a powerful tool for differentiating N- and S-oxidations from hydroxylations. Using M to represent the molecular mass of the neutral parent compound, each biotransformation yields a protonated M+16 species in conventional LC-MS, and the mass M+17 is detected. In deuterated solvents, the N- and S-oxides are detected as deuterated M+16, and the mass M+18 is detected. In deuterated solvent the hydroxyl OH will exchange to OD, and be detected as deuterated M+17, and the mass M+19 is detected. Thus N- and

S-oxides are detected one mass unit higher in deuterated solvents than regular solvents, while hydroxylations are detected two mass units higher [86].

### 6.3.4 *Selective analytical techniques*

An alternative to physical sample preparation is the use of a detection technique that shows optimal sensitivity for the desired metabolites whilst remaining essentially blind to background matrix constituents. Such a capability is offered by some of the more specialized modes of modern MS and NMR instrumentation. Thus, in effect, the selectivities of these detectors can be used as a surrogate for physical sample preparation procedures.

Triple quadrupole mass spectrometers are particularly useful for the detection of phase II conjugates in complex biological fluids. These species undergo facile and characteristic collision-induced fragmentations of either the entire conjugating moiety, or a part of it. These can be detected in either the parent ion or constant neutral loss scanning modes [77,78].

For instance sulfate conjugates tend to lose the $SO_3^-$ ion, *m*/*z* 80, and these can be detected using parent ion scanning. In this mode the first mass analyzing quadrupole of the mass spectrometer operates in a scanning mode, whilst the second mass analyzer is fixed to detect *m*/*z* 80. With collision gas in the intervening gas cell, any sulfate will be detected as a signal, and the electronics of the instrument are such that mass spectral data will reveal the mass of the intact metabolite which generated the *m*/*z* 80 ion that was actually detected.

Other conjugates have a characteristic neutral loss. Glucuronides fragment with loss of the glucuronic acid moiety (*m*/*z* 176). In this case it is not this fragment that is detected by MS, but the remaining part of the metabolite. Thus in constant neutral loss scanning for glucuronides, both mass analyzers operate in scanning mode, but with the mass range of the second analyzer displaced from the first by 176 Da. With gas in the collision cell signals are obtained when an analyte loses 176 Da in the collision cell, and again the electronics of the instrument are such that mass spectral data will reveal the mass of the intact metabolite.

These scanning modes can also be useful when analyzing complex biological matrices for drug related material. Generally, full scan MS data acquired on complex biological samples with unit resolution instruments provide no useful information, because drug related signals are swamped by signals from the endogenous background. If characteristic fragmentations are known to occur for the compound of interest, then constant neutral loss or parent scanning detection can yield LC-MS traces containing data only on the compound of interest. Thus these techniques perform the same function as a physical sample preparation, i.e. they lead to production of cleaner LC-MS traces. Each approach has caveats. Physical sample preparation assumes that the metabolites will extract under the conditions employed, and also that they will not be degraded. Constant neutral loss or parent scanning methods are used with the implicit assumption that MS fragmentation pathways for a metabolite will produce ions related to those produced by fragmentation of the parent compound.

High resolution mass spectrometers offer further possibilities for highly selective detection. The mass spectrometers traditionally used to collect high-resolution data were large, complex and expensive. The recent commercialization of time-of-flight instrumentation allows these data to be obtained in a more routine manner, and, with a suitable means of introducing a calibrant of known molecular mass, accurate masses within 5 mDa can be achieved. In the field of metabolite identification this allows unique empirical formulae to be assigned to nominally isobaric molecular ions and fragments, greatly aiding assignment of structures. These instruments are also useful for initial profiling of metabolism samples containing large amounts of endogenous material [87]. The software supplied with the instrumentation is capable of using accurate mass measurements to eliminate all signals which are not plausible metabolites of the parent compound, thus providing the analyst with a greatly simplified dataset with which to work.

Possibly the epitome of selectivity is the use of techniques that detect based on the presence of an atom or functional group not found in biological systems. Inductively coupled plasma mass spectrometry (ICPMS) initiates a robust ionization that extends common fragmentation to complete atomization of a proportion of the analyte. Detection purely based on the presence of a bromine atom has been demonstrated. The resulting chromatogram represents purely those compounds that retain the bromine of a parent molecule [88]. Thus, the metabolic profile is displayed with the minimum of sample preparation. The molecular ion for each of the visible components can be discerned, thereby accomplishing the first stages of metabolite identification. This technique has also been shown to be applicable to sulfur and phosphorous, though these are of less use because of their presence in endogenous matrices.

Another use of the non-endogenous atomic form of detection utilizes the NMR properties of fluorine [89,90]. This element is almost as sensitive as protons to NMR, and importantly is almost completely absent from *in vivo* systems. If sufficient metabolite levels exist, on-flow $^{19}$F LC-NMR can be used to separate components from a mixture that requires only rudimentary clean-up. A chromatogram is generated of the NMR signal against elution time to characterize the metabolic profile. Heteronuclear $^{1}$H–$^{19}$F coupling correlations (COSY and TOCSY) can be used to provide insight into the structure of the metabolites. As the complexity of the NMR experiment increases, stop-flow, peak parking or peak trapping experiments become necessary to obtain the required signal-to-noise ratio.

Analysis of fragmentation data is a powerful means by which an experienced mass spectrometrist can assign metabolic transformations to specific regions of a molecular structure. Even with impure samples, tandem mass spectrometers can generate daughter ion spectra specific to the analytes of interest, negating the need for extensive sample preparation. While standard triple quadrupole instruments are capable of generating these data, ion trap instruments offer the advantage of $MS^n$ data [73,75], and hybrid quadrupole time-of-flight instruments can generate accurate mass fragmentation data [76]. Instruments are now supplied with software packages which perform many of the necessary MS experiments in an automated manner [91,92].

## 6.4 Conclusion

We have shown a number of different approaches to obtaining metabolite information from *in vitro* and *in vivo* samples. Both physical separation and isolation techniques, and selective analytical methods, can be used to generate the data required for structural elucidation. In practice, the particular method, or combination of methods, used will depend on the nature of the samples, the complexity of the metabolic profiles present and the absolute amounts of metabolite present. Additionally, many laboratories do not have access to the advanced instrumentation which negates the need for physical sample preparation, so physical separation and isolation techniques will retain their importance.

## References

1. Weller, H.N., Young, M.G., Michalczyk, S.J., Reitnauer, G.H., Cooley, R.S., Rahn, P.C., Loyd, D.J., Fiore, D. & Fischman, S. J. (1997) High throughput analysis and purification in support of automated parallel synthesis. *Mol. Diversity*, **3**, 61–70.
2. Yu, K. & Balogh, M. (2001) A protocol for high-throughput drug mixture quantitation: fast LC-MS or flow injection analysis-MS? *LC-GC*, **19**, 60–72.
3. Mutton, I.M. (1998) Use of short columns and high flow rates for rapid gradient reversed-phase chromatography. *Chromatographia*, **47**, 291–298.
4. Leroy, F., Presle, B., Verillon, F. & Verette, E. (2001) Fast generic-gradient reverse-phase high-performance liquid chromatography using short-bore columns packed with small nonporous silica particles for the analysis of combinatorial libraries. *J. Chromatogr. Sci.*, **39**, 487–490.
5. Lubda, D., Cabrera, K., Kraas, W., Schaefer, C. & Cunningham, D. (2001) New developments in the application of monolithic HPLC columns. *LC-GC*, **19**, 1186–1191.
6. Cheng, X. & Hochlowski, J. (2002) Current application of mass spectrometry to combinatorial chemistry. *Anal. Chem.*, **74**, 2679–2690.
7. McCrossen, S.D., Bryant, D.K., Cook, B.R. & Richards, J.J. (1998) Comparison of LC detection methods in the investigation of non-UV detectable organic impurities in a drug substance. *J. Pharm. Biomed. Anal.*, **17**, 455–471.
8. Taylor, E.W., Qian, M.G. & Dollinger, G.D. (1998) Simultaneous on-line characterization of small organic molecules derived from combinatorial libraries for identity, quantity and purity by reversed-phase HPLC with chemiluminescent nitrogen, UV and mass spectrometric detection. *Anal. Chem.*, **70**, 3339–3347.
9. Nussbaum, M.A., Baertschi, S.W. & Jansen, P.J. (2002) Determination of relative UV response factors for HPLC by use of chemiluminescent nitrogen-specific detector. *J. Pharm. Biomed. Anal.*, **27**, 983–993.
10. Lewis, K., Phelps, D. & Sefler, A. (2000) Automated high-throughput quantification of combinatorial arrays. *Am. Pharm. Rev.*, **3**, 63–68.
11. For example, see www.chiral.fr.
12. Hochlowski, J., Cheng, X., Sauer, D. & Djuric, S. (2003) Studies of the relative stability of TFA adducts vs non-TFA analogues for combinatorial chemistry library members in DMSO in a repository compound collection. *J. Combinatorial Chem.*, **5**, 345–349.
13. Zeng, L., Burton, L., Yung, K., Shushan, B. & Kassel, D.B. (1998) Automated analytical/preparative high-performance liquid chromatography-mass spectrometry system for the rapid characterisation and purification of compound libraries. *J Chromatogr. A*, **794**, 3–13.
14. Zeng, L. & Kassel, D.B. (1998) Developments of a fully automated parallel HPLC/mass spectrometry system for the analytical characterization and preparative purification of combinatorial libraries. *Anal. Chem.*, **70**, 4380–4388.
15. Diggelmann, M., Sporri, H. & Gassmann, E. (2001) Preparative LC/MS technology: a key component of the existing high speed synthesis platform at Syngenta. *Chimia*, **55**, 23–25.
16. Kibbey, C.E. (1997) An automated system for the purification of combinatorial libraries by preparative LC/MS. *Lab. Robotics Automation*, **9**, 309–321.
17. Kiplinger, J.P., Cole, R.O., Robinson, S., Roskamp, E.J., Wane, R.S., O'Connell, H.J., Brailsford, A. & Batt, J. (1998) Structure-controlled automated purification of parallel synthesis products in drug discovery. *Rapid Commun. Mass Spectrom.*, **12**, 658–664.
18. Xu, R., Wang, T., Isbell, J., Cai, Z., Sykes, C., Brailsford, A. & Kassel, D.B. (2002) High-throughput mass-directed parallel purification incorporating a multiplexed single quadrupole mass spectrometer. *Anal. Chem.*, **74**, 3055–3062.
19. Liu, J. (2002) Purification and analysis of reaction arrays using a 4-channel prep-HPLC-MS system, 224 th ACS National Meeting, Boston, MA. August 2002.

20. Blom, K.F. (2002) Two-pump at-column-dilution configuration for preparative liquid chromatography–mass spectrometry. *J. Combinatorial Chem.,* **4**, 295–301.
21. Cai, H, Kiplinger, J.P., Goetzinger, W.K., Cole, R.O., Laws, K.A., Foster, M. & Schrock, A. (2002) A straightforward means of coupling preparative high-performance liquid chromatography and mass spectrometry. *Rapid Commun. Mass Spectrom.,* **16**, 544–554.
22. Leister, W., Strauss, K., Wisnoski, D., Zhao, Z. & Lindsley, C. (2003) Development of a custom high-throughput preparative liquid chromatography/mass spectrometer platform for the preparative purification and analytical analysis of compound libraries. *J. Combinatorial Chem.*, **5**, 322–329.
23. For a review, see: Chester, T.L. & Pinkston, J.D. (2002) Supercritical fluid and unified chromatography. *Anal. Chem.*, **74**, 2801–2811.
24. Ramalakshmi, K. & Raghavan, B. (1999) Caffeine in coffee: its removal. Why and how? *Crit. Rev. Food Sci. Nutr.,* **39**, 441–456.
25. Berger, T.A., Fogleman, K., Staats, T., Bente, P., Crocket, I., Farrell, W. &. Osonubi, M., (2000) The development of a semi-preparatory scale supercritical fluid chromatograph for high-throughput purification of 'combi–chem' libraries. *J. Biochem. Biophys. Methods.,* **43**, 87–111.
26. Wang, T., Barber, M., Hardt, I. & Kassel, D.B. (2001) Mass directed fractionation and isolation of pharmaceutical compounds by packed-column supercritical fluid chromatography/mass spectrometry. *Rapid Commun. Mass Spectrom.*, **15**, 2067–2075.
27. Hochlowski, J., Olson, J., Pan J., Sauer, D., Searle, P. & Sowin, T. (2003) Purification of HTOS libraries by supercritical fluid chromatography. *J. Liquid Chromatogr. Relat. Technol.*, **26**, 333–354.
28. Ripka, W.C., Barker, G. & Krakover, J. (2001) High-throughput purification of compound libraries. *Drug Discovery Today*, **6**, 471–477.
29. Whatley, J. (1995) Enantiomeric separation by packed column chiral supercritical fluid chromatography. *J. Chromatogr. A*, **697**, 251–255.
30. Hopfgartner, G. & Bourgogne, E. (2003) Quantitative high-throughput analysis of drugs in biological matrices by mass spectrometry. *Mass Spectrom. Rev.*, **22**, 195–214.
31. Ackermann, B.L., Berna, M.J. & Murphy, A.T. (2002) Recent advances in use of LC/MS/MS for quantitative high-throughput bioanalytical support of drug discovery. *Curr. Topics Med. Chem.*, **2**, 53–66.
32. Brockman, A.H., Hiller, D.L. & Cole, R.O. (2000) High-speed HPLC/MS/MS analysis of biological fluids: a practical review. *Curr. Opin. Drug Discovery Devel.*, **3**, 432–438.
33. Fox, S.I. (1993) *Human Physiology*, 4th edn., p. 332. Wm C. Brown, Dubuque, Iowa.
34. Lemière, F. (2001) Mass analysers for LC-MS. *LC-GC Europe Guide to LC-MS*, pp. 22–28. Advanstar Communications, Chester.
35. Lemière, F. (2001) Interfaces for LC-MS. *LC-GC Europe Guide to LC-MS*, pp. 29–35. Advanstar Communications, Chester.
36. Bonfiglio, R., King, R.C., Olah, T.V. & Merkle, K. (1999) The effects of sample preparation methods on the variability of the electrospray ionization response for model drug compounds. *Rapid Commun. Mass Spectrom.*, **13**, 1175–1185.
37. Nelson, M.D. & Dolan, J.W. (2002) Ion suppression in LC-MS-MS – a case study. *LC-GC*, **20**, 24–32.
38. Shou, W.Z., & Naidong, W. (2003) Post-column infusion study of the 'dosing vehicle effect' in liquid chromatography/tandem mass spectrometric analysis of discovery pharmacokinetic samples. *Rapid Commun. Mass Spectrom.*, **17**, 589–597.
39. Mei, H., Hsieh, Y., Nardo, C., Xu, X., Wang, S., Ng, K. & Korfmacher, W.A. (2003) Investigation of matrix effects in bioanalytical high-performance liquid chromatography/tandem mass spectrometric assays: application to drug discovery. *Rapid Commun. Mass Spectrom.*, **17**, 97–103.
40. Matuszewski, B.K., Constanzer, M.L., & Chavez-Eng, C.M. (1998) Matrix effect in quantitative LC/MS/MS analyses of biological fluids: a method for determination of finasteride in human plasma at picogram per milliliter concentrations. *Anal. Chem.*, **70**, 882–889.
41. McCauley-Myers, D.L., Eichhold, T.H., Bailey, R.E., Dobrozsi, D.J., Best, K.J., Hayes, J.W., II & Hoke, S.H., II (2000) Rapid bioanalytical determination of dextromethorphan in canine plasma by dilute-and-shoot preparation combined with one minute per sample LC-MS/MS analysis to optimize formulations for drug delivery. *J. Pharm. Biomed. Anal.* **23**, 825–835.
42. Polson, C., Sarkar, P., Incledon, B., Raguvaran, V. & Grant, R. (2003) Optimization of protein precipitation based upon effectiveness of protein removal and ionization effect in liquid chromatography–tandem mass spectrometry. *J. Chromatogr. B*, **785**, 263–275.
43. Majors, R.E. (2001) New designs and formats in solid-phase extraction sample preparation. *LC-GC Europe,* **14**, 746–751.
44. Wells, D.A. (2003) *High-throughput Bioanalytical Sample Preparation.* Elsevier, Amsterdam.
45. Sadagopan, N.P., Li, W., Cook, J.A., Galvan, B., Weller, D. L., Fountain, S.T. & Cohen, L.H. (2003) Investigation of EDTA anticoagulant in plasma to improve the throughput of liquid chromatography/tandem mass spectrometric assays. *Rapid Commun. Mass Spectrom.*, **17**, 1065–1070.

46. Palmer, D.S., Rosborough, D., Perkins, H., Bolton, T., Rock, G. & Ganz, P.R. (1993) Characterization of factors affecting the stability of frozen heparinized plasma. *Vox Sanguinis,* **65**, 258–270.
47. Berna, M., Murphy, A.T., Wilken, B. & Ackermann, B. (2002) Collection, storage and filtration of *in vivo* study samples using 96-well filter plates to facilitate automated sample preparation and LC/MS/MS analysis. *Anal. Chem.,* **74**, 1197–1201.
48. Weaver, R., Graham, K.S., Beattie, I.G. & Riley R.J. (2003) Cytochrome P450 inhibition using recombinant proteins and mass spectrometry/multiple reaction monitoring technology in a cassette incubation. *Drug Metab. Disposition,* **31**, 955–966.
49. Watt, A.P., Morrison, D., Locker, K.L. & Evans, D.C. (2000) Higher throughput bioanalysis by automation of a protein precipitation assay using a 96-well format with detection by LC-MS/MS. *Anal. Chem.*, **72**, 979–984.
50. O'Connor, D., Clarke, D.E., Morrison, D. & Watt, A.P. (2002) Determination of drug concentrations in plasma by a highly automated, generic and flexible protein precipitation and liquid chromatography/tandem mass spectrometry method applicable to the drug discovery environment. *Rapid Commun. Mass Spectrom.*, **16**, 1065–1071.
51. Biddlecombe, R.A. & Pleasance, S. (1999) Automated protein precipitation by filtration in the 96-well format. *J. Chromatogr., B*, **734**, 257–265.
52. Zweigenbaum, J. & Henion, J. (2000) Bioanalytical high-throughput selected reaction monitoring-LC/MS determination of selected estrogen receptor modulators in human plasma: 2000 samples/day. *Anal. Chem.*, **72**, 2446–2454.
53. Eerkes, A., Shou, W.Z. & Naidong, W. (2003) Liquid/liquid extraction using 96-well plate format in conjunction with hydrophilic interaction liquid chromatography-tandem mass spectrometry method for the analysis of fluconazole in human plasma. *J. Pharm. Biomed. Anal.*, **31**, 917–928.
54. Peng, S.X., Branch, T.M. & King, S.L. (2001) Fully automated 96-well liquid–liquid extraction for analysis of biological samples by liquid chromatography with tandem mass spectrometry. *Anal. Chem.*, **73**, 708–714.
55. Wang, A.Q., Zeng, W., Musson, D.G., Rogers, J.D. & Fisher, A.L. (2002) A rapid and sensitive liquid chromatography/negative ion tandem mass spectrometry method for the determination of an indolocarbazole in human plasma using internal standard (IS) 96-well diatomaceous earth plates for solid-liquid extraction. *Rapid. Commun. Mass Spectrom.*, **16**, 975–981.
56. Peng, S.X., King, S.L., Bornes, D.M., Foltz, D.J., Baker, T.R. & Natchus, M.G. (2000) Automated 96-well SPE and LC-MS-MS for determination of protease inhibitors in plasma and cartilage tissues. *Anal. Chem.*, **72**, 1913–1917.
57. Shou, W.Z., Jiang, X., Beato, B.D. & Naidong, W. (2001) A highly automated 96-well solid phase extraction and liquid chromatography/tandem mass spectrometry method for the determination of fentanyl in human plasma. *Rapid Commun. Mass Spectrom.*, **15**, 466–476.
58. Mallet, C.R., Lu, Z., Fisk, R., Mazzeo., J.R. & Neue, U.D. (2003) Performance of an ultra-low elution-volume 96-well plate: drug discovery and development applications. *Rapid Commun. Mass Spectrom.,* **17**, 163–170.
59. Naidong, W., Shou, W.Z., Addison, T., Maleki, S. & Jiang, X. (2002) Liquid chromatography/tandem mass spectrometric bioanalysis using normal-phase columns with aqueous/organic mobile phases – a novel approach of eliminating evaporation and reconstitution steps in 96-well SPE. *Rapid Commun. Mass Spectrom.,* **16**, 1965–1975.
60. Zimmer, D., Pickard, V., Czembor, W. & Müller, C. (1999) Comparison of turbulent-flow chromatography with automated solid-phase extraction in 96-well plates and liquid–liquid extraction used as plasma sample preparation techniques for liquid chromatography–tandem mass spectrometry. *J. Chromatogr. A,* **854**, 23–35.
61. Chassaing, C., Luckwell, J., Macrae, P., Saunders, K., Wright, P. & Venn, R. (2001) Direct analysis of crude plasma samples by turbulent flow chromatography/tandem mass spectrometry. *Chromatographia*, **53**, 122–130.
62. Herman, J.L. (2002) Generic method for on-line extraction of drug substances in the presence of biological matrices using turbulent flow chromatography. *Rapid Commun. Mass Spectrom.,* **16**, 421–426.
63. Hsieh, Y., Bryant, M.S., Brisson, J.-M., Ng, K. & Korfmacher, W.A. (2002) Direct cocktail analysis of drug discovery compounds in pooled plasma samples using liquid chromatography–tandem mass spectrometry. *J. Chromatogr. B*, **767**, 353–362.
64. Hsieh, Y., Wang, G., Wang, Y., Chackalamannil, S. & Korfmacher, W.A. (2003) Direct plasma analysis of drug compounds using monolithic column liquid chromatography and tandem mass spectrometry. *Anal. Chem.*, **75**, 1812–1818.
65. Koster, E. & Ooms, B. (2001) Recent developments in on-line SPE for HPLC and LC-MS in bioanalysis. *LC-GC Europe Guide to LC-MS,* pp. 55–57. Advanstar Communications, Chester.
66. Koster, E., Ringeling, P. & Ooms, B. (2002) Denaturing solid-phase extraction for reduced protein interference in bioanalytical SPE-LC-MS. *LC-GC Europe* (*The Applications Book*), pp. 28–32.
67. Needham, S.R. (2000) Directions in discovery: fast, efficient separations in drug discovery-LC-MS analysis using column switching and rapid gradients *LC-GC*, **18**, 1156–1161.

68. Romanyshyn, L., Tiller, P.R., Alvaro, R., Pereira, A. & Hop, C.E.C.A. (2001) Ultra-fast gradient vs. fast isocratic chromatography in bioanalytical quantification by liquid chromatography/tandem mass spectrometry. *Rapid Commun. Mass Spectrom.*, **15**, 313–319.
69. Jemal. M. & Xia, Y.-Q. (1999) The need for adequate chromatographic separation in the quantitative determination of drugs in biological samples by high performance liquid chromatography with tandem mass spectrometry. *Rapid Commun. Mass Spectrom.*, **13**, 97–106.
70. King, R.C., Miller-Stein, C., Magiera, D.J. & Brann, J. (2002) Description and validation of a staggered parallel high performance liquid chromatography system for good laboratory practice level quantitative analysis by liquid chromatography/tandem mass spectrometry. *Rapid Commun. Mass Spectrom.*, **16**, 43–52.
71. Yang, L., Mann, T.D., Little, D., Wu, N., Clement, R.P. & Rudewicz, P.J. (2001) Evaluation of a four-channel multiplexed electrospray triple quadrupole mass spectrometer for the simultaneous validation of LC/MS/MS methods in four different preclinical matrixes. *Anal. Chem.*, **73**, 1740–1747.
72. Deng, Y., Wu, J.-T., Lloyd, T.L., Chi, C.L., Olah, T.V. & Unger, S.E. (2002) High-speed gradient parallel liquid chromatography/tandem mass spectrometry with fully automated sample preparation for bioanalysis: 30 seconds per sample from plasma. *Rapid Commun. Mass Spectrom.*, **16**, 1116–1123.
73. Lim, H.K., Chan, K.W., Sisenwine, S. & Scatina, J.A. (2001) Simultaneous screen for microsomal stability and metabolite profile by direct injection turbulent–laminar flow LC-LC and automated tandem mass spectrometry. *Anal. Chem.*, **73**, 2140–2146.
74. Hop, C.E.C.A., Tiller, P.R. & Romanyshyn, L. (2002) *In vitro* metabolite identification using fast gradient high performance liquid chromatography combined with tandem mass spectrometry. *Rapid Commun. Mass Spectrom.*, **16**, 212–219.
75. Cai, Z., Han, C., Harrelson, S., Fung, E. & Sinhababu, A.K. (2001) High throughput analysis in drug discovery: application of liquid chromatography/ion-trap mass spectrometry for simultaneous cassette analysis of α-1a antagonists and their metabolites in mouse plasma.*Rapid Commun. Mass Spectrom.*, **15**, 546–550.
76. Tiller, P.R. & Romanyshyn, L.A. (2002) Liquid, chromatography/tanden mass spectrometric quantification with metabolite screening as a strategy to enhance the early drug discovery process. *Rapid Commun. Mass Spectrom.*, **16**, 1225–1231.
77. Clarke, N.J., Rindgen, D., Korfmacher, W.A. & Cox, K.A. (2001) Systematic LC/MS metabolite identification in drug discovery. *Anal. Chem.*, **73**, 430A–439A.
78. Kostiainen, R., Kotiaho, T., Kuuranne, T. & Auriola, S. (2003) Liquid chromatography/atmospheric pressure ionization–mass spectrometry in drug metabolism studies. *J. Mass Spectrom.*, **38**, 357–372.
79. Rushmore, T.H., Reider, P.J., Slaughter. D., Assang, C. & Shou, M. (2000) Bioreactor systems in drug metabolism: synthesis of cytochrome P450-generated metabolites. *Metabolic Eng.*, **2**, 115–125.
80. Ulrich, S. (2000) Solid-phase microextraction in biomedical analysis. *J. Chromatogr. A*, **902**, 167–194.
81. Plumb, R.S., Ayrton, J., Dear, G.J., Sweatman, B.C. & Ismail, I.M. (1999) The use of preparative high-performance liquid chromatography with tandem mass spectrometric directed fraction collection for the isolation and characterization of drug metabolites in urine by nuclear magnetic resonance spectroscopy and liquid chromatography/sequential mass spectrometry. *Rapid Commun. Mass Spectrom.*, **13**, 845–854.
82. Gerhard, U., Thomas, S. & Mortishire-Smith, R. (2003) Accelerated metabolite identification by 'Extraction-NMR'. *J. Pharm. Biomed. Anal.*, **32**, 531–538.
83. Peng, S.X. (2000) Hyphenated HPLC-NMR and its application in drug discovery. *Biomed. Chromatogr.*, **14**, 430–441.
84. Dear, G.J., Mallett, D.N., Higton, D.M., Roberts, A.D., Bird, S.A., Young, H., Plumb. R.S. & Ismail, I. M. (2002) The potential of serially coupled alkyl-bonded silica monolithic columns for high resolution separations of pharmaceutical compounds in biological fluids. *Chromatographia*, 55, 177–184.
85. Yokoyama, Y., Kishi, N., Tanaka, M. & Asakawa, N. (2000) On-line sample preparation system using column-switching HPLC for the structure elucidation of compounds in mixtures by NMR. *Anal. Sci.*, **16**, 1183–1188.
86. Liu, D.Q., Hop, C.E.C.A., Beconi, M.G., Mao, A. & Chiu, S.-H.L. (2001) Use of on-line hydrogen–deuterium exchange to facilitate metabolite identification. *Rapid Commun. Mass Spectrom.*, **15**, 1832–1839.
87. Zhang, N., Fountain, S.T., Bi, H. & Rossi, D.T. (2000) Quantification and rapid metabolite identification in drug discovery using API time-of-flight LC/MS. *Anal. Chem.*, **72**, 800–806.
88. Nicholson, J.K., Lindon, J.C., Scarfe, G.B., Wilson, I.D., Abou-Shakra, F., Sage., A.B. & Castro-Perez, J. (2001) High-performance liquid chromatography linked to inductively coupled plasma mass spectrometry and orthogonal acceleration time-of-flight mass spectrometry for the simultaneous detection and identification of metabolites of 2-bromo-4-trifluoromethyl-[$^{13}$C]-acetanilide in rat urine. *Anal. Chem.*, **73**, 1491–1494.
89. Corcoran, O., Lindon, J.C., Hall, R., Ismail, I.M, & Nicholson, J.K. (2001) The potential of $^{19}$F NMR spectroscopy for rapid screening of cell cultures for models of mammalian drug metabolism. *Analyst*, **126**, 2103–2106.
90. Martino, R., Malet-Martino, M. & Gilard, V. (2000) Fluorine nuclear magnetic resonance, a privileged tool for metabolic studies of fluoropyrimidine drugs. *Current Drug Metab.*, **1**, 271–303.

91. Decaestecker, T.N., Clauwaert, K.M., Van Bocxlaer, J.F., Lambert, W.E., Van den Eeckhout, E.G., Van Peteghem, C.H. & De Leenheer, A.P. (2000) Evaluation of automated single mass spectrometry to tanden mass spectrometry function switching for comprehensive drug profiling analysis using a quadrupole time-of-flight mass spectrometer. *Rapid Commun. Mass Spectrom.*, **14**, 1787–1792.
92. Ramanathan, R., McKenzie, D.L., Tugnait, M. & Siebenaler, K. (2002) Application of semi-automated metabolite identification software in the drug discovery process for rapid identification of metabolites and the cytochrome P450 enzymes responsible for their formation. *J. Pharm. Biomed. Anal.*, **28**, 945–951.

# 7 Environmental organic analytes

MIRA PETROVIC and DAMIÀ BARCELÓ

## 7.1 Introduction

The analysis of organic microcontaminants in environmental samples constitutes a difficult task, first because of the complexity of the matrices, and second because of the normally very low concentrations of the target compounds. Additionally, a variety of chemical and physical characteristics of contaminants, often present in very complex mixtures, 'chemical cocktails,' requires a comprehensive approach for their analysis.

The growing number of samples to be analyzed in laboratories carrying out monitoring studies requires employment of high-throughput and fully automated analytical techniques. For these reasons, great efforts are going into the development of cost-effective sample handling techniques characterized by efficiency and simplicity of operations and devices. In general, the objective is to develop analytical methods that are faster, simpler and more sensitive. Impressive improvements in detection limits for organic contaminants have pushed the target concentrations from the microgram per liter to the nanogram or picogram per liter range. Such progress is mostly due to the development of hyphenated chromatography–mass spectrometry (MS) techniques, which are today the methods of choice for the determination of trace organic analytes in environmental samples. Some of techniques that have advanced significantly in recent years are tandem (MS-MS) systems, time-of-flight MS (TOF-MS), quadrupole-time-of-flight (MS-TOF-MS) and membrane introduction MS (MIMS). However, instrumental analysis of environmental samples is often hampered by the complexity of matrices and negative effects resulting from the presence of high concentrations of interfering substances. Therefore, the analysis of trace components in complex environmental matrices generally requires multistep sample preparation. The general problem in analysis of complex environmental samples is that the extract obtained by exhaustive extraction techniques typically contains a large number of matrix components which may co-elute with the analytes and disturb the quantitative analysis. The choice of an appropriate sampling and sample preparation strategy has to be adjusted to the individual case and optimized depending on the target pollutants, number of samples, sampling locations and subsequent analytical method.

The presence of interfering substances demands either very selective detection or tedious extract clean-up, or even both. Generally, multistep sample pretreatment aimed at reducing the matrix content and enriching the target compounds remains the most direct means of obtaining maximum sensitivity. However, extraction and clean-up protocols used are time- and labor-consuming, and they often constitute the bottleneck of an analytical method.

This chapter attempts to survey current state-of-the-art sample preparation techniques used for the analysis of organic contaminants in the environment. Nowadays, a wide range of man-made chemicals, designed for use in industry, agriculture and consumer goods, and chemicals formed unintentionally, or produced as by-products of industrial processes or combustion, are potentially of environmental concern. Consequently, the research in such a broad field is very active and results in a huge number of papers being published every year. Because, it is impossible for a single chapter to cover comprehensively all analytes of interest, we have focused our attention on several important classes of organic contaminants, mainly from the group of new, so-called emerging contaminants.

## 7.2 Initial considerations

### 7.2.1 *Sampling of aqueous and solid environmental samples*

The sampling strategy for water samples often includes just simply placing the water of interest in a bottle. For surface waters (river, lake or sea), discrete (grab) sampling has frequently been employed, whereas for wastewaters, composite sampling over periods of between 6 h and several days is often used. One of major drawbacks of conventional 'grab' sampling is that it yields samples representing one point in time and may fail to detect episodic contamination. Long-term monitoring, performed with the objective of establishing the quality of the aquatic environment, requires a large number of samples to be taken from one sampling point over the entire sampling period; this is usually time-consuming and costly. To overcome this problem modern sampling strategies for long-term monitoring combine sampling, analyte isolation and preconcentration into a single step. An example of this strategy is the application of passive samplers, mainly for water and air analysis.

When analyzing solid environmental samples, because of the heterogeneity of solid matrices, special care must be taken that the samples and subsamples analyzed are representative of the system studied. In the case of sediment or soil samples, depending on the objective of the study (determination of vertical distribution profiles, or concentrations in a surface layer), one should decide whether to take core samples, or grab samples, respectively.

When studying accumulation of organic pollutants in biota (e.g. fish, mussels and snails) one should have in mind that the selection of species to be analyzed should be based on considerations such as availability at sampling sites, size, migratory behavior and their position in the food chain [1]. For example, if fish had not been caged, samples from a particular site would actually represent a segment of water that could range several kilometers upstream or downstream from the sampling point; thus, when studying the impact of sewage treatment plant (STP) effluents in receiving rivers, fewer migratory species are preferred, and also species that reside primarily in the middle depths of the water column.

### 7.2.2 *Sample storage, homogenisation and preservation*

After collection of water samples, their further pretreatment, including preservation, is highly dependent on the type of analysis and analytes, and their stability within the sample matrix during its transport and storage before extraction of the target analytes.

The stability of target compounds within sample matrix during its transport and storage is the key issue regarding quality assurance parameters. Loss of sample integrity for some compounds during transportation, because of hydrolysis or adverse shipping conditions (high temperature and/or humidity), may compromise the reliability of the results obtained. Several comprehensive stability studies have been performed comparing standard and alternative preservation methods for different organic pollutants in water matrices: pesticides [2], phenolic compounds [3], benzene and naphthalenesulfonates [4,5], ionic and nonionic surfactants [6,7] and estrogens [8]; all these showed that conventional methods of chemical preservation are not always appropriate and that significant losses of some compounds occur when samples are kept in a water matrix, even when preserved with the chosen additive. The use of solid phase extraction (SPE) materials (disks, cartridges and disposable precolumns) for the stabilization and storage of analytes preconcentrated from water samples was found to be more suitable.

For solid matrices, the common practice is to remove water from the solid matrix and to store dry samples under cooling conditions. Removal of water from sediment before extraction was found to be crucial in obtaining good recoveries [9]. Freeze-drying is an accepted and commonly used procedure for drying sediments and sludges; however, it is not known whether it could affect the levels of target compounds measured, especially some relatively volatile compounds.

The most common pretreatment procedure used for fresh soft tissue (for mussels) or fish muscle, liver or whole fish (or cross-section), is homogenization with anhydrous $Na_2SO_4$ and immediate extraction. Some protocols include freezing at $-20°C$ in a freezer or cutting into small pieces and subsequent freezing in liquid nitrogen. Before analysis frozen samples were thawed and then homogenized with an organic solvent (methanol, acetonitrile) or placed directly in a mortar filled with liquid nitrogen and crushed.

## 7.3 Extraction of target compounds

### 7.3.1 *Extraction from aqueous samples*

Extraction of organic contaminants from environmental and wastewater samples is usually performed by SPE, solid phase microextraction (SPME), or liquid-liquid extraction (LLE). The main practical and functional attributes of these methods, routinely used in modern laboratory practice, are shown in Table 7.1.

7.3.1.1 *Solid phase extraction (SPE)* SPE was initially developed as a complement or replacement for LLE, but it has become the most common sample preparation technique now used in environmental analysis. Since the 1990s, SPE has undergone considerable development, with many improvements in sampling formats and sorbent materials. The growing number of samples to be analysed in laboratories carrying out monitoring studies requires development of high-throughput and fully automated analytical techniques. One of the well established and robust options is application of on-line coupling of SPE and LC, utilizing special sample preparation units, e.g. PROSPEKT (Spark, Holland) or OSP-2 (Merck, Germany) and disposable extraction

**Table 7.1** Comparison of extraction methods for liquid samples.

| LLE | SPE | SPME |
|---|---|---|
| Use of an extractant immiscible with the sample | Wide choice of solid phase | Non-solvent technique |
| Large volume of organic solvent | High selectivity | Equilibrium (nonexhaustive) extraction |
| Diluted extracts | Small volume of organic solvent | Wide choice of fibers |
| Difficulty in extracting polar and ionic compounds from water | High concentration factor | Applicable to volatile compounds |
| Difficult to automate | Amenable to automation | On-line coupling to GC and LC |
| | On-line coupling to chromatographic systems | Portable field samplers available |

cartridges. On-line SPE-LC coupling techniques have been successfully applied to the analysis of pesticides, polycyclic aromatic hydrocarbons (PAHs), antifouling agents and endocrine disrupting compounds in water, providing a significant time benefit, accompanied by increased sensitivity and decreased amounts of sample and solvent required [10]. Similarly, on-line coupling of SPE to GC is a promising approach with good prospects for the future [11].

Applied in an off-line or on-line mode, SPE offers, respectively, reduced processing time, or higher selectivity due to wide choice of sorbent materials, and significant solvent economy. Today, SPE is the most routinely applied extraction method and is applicable to virtually all nonvolatile and semivolatile compounds in aqueous and organic matrices. Although this extraction technique has reached an advanced stage of maturity, various aspects of SPE continue to be studied. The basic principles of classical off-line SPE have not changed substantially; however, a high degree of commercialization together with the development of vacuum manifolds, microprocessors and robotic systems, has permitted a remarkable increase in sample throughput [12].

For the isolation of organic contaminants from water samples (the most common application of SPE in environmental analysis), two general approaches may be applied: the use of low specificity sorbents for rather nonspecific extraction of different classes of organic compounds; the use of compound-specific or class-specific sorbents for more targeted extraction. For the first group of sorbents, the most popular for environmental applications are chemically bonded silicas (e.g. containing polar or apolar functional groups such as cyanopropyl, aminopropyl, short and long alkyl chains), porous polymers (e.g. copolymers of styrene and divinylbenzene, macroporous polypolymers, sulfonated polymers and vinylpyrrolidine copolymers) and carbon (e.g. graphitized carbon black and porous graphitic carbon).

The introduction of several new materials for class specific sorption, such as immunosorbents, molecularly imprinted polymers (MIP) and restricted access materials (RAM) undoubtedly improved cost-effective analysis of organic contaminants and enhanced the selectivity of sample preparation.

The SPE immunosorbents with polyclonal antibodies immobilized on the packing of a cartridge or precolumn (i.e. silica based supports, activated Sephadex gels, synthetic polymers, sol–gel materials, cyclodextrins and RAM) have enabled rapid and highly selective enrichment of trace analytes [13]. New developments and applications of immuno-based sample preparation for trace analysis recently reported have been

**Table 7.2** Application of molecularly imprinted polymers to the analysis of environmental samples (MIP-SPE).

| Analyte | Sample | Reference |
|---|---|---|
| Chlorinated phenoxyacids | River water | 98 |
| Triazines and chlorotriazines | River water | 28,99 |
| | Water and fruit extracts | 100 |
| | Tap water | 101 |
| | Water | 102 |
| | Natural waters and sediments | 103 |
| Carbamates | Tap water, spring water, river water and seawater | 104 |
| Nitrophenols | Surface waters | 105,106 |

reviewed [14] with emphasis on (i) antigen–antibody interactions and (ii) their importance for the properties and use of immunosorbents, (iii) multiresidue extractions, (iv) on-line coupling to chromatographic or electrophoretic separations, and (v) the high potential for improving MS detection. The first applications of immunosorbents involved selective extraction of triazine and phenylurea herbicides [15–18] in environmental waters; since then immunoaffinity extraction has been successfully applied to the analysis of different classes of organic contaminants, such as PAHs in groundwaters, wastewaters and industrial effluents [19–22], benzene derivatives (BTEX) in gasoline-contaminated waters [23], and steroid sex hormones (E2, E1) in wastewater [24].

Current research in the molecular imprinting field is concentrated on MIP-SPE; this seems to be one of the most promising applications for MIPs today and also the application that is closest to commercialization [25]. MIP-SPE has mainly been used in bioanalysis for extraction of target analytes from blood, plasma, serum, urine, liver extracts, bile, etc. [26], whereas its application in the environmental field is still rare. Environmental applications of MIPs are summarized in Table 7.2. MIPs are synthetic antibodies developed for class-specific or compound-specific extraction. Porous MIPs that have antibody-like affinities toward an imprinted target analyte could be prepared by noncovalent or covalent imprinting. Selective binding (molecular recognition) of a given target compound occurs in an aprotic, low polarity organic solvent (i.e. a solvent used for the polymerization), which results in specific hydrogen bonds between analyte and imprinted polymer. MIPs offer important advantages, such as the possibility of synthesizing polymers with a predetermined selectivity for a particular analyte, they are stable in different environments, reusable, and have higher stability and shorter preparation times than antibodies. However, in some cases MIPs tend to exhibit class-specific rather than analyte-specific behavior in combination with some non-specific binding interactions with non-target analytes, and their affinities and capacities are in most cases lower than of those of immunoaffinity cartridges or columns. Nevertheless, the technique is still under development and the use of MIP sorbents in environmental applications is expected to increase.

SPE sorbents, based on RAMs, are specifically tailored to fractionate samples into macromolecular matrix components and low molecular target analytes. They have been successfully applied for direct extraction and enrichment of hydrophobic low molecular mass analytes from biological fluids carrying a high load of proteins (plasma, blood, urine, saliva, supernatants of cell cultures and tissue) and from food samples (milk, food homogenates) [27]. In environmental analysis, RAMs have been successfully applied for the separation of humic substances interfering in the analysis of triazines [28], acidic herbicides [29] and polar fungicides [30] in river water and groundwater. They have also been applied in on-line clean-up of complex samples and for high-throughput flow immunoassay of triazines [31]. This separation of high molecular mass impurities is based on size-exclusion of unwanted matrix components as a result of topographical restriction achieved either by a physical diffusion barrier (small pore diameter), or by a chemical diffusion barrier (chemically bonded phase). Low molecular mass analytes able to access active adsorption centres (alkyl chains or ion-exchange groups) at the inner pore surface, are retained by a reversed-phase, affinity, or ion-pair mechanism, respectively [32].

One example of the environmental applications of RAM and MIP sorbents is shown schematically in Fig. 7.1. The two sorbents coupled in line were used for selective extraction of triazine pesticides from water samples [28]. The procedure, the so called Six-SPE (size-selective sample separation and solvent switch), involves three different chromatographic processes. First, a RAM precolumn eliminates macromolecular matrix components by size-exclusion chromatography (SEC). The low molecular mass components and the analytes are enriched in a second process by adsorption chromatography (reversed-phase active sites in an inner pore surface of RAM). Finally, after a solvent switch in the third process, the target analytes are selectively adsorbed by tailor-made MIP sorbents during affinity chromatography.

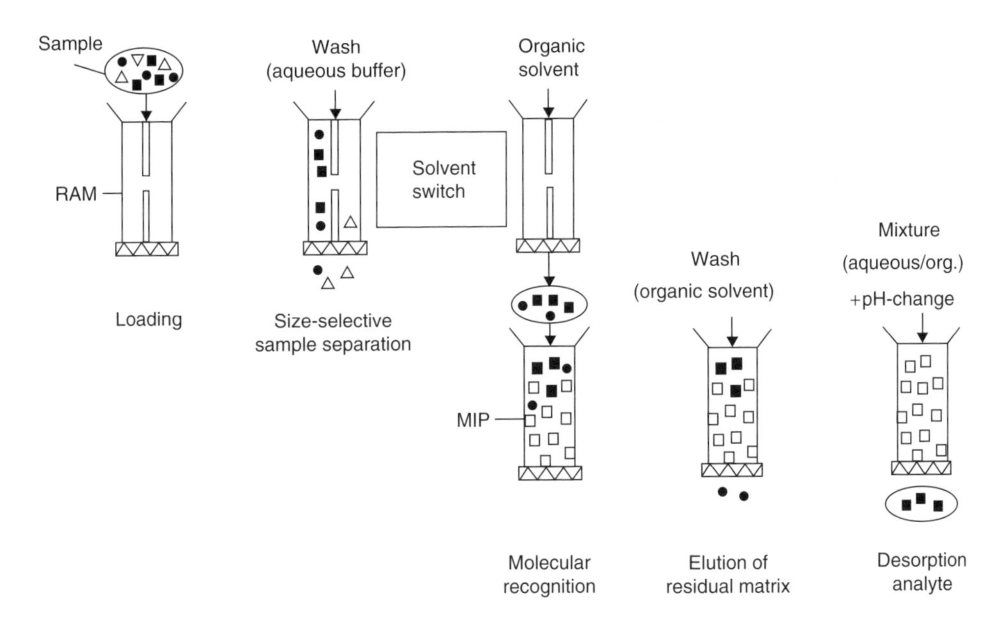


**Fig. 7.1** Schematic representation of the working principle of the RAM-MIP coupling procedure, i.e. six-SPE. The upper left row schematically shows the process in the RAM column, the lower right, the process in MIP column. Reprinted with permission from [28]. Copyright 2001, American Chemical Society.

7.3.1.2 *Solid phase microextraction (SPME)* One of the prevailing trends in environmental analysis is the use of solvent-free extraction techniques that combine efficiency, cheapness and a high level of automation. SPME fulfils these requirements and is considered to be a universal extraction technique that allows the determination of a wide range of volatile, semivolatile and nonvolatile organic contaminants and is particularly suited for field analyses and for rapid response applications. Depending on the analyte and matrix, SPME of water samples can be performed under three different modes: direct immersion extraction (for less volatile compounds and relatively clean samples), headspace extraction (for more volatile compounds and dirtier samples) and membrane protected SPME (for extraction of analytes in very polluted samples).

The major part of SPME applications has been developed for GC due to easy coupling of these two techniques, and numerous papers have described its application in environmental analysis. (Approximately 40% of more than 1200 references on SPME found in the Chemical Abstract Service, and published since the first SPME device was described in 1990, are in the field of environmental analysis.) However, only a small percentage of all SPME applications use HPLC, because the coupling is more complex and requires specially designed interfaces to desorb the analytes from the fiber. The first automated in-tube SPME-HPLC system was successfully applied to the analysis of polar thermally labile analytes by Eisert & Pawliszyn [33]. This promising method has received increased attention over the past few years and has been termed 'in-tube' because the extraction phase is not coated on the surface of a fused silica rod as in syringe-like SPME devices, but instead is coated inside a section of fused silica tubing (i.e. an open tubular capillary GC column, or micro-LC capillary column). The method has been successfully applied to the analysis of different organic contaminants, such as chlorinated phenoxy acid herbicides [34], carbamates [35] and phthalates [36], and its application is steadily growing.

7.3.1.3 *Extraction using passive samplers* To overcome the problem of time-consuming and costly long-term monitoring requiring a large number of samples to be taken from one sampling point over an entire sampling period, modern sampling strategies combine sampling, analyte isolation and preconcentration into a single step. An example of this strategy applied to environmental water samples is the use of passive samplers to measure time-averaged or integrated dissolved water concentrations, not affected by short-term fluctuations in analyte concentrations. The sampling/extraction is based on free flow of analyte molecules from the sampled medium to a collecting medium, as a result of a difference in chemical potential of the analyte between the two media [37]. Different designs have been proposed for passive samplers, with semipermeable membrane devices (SPMDs) and solvent- or sorbent-filled devices being the most often used. SPMDs [38–40] consist of a sealed tubular low-density polyethylene flat membrane (25–250 μm thick) filled with a high molecular mass lipid (typically high purity synthetic triolein). After a typical sampling time of 30 days, the SPDMs are removed from the aquatic environment and the analytes are recovered through dialysis using nonpolar solvents. The procedure required to recover analytes from the trioleine collecting medium and subsequent purification of analytes is quite tedious and

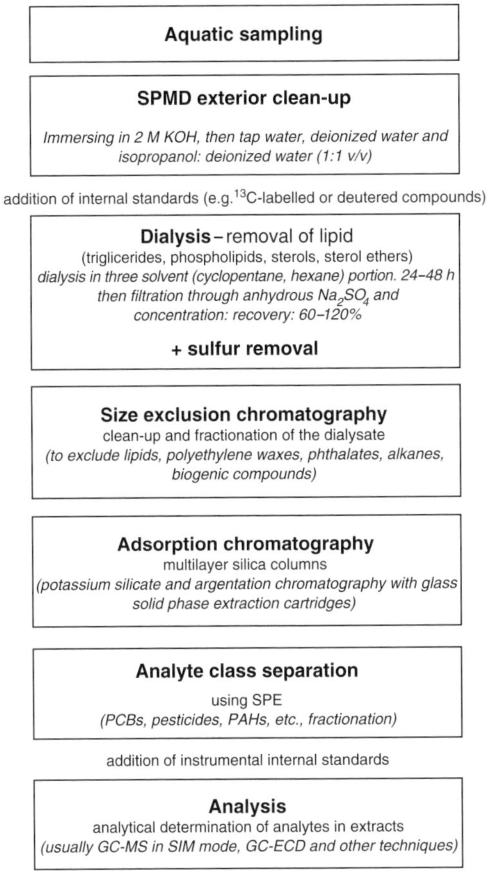

**Aquatic sampling**

**SPMD exterior clean-up**

*Immersing in 2 M KOH, then tap water, deionized water and isopropanol: deionized water (1:1 v/v)*

addition of internal standards (e.g.$^{13}C$-labelled or deutered compounds)

**Dialysis** - removal of lipid
(triglicerides, phospholipids, sterols, sterol ethers)
*dialysis in three solvent (cyclopentane, hexane) portion. 24–48 h then filtration through anhydrous $Na_2SO_4$ and concentration: recovery: 60–120%*

**+ sulfur removal**

**Size exclusion chromatography**
clean-up and fractionation of the dialysate
*(to exclude lipids, polyethylene waxes, phthalates, alkanes, biogenic compounds)*

**Adsorption chromatography**
multilayer silica columns
*(potassium silicate and argentation chromatography with glass solid phase extraction cartridges)*

**Analyte class separation**
using SPE
*(PCBs, pesticides, PAHs, etc., fractionation)*

addition of instrumental internal standards

**Analysis**
analytical determination of analytes in extracts
*(usually GC-MS in SIM mode, GC-ECD and other techniques)*

**Fig. 7.2** Schematic procedure for preparation of analytes from SPMDs before final analyses. Reprinted with permission from [40]. Copyright 2000, Elsevier.

represents the main disadvantage of the technique. The sequence of clean-up steps is schematically shown in Fig. 7.2. Passive samplers have found numerous applications in the analysis of PAHs [41–43], polychlorinated biphenyls (PCBs), dioxins and furans [44,45], organochlorine pesticides [44], organotin compounds [46], chlorophenols [47] and polar biocides (diuron, irgarol) [48] in surface waters, groundwater, agricultural run-offs and industrial effluents.

### 7.3.2 *Extraction from solid matrices samples (soil, sediment and sludge)*

Demands for new 'low-solvent, low-time, low-cost' extraction techniques amenable to automation, have led to accelerated development of a number of techniques that meet the above criteria. Thus, Soxhlet extraction and steam distillation, used almost exclusively in the 1980s and 1990s, have been partially replaced by more versatile extraction

**Table 7.3** Comparison of methods for extraction of organic pollutants from solid samples.

| Method | Advantages | Drawbacks |
|---|---|---|
| Soxhlet | Easy to handle<br>Inexpensive equipment<br>No filtration required<br>High matrix capacity | Demands large volumes of highly purified solvents (up to 200 ml)<br>High costs of purchase and disposal of solvents<br>Long extraction times (up to 48 h)<br>Generates dirty extracts<br>Not automatable |
| SFE [a] | Able to solvate a wide range of organic compounds<br>Rapid extraction<br>Low solvent requirements (low toxicity and low costs)<br>Mild extraction conditions reduce the risk of thermal degradation | Not suitable for polar organic solutes<br>Possible losses upon trapping of the analytes<br>Fastidious optimization procedures<br>High cost of equipment<br>Problems for matrices with a high water content |
| PLE | Simple optimization procedure (easy to transfer an existing Soxhlet or sonication method)<br>Fast extraction (ca. 10–20 min). Low solvent consumption (20–30 ml)<br>No filtration required<br>High level of automation. Easy to use | High investment cost of the commercialized systems<br>Elevated temperatures may degrade thermolabile analytes |
| MAE | Several extractions can be performed simultaneously (up to 12)<br>High level of automation<br>Fast extraction (*ca.* 10–30 min)<br>Low solvent consumption (10–70 ml)<br>Moderate investment | Requires further filtration of the extracts<br>Solvent must absorb microwaves (unless water is present in the matrix)<br>Non-homogenous field inside the cavity |

[a]Using supercritical $CO_2$.

systems, such as sonication, microwave assisted extraction (MAE), pressurized liquid extraction (PLE) and supercritical fluid extraction (SFE). Methods based on SPE (e.g. headspace SPME) and purge-and-trap (P&T) have been developed for volatile and semivolatile compounds. An overview of the most representative methods used to extract selected organic pollutants from solid environmental samples (sediment, soil and sludge) is shown in Table 7.3. Several complementary reviews were recently published discussing potentials and pitfalls of modern extraction techniques (SFE, PLE and MAE and subcritical water extraction (SWE)) for solid environmental matrices, summarizing the basic set-ups and relevant experimental parameters for efficient extraction of organic trace pollutants in environmental samples [49–52].

Extraction at elevated temperatures and pressures, which results in an improved mass transfer of the analytes, is employed in several techniques such as SFE, MAE and PLE. SFE with solid-phase trapping has been tested and used for different groups of organic pollutants. Although excellent results and unique elevated selectivity were obtained for selected applications, the method did not reach the status of a widely applicable technology because of the high dependence of the extraction conditions on the sample, leading to fastidious optimization procedures [53]. The lower interest is also partly due to the development of PLE that has become a widely accepted extraction technique. PLE, also termed accelerated solvent extraction (ASE) or pressurized fluid extraction (PFE), has been found to be a good alternative to the more traditional and laborious extraction methods because it offers great reduction in solvent consumption coupled with faster sample processing and a high level of automation. The considerable saving in extraction time (up to 30 min) and solvent consumption (30–60 ml) has made this technique a very attractive alternative to the conventional Soxhlet extraction method. Additionally, method development is easy because only the extraction time and temperature have to be optimized while the solvent chosen can be the same as that used in the Soxhlet or sonication extraction.

A modification of the above mentioned technique, extraction with pressurized (supercritical) hot water, and also extraction with hot water under subcritical conditions, have also been used for the analysis of polar alkylphenolic compounds (e.g. carboxylates) [54], PAHs [55–59], polychlorinated and polybrominated organic compounds [60,61] and brominated flame retardants [62].

### 7.3.3 *Biota*

Isolation of target organic compounds from biological tissues is a complicated and laborious task because of the nature of the matrix. Disruption of the cellular structure of biological samples results in high abundance of lipids and proteins. Classical methods such as Soxhlet often yield high concentrations of co-extracted lipids and therefore require further manipulation in many steps (i.e. GPC, multiple LL partitioning), which may lead to low recovery or poor reproducibility.

Another approach for carrying out simultaneous disruption and extraction of solid and semisolid samples involves matrix solid-phase dispersion (MSPD), a technique that combines extraction, concentration and clean-up in one step [63,64]. Its main characteristic is the ability to handle directly solid, semisolid and viscous samples, and to simultaneously disrupt and disperse a sample over a bonded-phase solid support that may subsequently be used as column packing. A small amount of sample tissue is blended together with the selected solid phase (e.g. $C_{18}$ sorbent). This mixture is then used as column packing placed above an aluminum oxide layer used as clean-up adsorbent. According to the solubility characteristics of the target compounds, a proper solvent, or solvent sequence, is used to elute the MSPD column. The processes of blending and preparing a column for MSPD extraction have proven to be quite generic, with the same general approach proving to be applicable to a wide range of matrices and analytes. MSPD has found wide application in the analysis of different organic pollutants from animal tissues, fruits, vegetables and other matrices. In environmental analysis, the method was successfully applied to extract analytes from fish and mussel samples. Pollutants studied included penicillins, sulfonamides and tetracycline antibiotics [65,66], ionic [67,68] and nonionic surfactants [69], organochlorine pesticides, PCB and PAH [70,71]. The major advantage of the MSPD technique compared to classical solid–liquid extraction methods is that isolation and purification are combined and accomplished within one sequential step. Thus, solvent consumption is greatly reduced, as is the total pretreatment time. However, the limitations are manual packing of MSPD columns and manual control of elution speed, which can significantly affect reproducibility and optimal recovery.

## 7.4 Extract clean-up

Because of the complexity of samples and the low selectivity of exhaustive extraction techniques applied, substantial amounts of interfering substances are found in crude extracts, and subsequent clean-up and fractionation is indispensable. The conventional approach for extract clean-up is based on solid–liquid adsorption chromatography either using long open columns, or commercially disposable cartridges packed with different adsorbents (Florisil, alumina, different types of carbon, etc.). Purification and fractionation could be also performed using off-line SPE on cartridges packed with $C_{18}$, $NH_2$, or CN modified silica, using reversed-phase, or normal-phase LC, gel permeation chromatography (GPC), or size-exclusion chromatography (SEC). Recently, a column-switching system using precolumns packed with RAM was successfully employed for the simultaneous analysis of alkyphenolic compounds and steroid sex hormones in sediments [72].

For the purification of extracts of biological samples, which usually contain high concentrations of high-molecular mass impurities, treatment with concentrated sulfuric acid is frequently used for the removal of lipids. However, it may change sample integrity and destroy some of the compounds. In this case, adsorption chromatography and GPC, or a combination of both, offers less harsh treatment.

## 7.5 Sample preparation for specific compound classes

### 7.5.1 *Polychlorinated and polybrominated compounds*

For persistent polychlorinated (PCBs, polychlorinated dibenzo-*p*-dioxins (PCDDs), polychlorinated dibenzofurans (PCDFs)) and polybrominated compounds (polybrominated diphenyl ethers, PBDEs), Soxhlet extraction is still the most commonly used technique. However, as described above, the main drawbacks of this technique are the amounts of time and solvent required. Therefore, a number of approaches have been tried to reduce solvent consumption and to speed up extraction time. SFE has been tested, but the major disadvantage of this method is that extraction efficiencies are not as robust as those obtained by Soxhlet extraction. MAE has been tested by a number of laboratories for the extraction of toxic organics such as polycyclic aromatic hydrocarbons (PAHs), chlorinated pesticides and other semivolatile contaminants, and PCDDs and PCDFs [73,74]. PLE has also been developed for extraction of persistent organic micropollutants from solids and has the robustness of conventional Soxhlet extractions.

The analysis of trace levels of persistent organic pollutants (POPs) requires laborious clean-up procedures that involve multiple column elutions and consume large quantities of potentially hazardous solvents. An analytical protocol to determine PCDDs, PCDFs, PCBs and PBDEs in sediment samples includes the steps shown in Fig. 7.3. The clean-up step is based on solid–liquid adsorption chromatography in open columns using a combination of different adsorbents, such as modified silica, Florisil, alumina and different types of carbon (Amoco PX-21, Carbosphere, Carbopack). Automated clean-up systems have also been developed based on the use of pressurized column chromatographic procedures [75].

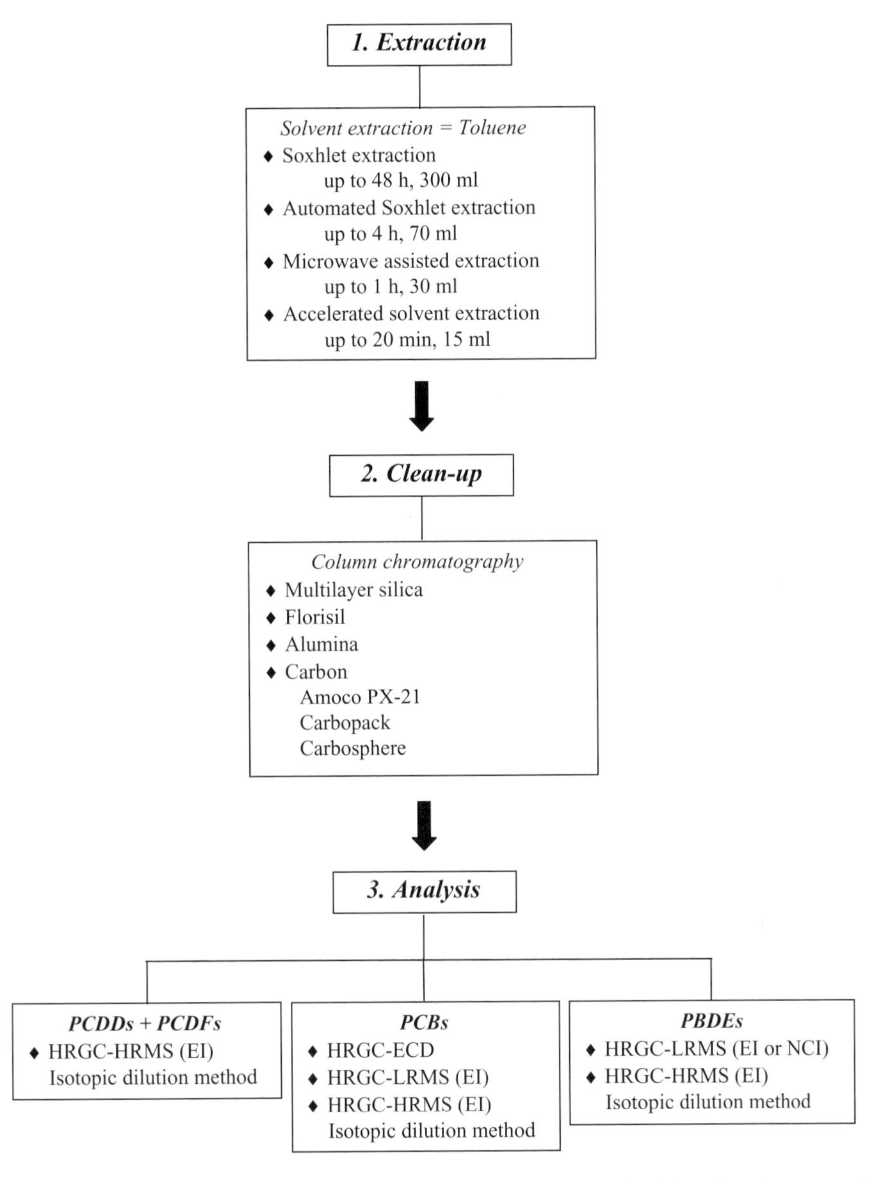


**Fig. 7.3** Analytical methods used for the analysis of polychlorinated and polybrominated compounds in sediments. Reprinted with permission from [95]. Copyright 2001, Elsevier.

Another approach, based on the use of an antibody-based affinity column has been explored as a means to shorten the length of time needed for the dioxin analysis and decrease the amount of solvent consumption [76–78]. When compared to classical clean-up and isolation methods, the immunochromatographic methods are more than 20 times faster and use 100 times less organic solvents, and their selectivity is enormously enhanced. Immunological methods, because of the specificity of antibody–antigen recognition, are known to be highly selective. In immunoaffinity chromatography methods, antibodies are bound to the packing of a chromatographic column, into which the sample containing the antigen is injected. The antigen–antibody binding is a reversible reaction that allows the use of immunoaffinity chromatography as an isolation and concentration step for the analytes of interest present in the sample.

### 7.5.2 *Surfactants*

Among different surfactant classes, this chapter focuses on nonionic surfactants: alkylphenol ethoxylates (APEO) deserve particular attention because of their ecotoxicological relevance, massive use and ubiquity in the environment. In the 1990s concern was raised regarding the environmental safety of APEOs and particularly of their lipophilic metabolites, short ethoxy-chain oligomers and alkylphenols (APs) [79,80].

For preconcentration of alkylphenolic surfactants from aqueous samples, SPE is considered the most appropriate technique, in terms of its speed, selectivity and percentage of recovery, and is preferred over conventional methods (e.g. LLE). Octadecyl ($C_{18}$) bonded silica has been the SPE material most widely employed for extraction of both neutral and acidic alkylphenolic compounds [81–84] with the efficiency of extraction from wastewater and surface water being higher than 80% for all compounds investigated. Strong anionic exchange (SAX) [85] and graphitized carbon black (GCB) have also been employed [86].

A review on the sample preparation methods for the determination of APEOs and their degradation products in solid environmental matrices has recently been published by Petrovic and Barceló [87]. The common approach that permits simultaneous extraction of nonpolar compounds (APEOs and alkylphenols (APs)) and moderately polar degradation products (alkylphenolethyl carboxylates, APECs), is either sonication or PLE. Typical solvents for the extraction of APEOs and APs have been methanol, dichloromethane, dichloromethane/hexane, hexane/acetone or hexane/isopropanol mixtures [87]. All these solvents yielded satisfactory extraction efficiency and maintained the integrity of the oligomeric distribution while allowing the preconcentration of APEOs.

Extraction with pressurized (subcritical) hot water has also been used for the analysis of polar APs (e.g. carboxylates) in sewage sludge. Water, under subcritical conditions, efficiently extracts both polar and nonpolar compounds from solid matrices [55,56]. Field and Reed [54] evaluated subcritical (hot) water extraction of nonylphenol polyethoxy carboxylates (NPECs) from 25 to 100°C at a pressure of 350 bar. Ethanol-modified hot water (30% ethanol) yielded quantitative recovery of native NPECs from sludge. Analytes were concentrated by a SAX Empore disk, simultaneously eluted with acetonitrile and derivatized to their methyl esters and finally analysed by GC-MS.

SFE with solid-phase trapping has been tested and used by some groups [88–91]. The native nonylphenols (NPs) and octylphenols (OPs) from STP sludge were extracted by combined static and dynamic SFE with pure $CO_2$ and *in situ* derivatization to corresponding acetyl derivatives in the presence of acetic anhydride and triethylamine. To improve recoveries of more polar, long chain ethoxylates (up to 17 ethoxylate units) and NPECs, supercritical $CO_2$ was modified with water [91].

For the extract clean-up, beside conventional SPE protocols, the column-switching LC using a RAM precolumn was proposed for efficient separation of high molecular

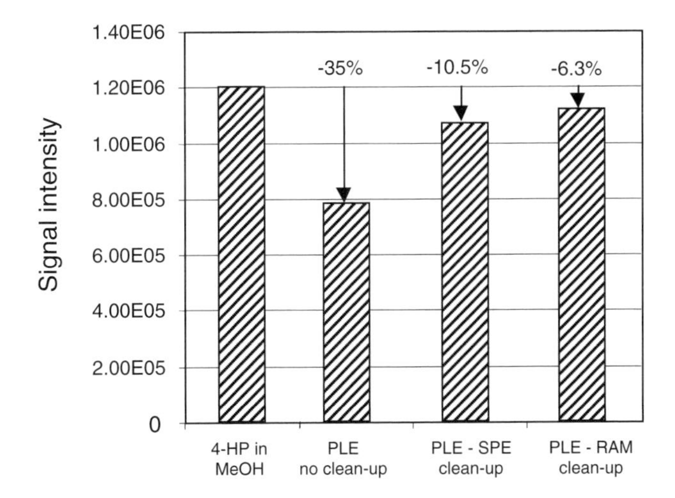


**Fig. 7.4** Comparison of signal intensity of 4-heptylphenol (HP; internal standard) in a standard solution and in sediment extracts obtained by pressurized liquid extraction (PLE) and analyzed using LC-ESI-MS without any clean-up, applying off-line SPE clean-up and on-line RAM, respectively. Conditions of PLE: methanol: acetone (1:1, v/v), 50°C, 1500 psi, 2 static cycles, extraction cell 11 ml. RAM column LiChrospher ADS C4 (Merck, Darmstadt, Germany). Reprinted with permission from [87]. Copyright 2002, Friedr. Viewig & Sohn.

mass matrix components (humic substances), polar impurities, and inorganic salts [72]. This approach helped to alleviate the problems associated with the ion suppression effect of LC-ESI-MS analysis, and allowed cleaner extracts to be analyzed providing better sensitivity. It has been shown that a restricted access precolumn efficiently separates high molecular mass matrix components (humic substances), polar impurities, and inorganic salts, thus reducing significantly ion suppression effects in ESI-MS detection (Fig. 7.4) and improving the resolution between the analytes and impurities. As a consequence, limits of detection were approximately one order of magnitude lower than those obtained with SPE clean-up.

### 7.5.3 *Hormones and drugs*

The published analytical methods for the analysis of estrogens and progestogens in water samples have been recently reviewed by López de Alda *et al.* [92]. Extraction of both estrogens and progestogens from water has been usually carried out by off-line SPE with either disks or, most frequently, cartridges. Attempts to develop fully automated methodologies for the on-line SPE and analysis of estrogens in water have also been made on a few occasions [93,94].

Analytical methods described so far for the determination of estrogens and progestogens in fresh water sediments [95] and in solid environmental samples in general [96] have been reviewed in two recently published articles. It should be mentioned that most of the environmental programs carried out to assess the presence and impact of natural and synthetic estrogens and progestogens in the aquatic environment have focused on the investigation of environmental waters and, to a lesser extent, of sewage sludge, whereas soils and sediments have seldom been analyzed. Extraction of both

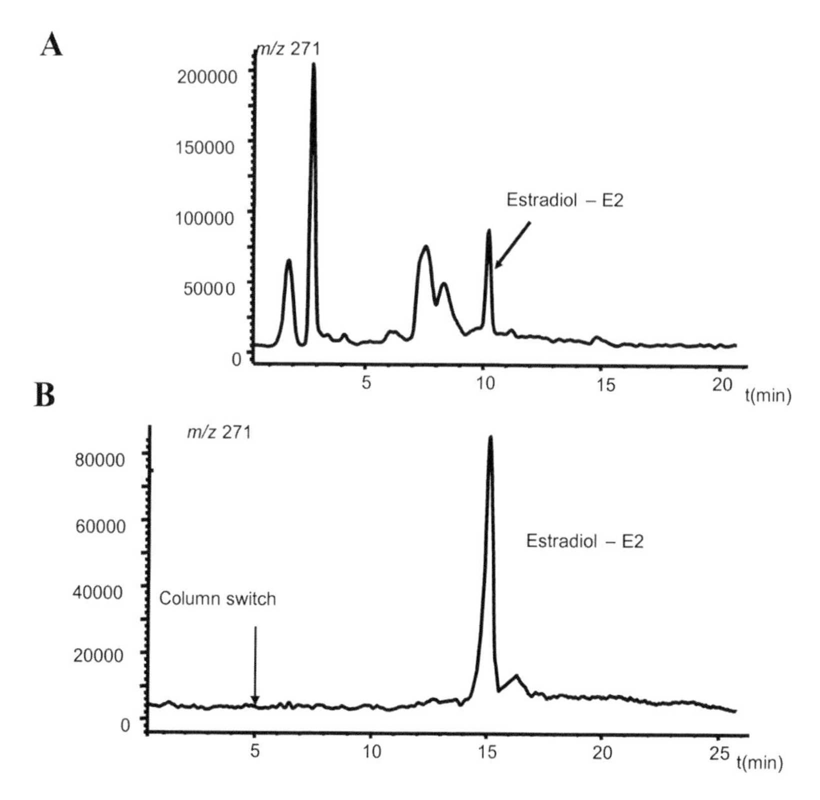


**Fig. 7.5** Reconstructed ion chromatograms of SIM channel $m/z$ 271 corresponding to $[M-H]^-$ of estradiol obtained under NI conditions of spiked river sediment. (A) PLE-SPE-LC-MS; (B) PLE-RAM-LC-MS. Reprinted with permission from [72]. Copyright 2002, Elsevier.

natural and synthetic estrogens from river sediments has always been performed with acetone:methanol 1:1, using either sonication or PLE. Clean-up of the extracts, when performed, has been carried out by SPE, LLE, GPC and semi-preparative LC. SPE has been preferred in most instances because it is fast, requires a low volume of organic solvent, presents low contamination risk, and can be used on-line.

Recently, an advanced sample preparation strategy was developed for the determination of steroid sex hormones in solid environmental samples [72]. A column switching LC-MS using LiChrospher ADS® RAM precolumns (Merck) is described for the integrated sample clean-up and analysis of estrogenic compounds in sediment. The use of RAM precolumns enables the LC-MS determination of steroid sex hormones and other pollutants, such as alkylphenolic compounds and bisphenol A, in sediments at very low levels (limits of detection (LODs) $= 0.5-5$ ng/g) due to the efficient removal of co-extracted matrix components, as shown in Fig. 7.5 for the analysis of 17β-estradiol from spiked sediment. With the methodology described, a complete analysis, including PLE of target compounds, on-line clean-up, chromatographic separation and MS detection takes approximately 2 h, which is a significant improvement compared to methods previously reported.

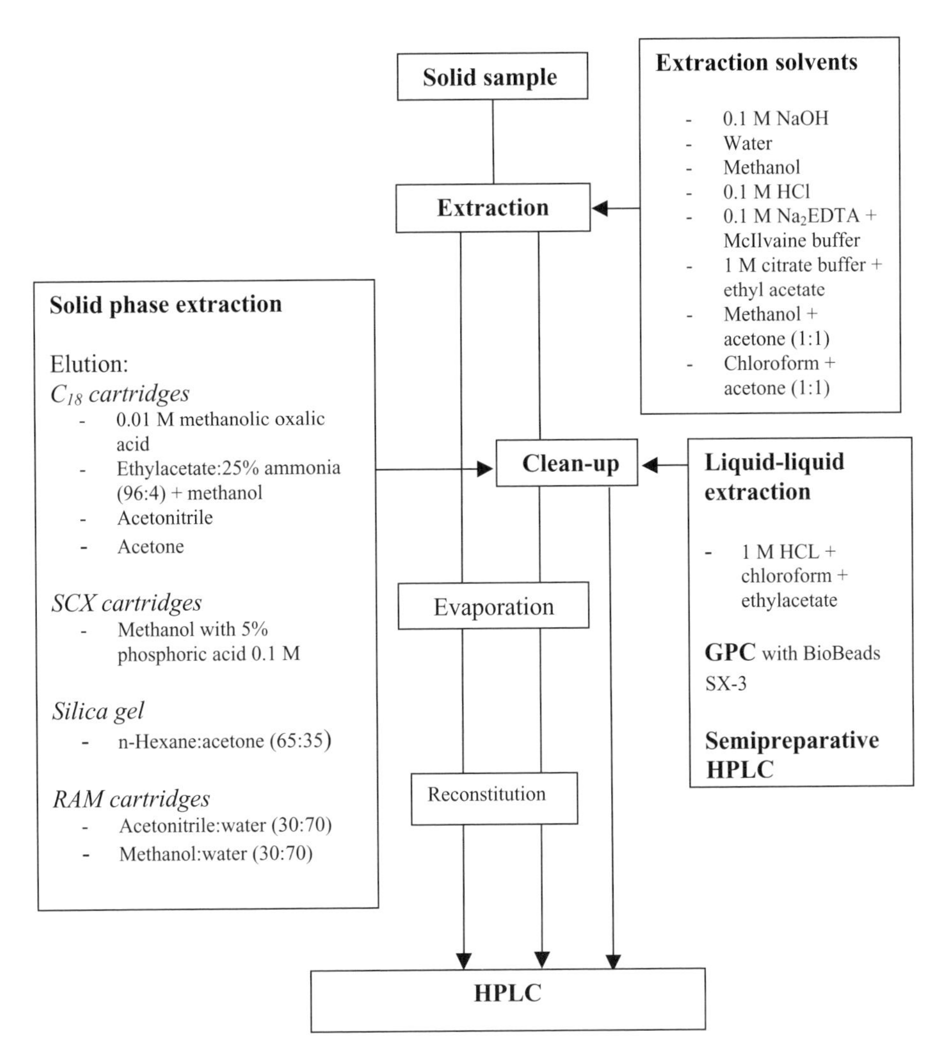


**Fig. 7.6** Scheme of the sample pre-treatment procedures applied to the determination of drugs in soil and sediments before HPLC analysis. Reprinted with permission from [96]. Copyright 2002, Elsevier.

Diaz-Cruz *et al.* [96] summarized the methodologies developed thus far for determination of pharmaceuticals in real and synthetic sediment, sludge and soil samples, with indication of the target analytes, type of sample, and the main experimental conditions used in the analysis by LC or GC. Another comprehensive review [97] summarized those analytical methods which allow for the determination of acidic drugs, β-blockers, neutral pharmaceuticals, antibiotics and iodinated X-ray contrast media in different aqueous matrices down to the nanogram per liter range. Figure 7.6 shows a flowchart of the main sample pretreatment steps and experimental conditions applied in the cited literature to the determination of drugs in soils, sediments and sludge.

### 7.5.4 *Volatile organic compounds (VOCs)*

VOCs are a class of compounds with sufficiently low boiling points (most VOCs of concern have boiling points below 150°C) to give them appreciable vapor pressures at 1 atm pressure. The group includes compounds such as low molecular mass aromatics, hydrocarbons, halogenated hydrocarbons, ketones, acetates, ethers, sulfides, acrylates and nitriles, and their solubility in water varies from insoluble (e.g. hydrocarbons and substituted hydrocarbons) to soluble (e.g. oxygenated compounds).

Sample collection and preservation procedures for this class of compound should be carefully selected to minimize VOC losses before sample treatment and analysis. Generally, aqueous and solid samples for VOC analysis should be collected into a closed system to prevent constituent losses using appropriate devices (corers and water samplers), and after sampling be rapidly transferred to a properly sealed VOA vial to be used for analysis. Physical preservation (storage at 4°C for up to a maximum of 7 days) is recommended for all environmental aqueous samples, although the use of chemical preservatives (sodium bisulfate or HCl) may extend the sample storage time. However, in this case one should have in mind that some types of analytes are unstable at low pH values.

The most commonly used methods for the preparation of solid samples for VOC analysis are vapor partitioning and methanol extraction. Vapor partitioning involves direct analysis of the sample either by equilibrium headspace, which is well suited to the analysis of very light molecular mass volatiles, or by a purge-and-trap method that permits detection of compounds of higher molecular mass. Generally, the headspace technique is considered less sensitive than purge-and-trap.

The extraction of VOCs from solid material with methanol (typical methanol:soil ratios range from 1:1 to 10:1) is followed by either purge-and-trap or headspace analysis of diluted extracts. The method is applicable to a wide range of concentrations, so obtained extract can be stored and used for multiple determinations.

Generally, LODs achieved with vapor partitioning are lower than those obtained with methanol extraction methods that require extract dilution (typically a 50-fold dilution is used).

Among VOCs, aromatic hydrocarbons (BTEX) are a group of environmentally significant compounds that are routinely monitored in contaminated soil. Comparison of three commonly employed methods, purge-and-trap followed by GC-MS, headspace GC-MS, and dichloromethane (DCM) extraction followed by GC-FID, showed that the measurements obtained by P&T/GC-MS are more reliable than those obtained by the other two techniques tested. The magnitudes of the recovered concentrations were significantly higher by P&T/GC-MS than by DCM/GC-FID, which in turn were significantly higher than the magnitudes determined by headspace/GC-MS.

## 7.6 Conclusions and future trends

The introduction of advanced instrumental techniques, such as fast LC, fast GC, or GC × GC, and application of very selective and sensitive MS and MS-MS detectors, have improved the analysis of environmental samples. However, to achieve the sensitivity and selectivity necessary for analysis of toxicologically relevant concentrations of organic microcontaminants in aqueous and solid environmental matrices, laborious, time-consuming and complicated extraction and multistep purification procedures are often applied. Thus, sample preparation often constitutes the bottleneck of the analysis of environmental samples. Because of this great effort is going into the development of cost-effective sample handling techniques characterized by efficiency and simplicity of operations and devices. In this sense, the developments in preparation of aqueous and solid samples are directed toward general application of advanced extraction techniques, such as SPE, SPME, PLE, subcritical hot water extraction and continuous-flow sonication, integrated within completely automated, on-line systems. Further development of more selective sorbents, such as MIPs, RAMs and immunosorbents might greatly simplify the trace level determination of environmentally relevant compounds and also improve the selectivity of trace level analysis. Further development and application of dual-column systems for integrated purification of extracts and analysis, and on-line coupling of sample preparation units and chromatographic and detection systems, will increase sample throughput and reduce operating costs and contamination risks.

The application of advanced sample preparation and detection techniques will no doubt expand our knowledge about the presence, fate and persistence of known and newly identified environmental pollutants and their degradation products, and will help environmentalists to assess potential risks and propose quality criteria.

## Acknowledgements

This work has been supported by the EU Program Copernicus (EXPRESS-IMMUNOTECH) contract Number ICA2-CT-2001-10007, by the Spanish Ministerio de Ciencia y Tecnologia (PPQ 2002-10945-E and PPQ2001-1805-CO3-01). Mira Petrovic acknowledges the Ramon y Cajal contract from the Spanish Ministry of Science and Technology.

## References

1. Keith, T.L., Snyder, S.A., Naylor, C.G., Staples, C.A., Summer, C., Kannan, K. & Giesy, J.P. (2001) Identification and quantitation of nonylphenol ethoxylates and nonylphenol in fish tissues from Michigan. *Environ. Sci. Technol.* **35**, 10–13.
2. Lacorte, S., Ehresmann, N. & Barceló, D. (1995) Stability of organophosphorus pesticides on disposable solid-phase extraction precolumns. *Environ. Sci. Technol.* **29**, 2834–2841.
3. Castillo, M., Puig, D. & Barceló, D. (1997) Determination of priority phenolic compounds in water and industrial effluents by polymeric liquid–solid extraction cartridges using automated sample preparation with extraction columns and liquid chromatography use of liquid-solid extraction cartridges for stabilization of phenols. *J. Chromatogr. A*, **778**, 301–311.
4. Alonso, M.C. & Barceló, D. (2002) Stability study and determination of benzene- and naphthalenesulfonates following an on-line solid-phase extraction method using the new programmable field extraction system. *Analyst*, **127**, 472–479.
5. Alonso, M.C. & Barceló, D. (2000) Stability of sulfonated derivatives of benzene and naphthalene on disposable solid-phase extraction pre-columns and in an aqueous matrix. *J. Chromatogr. A*, **889**, 231–244.

6. Naylor, C.G., Mieure, J.P., Adams, W.J., Weeks, J.A., Castaldi, F.J., Ogle, L.D. & Romano, R.R. (1992) Alkylphenol ethoxylates in the environment. *J. Am. Oil Chem. Soc.* **69**, 695–703.
7. Petrovic, M. & Barceló, D. (2000) The stability of non-ionic surfactants and linear alkylbenzene sulfonates in a water matrix and on solid-phase extraction cartridges. *Fresenius J. Anal. Chem.*, **368**, 676–683.
8. Baronti, C., Curini, R., D'Ascenzo, G., Di Corcia, A., Centili, A. & Samperi, R. (2000) Monitoring natural and synthetic estrogens at activated sludge sewage treatment plants and in a receiving river water. *Environ. Sci. Technol.*, **34**, 5059–5066.
9. Shang, D.Z., Ikonomou, M.G. & Macdonald, R.W. (1999) Quantitative determination of nonylphenol polyethoxylate surfactants in marine sediment using normal-phase liquid chromatography–electrospray mass spectrometry. *J Chromatogr.* A, **849**, 467–482.
10. Barcelo, D., Petrovic, M., Eljarrat, E., Lopez de Alda, M.J. & Kampioti, A. (2004) Environmental analysis, In *Chromatography* Vol.VI, (ed. E. Heftmann), Elsevier, Amsterdam pp. 987–1036.
11. Santos, F. J. & Galceran, M.T. (2002) The application of gas chromatography to environmental analysis. *Trends Anal. Chem.*, **21**, 672–685.
12. Liška, I. (2000) Fifty years of solid-phase extraction in water analysis–historical development and overview. *J. Chromatogr.* A , **885**, 3–16.
13. Delaunay, N., Pichon, V. & Hennion, M. (2000) Immunoaffinity solid-phase extraction for the trace-analysis of low-molecular-mass analytes in complex sample matrices. *J. Chromatogr. B*, **745**, 15–37.
14. Hennion, M.C. & Pichon, V. (2003) Immuno-based sample preparation for trace analysis. *J. Chromatogr. A*, **1000**, 29–52.
15. Bou Carrasco, P., Escolà, R., Marco, M.P. & Bayona, J.M. (2001) Development and application of immunoaffinity chromatography for the determination of the triazinic biocides in seawater. *J. Chromatogr.* A, **909**, 61–72.
16. Houben, A., Meulenberg, E., Noij, T., Gronert, C. & Stoks, P. (1999) Immuno affinity extraction of pesticides from surface water. *Anal. Chim. Acta*, **399**, 69–74.
17. Dallüge, J., Hankemeier, T. Vreuls, R.J. & Brinkman, U.T. (1999) On-line coupling of immunoaffinity-based solid-phase extraction and gas chromatography for the determination of *s*-triazines in aqueous samples. *J. Chromatogr. A*, **830**, 377–386.
18. Martin-Esteban, A., Fernandez, P., Stevenson, D. & Camara, C. (1997) Mixed immunosorbent for selective on-line trace enrichment and liquid chromatography of phenylurea herbicides in environmental waters. *Analyst*, **122**, 1113–1118.
19. Miege, C., Bouzige, M., Nicol, S., Dugay, J., Pichon, V. & Hennion, M.C. (1999) Selective immunoclean-up followed by liquid or gas chromatography for the monitoring of polycyclic aromatic hydrocarbons in urban waste water and sewage sludges used for soil amendment. *J. Chromatogr. A*, **859**, 29–39.
20. Perez, S., Ferrer, I., Hennion, M.C. & Barceló, D. (1998) Isolation of priority polycyclic aromatic hydrocarbons from natural sediments and sludge reference materials by an anti-fluorene immunosorbent followed by liquid chromatography and diode array detection. *Anal. Chem.*, **70**, 4996–5001.
21. Perez, S. & Barceló, D. (2000) Evaluation of anti-pyrene and anti-fluorene immunosorbent clean-up for PAHs from sludge and sediment reference materials followed by liquid chromatography and diode array detection. *Analyst*, **125**, 1273–1280.
22. Bouzige, M., Machtalère, G. Legeay, P. Pichon, V. & Hennion, M.C. (1999) New methodology for a selective on-line monitoring of some polar priority industrial chemicals in waste water. *Waste Management*, **19**, 171–180.
23. Ouyang, S., Xu, Y., & Chen, Y.H., (1988) Selective determination of a group of organic compounds in complex sample matrixes by LC/MIMS with on-line immunoaffinity extraction. *Anal. Chem.*, **70**, 931–935.
24. Ferguson, P.L., Iden, C.R., McElroy, A.E. & Brownawell, B.J. (2001) Determination of steroid estrogens in wastewater by immunoaffinity extraction coupled with HPLC-electrospray-MS. *Anal. Chem.*, **73**, 3890–3895.
25. Haupt, K. (2001) Molecularly imprinted polymers in analytical chemistry. *Analyst*, **126**, 747–756.
26. Masqué, N., Marcé, R.M. & Borull, F. (2001) Molecularly imprinted polymers: new tailor-made materials for selective solid-phase extraction. *Trends Anal. Chem.*, **20**, 477–486.
27. Schafer, C. & Lubda, D. (2001) Alkyl diol silica: restricted access pre-column packings for fast liquid chromatography-integrated sample preparation of biological fluids. *J. Chromatogr. A*, **909**, 73–78.
28. Koeber, R., Fleischer, C., Lanza, F., Boos, K.S., Sellergren, B. & Barceló, D. (2001) Evaluation of a multidimensional solid-phase extraction platform for highly selective on-line cleanup and high-throughput LC-MS analysis of triazines in river water samples using molecularly imprinted polymers. *Anal. Chem.*, **73**, 2437–2444.
29. Hogendoorn, E.A., Dijkman, E., Baumann, B., Hidalgo, C., Sancho, J.V. & Hernandez, F. (1999) Strategies in using analytical restricted access media columns for the removal of humic acid interferences in the trace analysis of acidic herbicides in water samples by coupled column liquid chromatography with UV detection. *Anal. Chem.*, **71**, 1111–1118

30. Parilla Vazquez, P., Marinez Vidal, J.L. & Marinez Fernández, J. (2000) A restricted access medium column for the trace determination of triadimefon and tetraconazole in water with large-volume injection coupled-column RPLC with UV detection. *Analyst*, **125**, 1549–1554.
31. Önnerfjord, P., Barceló, D., Emnéus, J., Gorton, L. & Marko-Varga, G. (1996) On-line solid-phase extraction in liquid chromatography using restricted access pre-columns for the analysis of *s*-triazines in humic-containing waters. *J. Chromatogr.* A, **737**, 35–45.
32. Boos, K.S. & Grimm, C.H. (1997) High-performance liquid chromatography integrated solid-phase extraction in bioanalysis using restricted access precolumn packings. *Trends Anal. Chem.*, **18**, 175–180.
33. Eisert, R. & Pawliszyn, J. (1997) Automated in-tube solid-phase microextraction coupled to high-performance liquid chromatography. *Anal. Chem.*, **69**, 3145–3147.
34. Takino, M. Daishima, S. & Nakahara, T. (2001) Automated on-line in-tube solid-phase microextraction followed by liquid chromatography/electrospray ionization-mass spectrometry for the determination of chlorinated phenoxy acid herbicides in environmental waters. *Analyst*, **126**, 602–608.
35. Gou, Y. & Pawliszyn, J. (2000) In-tube solid-phase microextraction coupled to capillary LC for carbamate analysis in water samples. *Anal. Chem.*, **72**, 2774–2779.
36. Saito, Y., Nakao, Y., Imaizumi, M., Morishima, Y., & Kiso, Y. Jinno, K. (2002) Miniaturized solid-phase extraction as a sample preparation technique for the determination of phthalates in water. *Anal. Bioanal. Chem.*, **373**, 81–86.
37. Gorecki, T. & Namiesnik, J. (2002) Passive sampling. *Trends Anal. Chem.*, **21**, 276–291.
38. Huckins, J.N., Tubergen, M.W., & Manuweera, G.K. (1990) Semipermeable membrane devices containing model lipid: a new approach to monitoring the bioavailability of lipophilic contaminants and estimating their bioconcentration potential. *Chemosphere*, **20**, 533–552.
39. Bennett, E.R., Metcalfe, T.L,. & Metcalfe, C.D. (1996) Semi-permeable membrane devices (SPMDS) for monitoring organic contaminants in the Otonabee River, Ontario. *Chemosphere*, **33**, 363–375.
40. Kot, A., Zabiegala, B., & Namiesnik, J. (2000) Passive sampling for long-term monitoring of organic pollutants in water. *Trends Anal. Chem.*, **19**, 446–459.
41. Petty, J.D., Poulton, B.C., Charbonneau, C.S., Huckins, J.N., Jones, S.B., Caneron J.T. & Prest, H.F. (1998) Determination of bioavailable contaminants in the Lower Missouri River following the flood of 1993. *Environ. Sci Technol.*, **32**, 837–842.
42. Prest, H.F., Jacobson, L.A., & Wilson, M. (1997) Passive water sampling for polynuclear aromatic hydrocarbons using lipid-containing semipermeable membrane devices (SPMDs): application to contaminant residence times. *Chemosphere*, **35**, 3047–3063.
43. Huckins, J.N., Petty, J.D. Orazio, C.E., Lebo, J. A., Clark, R.C. Gibson, V.L. Gala W.R. & Echols, K.R. (1999) Determination of uptake kinetics (sampling rates) by lipid-containing semipermeable membrane devices (SPMDs) for polycyclic aromatic hydrocarbons (PAHs) in Water. *Environ. Sci. Technol.*, **33**, 3918–3923.
44. Bergovist, A., Strandberg, B., Ekelund, R., Rappe, C. & Granmo, A. (1998) Temporal monitoring of organochlorine compounds in seawater by semipermeable membranes following a flooding episode in Western Europe. *Environ. Sci. Technol.*, **32**, 3887–3892.
45. Lebo, J.A., Gale, R.W., Petty, J.D., Tillit, D.E., Huckins, J.N., Meadows, J.C., Orazio, C.E., Echols, K.R., Schroeder D.J. & Inmon, L.E. (1995) Use of the semipermeable membrane device as an *in situ* sampler of waterborne bioavailable PCDD and PCDF residues at sub-parts-per-quadrillion concentrations. *Environ. Sci. Technol.*, **29**, 2886–2892.
46. Folsvik, N., Brevik E. & Berge, J.A. (2000) Monitoring of organotin compounds in seawater using semipermeable membrane devices (SPMDs)–tentative results. *J. Environ. Monit.*, **2**, 281–284.
47. Granmo, A., Ekelund, R., Berggren, M., Brorstrom-Lunden, E. & Bergvist, P.A. (2000) Temporal trend of organochlorine marine pollution indicated by concentrations in mussels, semipermeable membrane devices, and sediment. *Environ. Sci. Technol.*, **34**, 3323–3329.
48. Kingston, J.K., Greenwood, R., Mills, G.A., Morrison, G.M. & Persson, L.B. (2000) Development of a novel passive sampling system for the time-averaged measurement of a range of organic pollutants in aquatic environments. *J. Environ. Monit.*, **2**, 487–495.
49. Ramos, L., Kristenson, E.M. & Brinkman, U.A.Th. (2002) Current use of pressurised liquid extraction and subcritical water extraction in environmental analysis. *J. Chromatogr. A*, **975**, 3–29.
50. Björklund, E., Nilsson, T. & Bøwadt, S. (2000) Pressurised liquid extraction of persistent organic pollutants in environmental analysis. *Trends Anal. Chem.*, **19**, 434–445.
51. Camel, V. (2001) Recent extraction techniques for solid matrices–supercritical fluid extraction, pressurized fluid extraction and microwave-assisted extraction: their potential and pitfalls. *Analyst*, **126**, 1182–1193.
52. Giergielewicz-Mozajska, H., Dabrowski, L. & Namienik, J. (2001) Accelerated solvent extraction (ASE) in the analysis of environmental solid samples–some aspects of theory and practice. *Crit. Rev. Anal. Chem.*, **31**, 149–165.
53. Camel, V. (2002) Extraction techniques. *Anal. Bioanal. Chem.*, **372**, 39–40.

54. Field, J.A., & Reed, R.L. (1999) Subcritical (hot) water/ethanol extraction of nonylphenol polyethoxy carboxylates from industrial and municipal sludges. *Environ. Sci. Technol.*, **33**, 2782–2787.
55. Hawthorne, S.B., Yang, Y. & Miller, D.J. (1994) Extraction of organic pollutants from environmental solids with sub- and supercritical water. *Anal. Chem.*, **66**, 2912–2920.
56. Yang, Y., Hawthorne, S.B. & Miller, D.J. (1997) Class-selective extraction of polar, moderately polar, and nonpolar organics from hydrocarbon wastes using subcritical water. *Environ. Sci. Technol.*, **31**, 430–437.
57. Fernandez-Perez, V. & Luque de Castro, M.D. (2000) Micelle formation for improvement of continuous subcritical water extraction of polycyclic aromatic hydrocarbons in soil prior to high-performance liquid chromatography–fluorescence detection. *J. Chromatogr.* A, **902**, 357–367.
58. Morales-Muñoz, S., Luque-García, J.L. & Luque de Castro, M.D. (2002) Pressurized hot water extraction with on-line fluorescence monitoring: a comparison of the static, dynamic, and static-dynamic modes for the removal of polycyclic aromatic hydrocarbons from environmental solid samples. *Anal. Chem.*, **74**, 4213–4219.
59. Hyötyläinen, T., Andersson, T., Hartonen, K., Kuosmanen, K. & Riekkola, M.L. (2000) Pressurized hot water extraction coupled on-line with LC-GC: determination of polyaromatic hydrocarbons in sediment. *Anal. Chem.*, **72**, 3070–3076.
60. Hartonen, K., Bowadt, S., Hawthorne, S.B. & Riekkola, M.L. (1997) Supercritical fluid extraction with solid-phase trapping of chlorinated and brominated pollutants from sediment samples. *J. Chromatogr.* A, **774**, 229–242.
61. van Bavel, B., Hartonen, K., Rappe, C. & Riekkola, M.L. (1999) Pressurised hot water/steam extraction of polychlorinated dibenzofurans and naphthalenes from industrial soil. *Analyst*, **124**, 1351–1354.
62. Kuosmanen, K., Hyötyläinen, T., Hartonen, K. & Riekkola, M.L. (2002) Pressurised hot water extraction coupled on-line with liquid chromatography–gas chromatography for the determination of brominated flame retardants in sediment samples. *J. Chromatogr. A*, **943**, 113–122.
63. Barker, S.A., Long, A.R., Hines, M.E. (1993) Disruption and fractionation of biological materials by matrix solid-phase dispersion. *J. Chromatogr.*, **629**, 23–34.
64. Barker, S.A. (2000) Matrix solid-phase dispersion. *J. Chromatogr. A*, **885**, 115–127.
65. Reimer, G.J. Suarez, A. (1991) Development of a screening method for five sulfonamides in salmon muscle tissue using thin-layer chromatography. *J. Chromatogr.*, **555**, 315–320.
66. Walker, C.C., Lott, H.M. & Barker, S.A. (1993) Matrix solid-phase dispersion extraction and the analysis of drugs and environmental pollutants in aquatic species. *J. Chromatogr.*, **642**, 225–242.
67. Tolls, J., Haller, M. & Sijm, D.T. (1999) Extraction and isolation of linear alcohol ethoxylates from fish. *J. Chromatogr. A*, **839**, 109–117.
68. Tolls, J., Haller, M., Sijm, D.T. (1999) Extraction and isolation of linear alkylbenzenesulfonate and its sulfophenylcarboxylic acid metabolites from fish samples. *Anal. Chem.*, **71**, 5242–5247.
69. Zhao, M., van der Wielen, F. & de Voogt, P. (1999) Optimization of a matrix solid-phase dispersion method with sequential clean-up for the determination of alkylphenol ethoxylates in biological tissues. *J. Chromatogr. A*, **837**, 129–138.
70. Ling, Y.C., Chang, M.Y. & Huang, I.P. (1994) Matrix solid-phase dispersion extraction and gas chromatographic screening of polychlorinated biphenyls in fish. *J. Chromatogr. A*, **669**, 119–124.
71. Crouch, M.D. & Barker, S.A. (1997) Analysis of toxic wastes in tissues from aquatic species: applications of matrix solid phase dispersion. *J. Chromatogr. A*, **774**, 287–309.
72. Petrovic, M., Tavazzi, S. & Barceló, D. (2002) Column-switching system with restricted access pre-column packing for an integrated sample cleanup and liquid chromatographic–mass spectrometric analysis of alkylphenolic compounds and steroid sex hormones in sediment. *J. Chromatogr. A*, **971**, 37–45.
73. López-Avila, V., Young, R., Beckert, W.F. (1994) Microwave-assisted extraction of organic compounds from standard reference soils and sediments. *Anal. Chem.*, **66**, 1097–1106.
74. Eljarrat, E., Caixach, J. & Rivera, J. (1998) Microwave vs. soxhlet for the extraction of PCDDs and PCDFs from sewage sludge samples. *Chemosphere*, **36**, 2359–2366.
75. Turner, W.E., Cash, T.P., DiPietro, E.S. & Patterson, D.G., Jr. (1998) Evaluation and applications of the Power-Prep$^{TM}$ Universal automated cleanup system for PCDDs, PCDFs, cPCBs, PCB congeners and chlorinated pesticides in biological samples. *Organohalogen Compd.*, **35**, 21–24.
76. Shelver, W.L., Larsen, G.L., & Huwe, J. K. (1998) Use of an immunoaffinity column for tetrachlorodibenzo-*p*-dioxin serum sample cleanup. *J. Chromotogr. B*, **705**, 261–268.
77. Concejero, M.A., Galve, R., Herradón, B., González, M.J. & de Frutos, M. (2001) Feasibility of high-performance immunochromatography as an isolation method for PCBs and other dioxin-like compounds. *Anal. Chem.*, **73**, 3119–3125.
78. Shelver, W.L., Huwe, J.K. Stanker, L.H. Patterson, D.G., Jr. & Turner, W.E. (2000) A monoclonal antibody based immunoaffinity column for isolation of PCDD/PCDF from serum. *Organohalogen Compd.*, **45**, 33–37.

79. Sonnenschein, C. & Soto, A.M. (1998) An update review of environmental estrogen and androgen mimics and antagonists. *J. Steroid. Biochem. Mol. Biol.*, **65**, 143–150.
80. Jobling, S., Sheahan, D., Osborne, J.A., Matthiessen, P. & Sumpter, J.P. (1996) Inhibition of testicular growth in rainbow trout (*Oncorhynchus mykiss*) exposed to estrogenic alkylphenolic chemicals. *Environ. Toxicol. Chem.*, **15**, 194–202.
81. Maki, H., Okamura, H., Aoyama, I. & Fujita, M. (1998) Halogenation and toxicity of the biodegradation products of a nonionic surfactant, nonylphenol ethoxylates. *Environ. Toxicol. Chem.*, **17**, 650–654.
82. Petrovic, M., Rodrigez Fernández-Alba, A.M., Borrull, F., Marce, R.M., González Mazo, E. & Barceló, D. (2001) Occurrence and distribution of nonionic surfactants, their degradation products and LAS in coastal waters and sediments in Spain. *Environ. Toxicol. Chem.*, **21**, 37–46.
83. Blackburn, M.A. & Waldock, M.J., (1995) Concentrations of alkylphenols in rivers and estuaries in England and Wales. *Wat. Res.*, **29**, 1623–1629.
84. Marcomini, A., Di Corcia, A., Samperi, R., & Capri S. (1993) Reversed-phase high-performance liquid chromatographic determination of linear alkylbenzene sulfonates, nonylphenol polyethoxylates and their carboxylic biotransformation products. *J. Chromatogr. A*, **644**, 59–71.
85. Field, J.A. & Reed, R.L. (1996) Nonylphenol polyethoxy carboxylate metabolites of nonionic surfactants in U.S. paper mill effluents, municipal sewage treatment plant effluents, and river waters. *Environ. Sci. Technol.*, **30**, 3544–3550.
86. Crescenzi, C., Di Corcia, A., Samperi, R. & Marcomini, A. (1995) Determination of nonionic polyethoxylates surfactants in environmental waters by liquid chromatography–electrospray mass spectrometry. *Anal. Chem.*, **67**, 1797–1804.
87. Petrovic, M. & Barceló, D. (2002) Review of advanced sample preparation methods for the determination of alkylphenol ehtoxylates and their degradation products in solid environmental matrices. *Chromatographia*, **56**, 535–544.
88. Kreisselmeier, A. & Dürbeck, H.W. (1997) Determination of alkylphenols, alkylphenolethoxylates and linear alkylbenezenesulfonates in sediments by accelerated solvent extraction and supercritical fluid extraction. *J. Chromatogr. A*, **775**, 187–196.
89. Lin, J.G., Arunkumar, R. & Liu, C.H. (1999) Efficiency of supercritical fluid extraction for determining 4-nonylphenol in municipal sewage sludge. *J. Chromatogr. A*, **840**, 71–79.
90. Lee, H.B. & Peart, T.E. (1995) Determination of 4-nonylphenol in effluent and sludge from sewage treatment plants. *Anal. Chem.*, **67**, 1976–1980.
91. Lee, H.B., Peart T.E., Bennie D.T. & Maguire R.J. (1997) Determination of nonylphenol polyethoxylates and their carboxylic acid metabolites in sewage treatment plant sludge by supercritical carbon dioxide extraction. *J. Chromatogr. A*, **785**, 385–394.
92. Lopez de Alda, M.J., Diaz-Cruz, S., Petrovic, M. & Barceló, D. (2003) Liquid chromatography—(tandem) mass spectrometry of selected emerging pollutants (steroid sex hormones, drugs and alkylphenolic surfactants) in the aquatic environment. *J. Chromatogr. A*, **1000**, 503–526
93. López de Alda, M.J. & Barceló, D. (2001) Use of solid-phase extraction in various of its modalities for sample preparation in the determination of estrogens and progestogens in sediment and water. *J. Chromatogr. A*, **938**, 145–153.
94. López de Alda, M.J. & Barceló, D. (2001) Determination of steroid sex hormones and related synthetic compounds considered as endocrine disrupters in water by fully automated on-line solid-phase extraction–liquid chromatography–diode array detection. *J. Chromatogr. A*, **911**, 203–210.
95. Petrovic, M., Eljarrat, E., López de Alda, M.J. & Barceló, D. (2001) Analysis and environmental levels of endocrine-disrupting compounds in freshwater sediments. *Trends Anal. Chem.*, **20**, 637–648.
96. Díaz-Cruz, S., López de Alda, M.J. & Barceló, D. (2002) Environmental behaviour and analysis of veterinary and human drugs in soils, sediments and sludge. *Trends Anal. Chem.*, **22**, 340–351.
97. Ternes, T.A. (2001) Analytical methods for the determination of pharmaceuticals in aqueous environmental samples. *Trends Anal. Chem.*, **20**, 419–434.
98. Baggiani, C., Giovannoli, C., Anfossi, L. & Tozzi, C. (2001) Molecularly imprinted solid-phase extraction sorbent for the clean-up of chlorinated phenoxyacids from aqueous samples. *J. Chromatogr. A*, **938**, 35–44.
99. Chapuis, F., Pichon, V., Lanza, F., Sellergren, S. & Hennion, M.C. (2003) Optimization of the class-selective extraction of triazines from aqueous samples using a molecularly imprinted polymer by a comprehensive approach of the retention mechanism. *J. Chromatogr. A*, **999**, 23–33.
100. Bjarnason, B., Chimuka, L. & Ramstrom. O. (1999) On-line solid-phase extraction of triazine herbicides using a molecularly imprinted polymer for selective sample enrichment. *Anal. Chem.*, **71**, 2152–2156.
101. Ferrer, I. & Barceló, D. (1999) Validation of new solid-phase extraction materials for the selective enrichment of organic contaminants from environmental samples. *Trends Anal. Chem.*, **18**, 180–192.
102. Matsui, J. Okada, M. Tsuruoka, M. & Takeuchi, T. (1997) Solid-phase extraction of a triazine herbicide using a molecularly imprinted synthetic receptor. *Anal. Commun.*, **34**, 85–87.

103. Ferrer, I., Lanza, F., Tolokan, A., Horvath, V., Sellergren, B., Horvai, G., & Barceló, D., (2000) Selective trace enrichment of chlorotriazine pesticides from natural waters and sediment samples using terbuthylazine molecularly imprinted polymers. *Anal. Chem.*, **72**, 3934–3941.
104. Mena, M.L., Martínez-Ruiz, P., Reviejo, A.J. & Pingarrón, J.M. (2002) Molecularly imprinted polymers for on-line preconcentration by solid phase extraction of pirimicarb in water samples. *Anal. Chim. Acta*, **451**, 297–304.
105. Masqué, N., Marcé, R.M., Borrull, F., Cormack, P.A. & Sherrington, D.C. (2000) Synthesis and evaluation of a molecularly imprinted polymer for selective on-line solid-phase extraction of 4-nitrophenol from environmental water. *Anal.Chem.*, **72**, 4122–4126.
106. Caro, E., Masqué, N., Marcé, R.M., Borrull, F., Cormack, P.A. & Sherrington, D.C. (2002) Non-covalent and semi-covalent molecularly imprinted polymers for selective on-line solid-phase extraction of 4-nitrophenol from water samples. *J. Chromatogr. A*, **963**, 169–178.

# 8 From cells to instrumental analysis

HOSSEIN AHMADZADEH and EDGAR A. ARRIAGA

## 8.1 Introduction

The heterogeneity of cells and multiplicity of their functions in a complex biological sample, such as a tissue, poses serious challenges for bioanalytical chemists that can only be addressed by studies based on single cell measurements. In recent years, single cell analyses have increasingly flourished in fields, such as medical diagnosis, forensic science, agriculture and biotechnology, that have sought to better understand and benefit from cellular heterogeneity. The scope of this chapter does not allow us to write an exhaustive review article on the subject of single cell analysis and we refer the interested reader to read the relevant literature [1–16]. Instead, we focus on the different instrumental techniques that can be used to extract information from single cells. Mainly, we discuss the challenges that analytical and bioanalytical chemists face when they sample a single cell with an instrument, and some of the cutting-edge strategies that promise to be useful for single cell analysis.

Typical mammalian cells are only 10–20 μm in diameter and were originally very difficult to manipulate because of their small size. Thus early single cell analysis was limited to relatively large cells like oocytes or neurons. For example, neuron cells from *Helix aspersa* were homogenized in a microvial and then analyzed by microcolumn capillary liquid chromatography (LC) and capillary electrophoresis (CE) [17,18]. These first single cell analyses proved that the information obtained from a single cell is more valuable than the average information obtained from a heterogeneous population of cells. Since then, both the types of single cells and the number of single cell chemical components that can be analyzed have been increasing rapidly thanks to advancements in analytical instrumentation, and there have been reports of the single cell analysis of proteins, DNA, and lipids. Even the specific enzymatic activity of a cell has been analyzed. Nonetheless, the total analysis and complete chemical inventory of a single cell, and the cell-by-cell monitoring of a chemical of interest, still remain the ultimate aim of biologists and biotechnologists. Recent advances that describe the subcellular localization of cellular components, achieve high throughput, or provide molecular structural information on the components of single cells are likely to bring this dream closer to reality. This chapter will touch on these recent advances in single cell analysis in an attempt to inspire other investigators in other fields to join those seeking more efficient alternatives for studying single cells.

One of the key aspects of single cell analysis is the need to adequately couple the sample defined by the single cell to the instrument that will extract the information from the cell. The instrument (or method) has to be able to handle dimensions in the micrometer range, sample volumes as low as picoliters, separate or detect the analytes in the presence of extremely complex cellular matrices, and detect as few as a hundred copies of molecules found within the cell. Only a handful of analytical techniques can adequately perform these challenging operations.

In this chapter, several of these techniques and the principles that make them useful for single cell analysis will be described. Flow cytometry will be discussed because this technique is widely accepted in the scientific community and has been considered the gold standard for whole cell sampling and analysis. However, the main emphasis of this chapter will be on CE because this technique is capable of separating the components of a single cell. The laser-induced fluorescence (LIF) detection scheme will be highlighted because of its high sensitivity and applicability to single cell analysis, in addition to providing structural information on the components of a single cell.

Levitating single cells will also be presented because this seems to be a very promising sampling technique, although less well known than optical tweezers and laser micropipets [19,20]. Methodologies for probing the cytoplasm and subcellular components of large cells (e.g. sampling with an etched capillary [21]) or for analyzing a mammalian neuronal process with a commercial capillary [4,22], will also be discussed because they represent the first attempts to obtain information. Mass spectrometry (MS) will be briefly discussed due to its potential for providing information about the compartmentalization of subcellular compounds. Last but not least, laser capture microdissection [23–27] combined with MS and direct tissue sampling by CE will also be described as this combination may serve to develop simpler approaches to single cell analysis.

## 8.2 Flow cytometry

Researchers in medicine, biology, biochemistry and biotechnology often need to analyze the biochemical and biophysical properties of single cells in a short period of time. Flow cytometry is currently their method of choice because of its high throughput and the statistical accuracy inherent in the measurement of large numbers of cells. Here, the selected properties of a cell such as its size and shape or its protein or DNA contents can be monitored individually. Before the invention of flow cytometry, individual cells were examined visually with an optical microscope. The task of examining 10 000 cells in order to determine their size was very cumbersome, if not impossible. Nowadays, the high throughput capability of flow cytometry has facilitated the analysis of cell populations and has proven to be extremely useful in the diagnosis of disease, e.g. in monitoring the DNA contents of cells for cancer prognosis [28–31]. It is not surprising that today there is at least one flow cytometer in almost every hospital.

Although flow cytometry was already in its infancy in the 1940s, its inception is usually dated to 1953 when Crossland-Taylor reported the use of a sheath-flow stream to count blood cells in a dark-field microscope [32]. The first flow cytometry device, the Coulter Counter, was used to measure cell volume in addition to counting cells. Coulter Counters are still routinely used in clinical chemistry. During the 1960s, researchers at Los Alamos Laboratory incorporated a laser into the instrumentation [33], creating the first generation of flow cytometers capable of sorting cells by isolating individual cells according to their volume. Later, the availability of cell staining techniques for specific molecules such as protein and DNA [34] led to the development of flow cytometers that could measure several cell properties simultaneously and sort the cells according to properties other than size.

A flow cytometer is made up of three main systems: fluidics, optics, and electronics. Sample introduction for flow cytometry, i.e. the transport of the cells from the cell suspension to the laser beam for excitation, is performed by the fluidics system. The optics system consists of the lasers, the source for the excitation of the fluorescent tags attached to the cells, and optical filters that direct the resulting fluorescence emission signals to the appropriate detectors. The third component of the flow cytometer, the electronics system, converts the detected fluorescence into electronic signals that can be processed by a computer.

When the flow cytometer is in use, cells in the suspension medium, either live or preserved by chemical fixation, are forced by hydrodynamic focusing to cross the laser beam. The cell suspension passes through a small chamber, an optically polished sheath-flow cuvette, in which the analysis sequence begins. Hydrodynamic focusing allows only one cell at a time to cross the laser beam for excitation. When a cell crosses the laser beam, it is excited, resulting in fluorescence emission. At the same time, a piezoelectric transducer is used to vibrate the sheath-flow cuvette at a high frequency (50 kHz or greater). The vibration causes a periodic production of droplets at the exit of the cuvette. When a droplet contains a cell, the computer triggers an electic field that induces an electrical charge onto the droplet that is then deflected by charge plates, making that particular droplet fall into a collection vial (Fig. 8.1). The collected cells can then be sorted for further analysis.

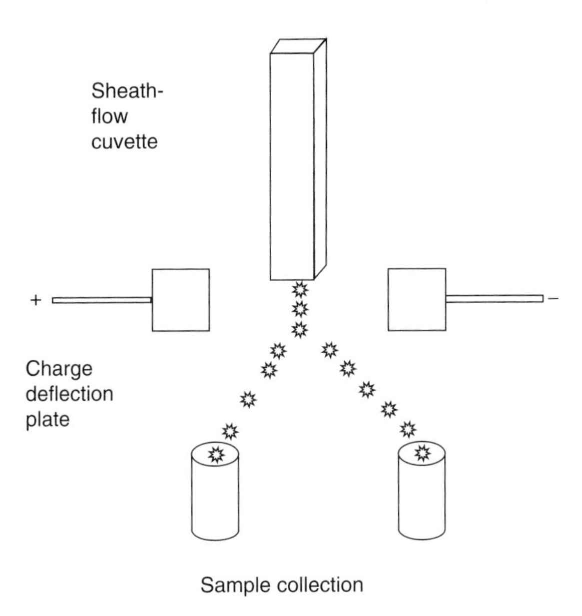


**Fig. 8.1** Block diagram of a flow cytometer. Sample introduction through the sheath-flow cuvette is followed by excitation with a laser and detection. After hydrodynamic focusing in the sheath-flow cuvette and fluorescence detection by a photomultiplier tube, the cells will be deflected in an electric field and then collected in sample collection vials.

The list of applications for flow cytometry in biology and medicine is very long, and it is not our objective to discuss these applications in this chapter. Instead, since flow cytometry is a well-established technique, we find it an ideal point of reference for comparison with new emerging technologies from an instrumental analysis perspective. Needless to say, the main advantage of flow cytometry is that it allows single cells to be analyzed with high throughput. The analysis of millions of cells, at rates of 1000–5000 cells per second, is routine.

Despite its widespread use, flow cytometry is not without limitations. A major disadvantage is that each separate component to be analyzed must have its own specific fluorescent tag. Another is that the fluorescence emission ranges for these tags must be sufficiently different for the tags to be distinguishable spectroscopically. Other techniques address these limitations by offering additional separation capabilities before detection of the various components found in single cells. CE is one of these techniques.

## 8.3 Capillary electrophoresis (CE)

Among the different techniques that probe single cells for information, none may have greater potential than single cell CE where an electric field is applied to a solution containing charged species. This electric field causes the charged species in the sample to migrate at rates that are dependent upon the electrical force (i.e. charge) and size (i.e. hydrodynamic drag force). High separation efficiencies can result when using electric fields around 400 V/cm. If this high separation capability is combined with LIF detection, CE becomes a very powerful tool in terms of the level of information that can be obtained from an individual cell [35,36].

CE is frequently used in bioanalytical chemistry, molecular biology, biochemistry and medicine. It has been applied to the separation and characterization of many different classes of molecules found inside a single cell, e.g. anions, cations, and neutral molecules [21,37].

The first step in single cell CE is sampling. Several modes of cell sampling have been used so far [17,18]: whole-cell sampling, cytoplasmic or subcellular sampling, cell-release or extracellular sampling. In whole-cell sampling, the entire cell is introduced into the separation capillary of the CE instrument. In cytoplasmic or subcellular sampling, a sharp capillary smaller than the cell is used to partially withdraw material from the cell. In cell-release or extracellular sampling, the CE instrument has a capillary that is very close to the cell surface to collect cellular secretions. In the following section, we describe whole-cell sampling (Fig. 8.2). Two other key components necessary for CE analysis, the single cell injector and the post-column LIF detector, are described in subsequent sections.

### 8.3.1 *Whole single cell analysis*

As shown in Fig. 8.2, whole single cell analysis includes: (i) cell selection using an inverted microscope, (ii) loading the cell into the capillary of the CE instrument, (iii) cell lysis inside the capillary, (iv) electrophoretic separation, and (v) detection.

First, siphoning or electro-osmotic flow is used to load the cell into a capillary attached to a macromanipulator with an X, Y and Z control that allows the capillary

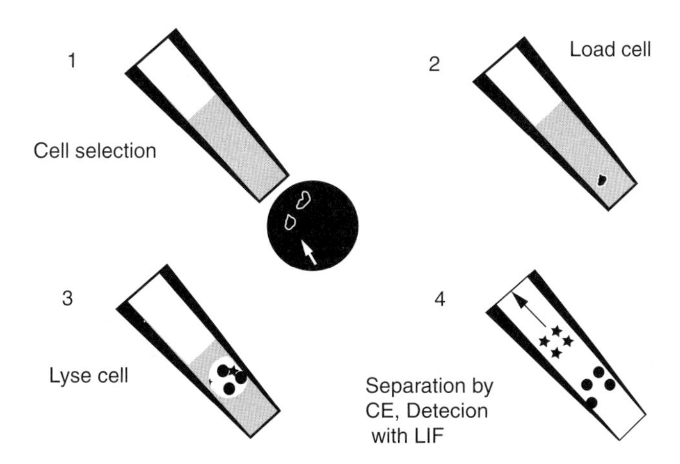


**Fig. 8.2** The sequence of events in single whole cell CE analysis. In step 1, the cell of interest is selected using an inverted microscope. Applying a negative pressure to the other side of the capillary will load the cell into the capillary (step 2). The loaded cell is lysed when it comes in contact with a detergent such as SDS (step 3). In step 4, high voltage is applied to the capillary to initiate the CE separation of the cellular components. LIF detection is then used to for the high sensitivity detection of those components that are fluorescent.

to be positioned precisely on top of the selected cell (Fig. 8.2, part 2). Either negative pressure is applied to the end of the capillary positioned away from the cell or a high voltage is delivered from a power supply. Applying this voltage to the capillary will introduce the cell into the capillary because of the resulting electro-osmotic flow (EOF) [38] or the electrophoretic mobility of the cell.

After the cell is loaded into the capillary, the presence of surfactants such as sodium dodecyl sulfate (SDS), either in the running buffer or added as a plug immediately after the cell is introduced, causes total cell lysis in 10–30 s (Fig. 8.2, part 3). Cell lysis then causes the dissolved cellular contents to be released. Cell lysis is most commonly produced by introducing a surfactant, especially when that surfactant can be an integral part of the run buffer, as in micellar electro kinetic chromatography (MEKC). Other techniques that have proven effective for cell disruption are ultrasound, the introduction of a strong magnetic field generated with a tesla coil, or exposure to an electric field [39,40]. Following lysis, high voltage is applied to the capillary and the contents of the single cell migrate towards the detector (Fig. 8.2, part 4).

The separation modes used in conventional CE can also be applied to single whole-cell CE analysis. For example, a run buffer containing a low percentage of water-soluble and replaceable polymers allows for the size-based separation of proteins [12]. Recently, Dovichi's group has used a low viscosity gel to fill the capillary, and after whole-cell injection, lysis and on-column fluorescent labeling, the protein content of a single cell has been analyzed [12]. This technique, capillary gel electrophoresis with LIF detection, has made it possible to map about 30 proteins in a single cell [12].

Despite its simplicity, on-column cell lysis will produce cellular debris that can be adsorbed onto the capillary surface, modifying its separation properties. When this occurs, the experiment is not easily reproduced. Cellular debris can be prevented from adsorbing onto the capillary surface by preconditioning the capillary in-between runs.

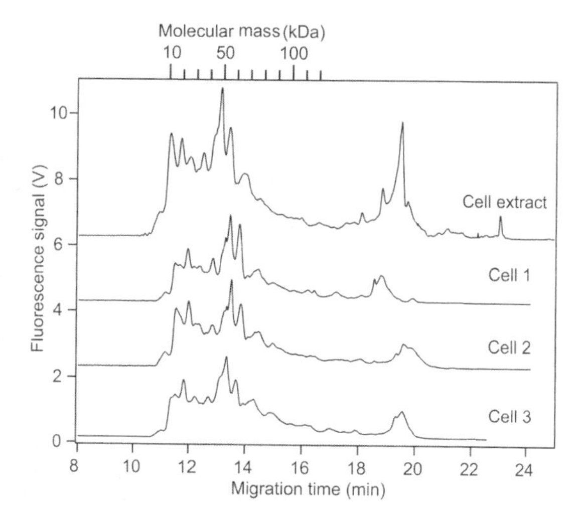


**Fig. 8.3** CE electropherograms of single HT29 cancer cells and HT29 cell extract obtained with LIF detection and the on-column labeling reaction. The steps in the experimental procedure are depicted in Fig. 8.2 and the experimental conditions can be obtained from the original publication [12]. Figure 8.3 is reproduced by courtesy of Professor N.J. Dovichi and permission from the publisher.

In addition, the capillary surface can be coated with a polymer such as acryloylamino propanol to reduce adsorption of cell debris onto the capillary wall [35].

Whole single cell analysis has many applications, one of which is given in Fig. 8.3 which shows the CE electropherograms of HT29 cancer cells and HT29 cell extract. For the detailed experimental conditions, the reader is encouraged to check the relevant publication of Dovichi and co-workers [12].

### 8.3.2 *Single-cell injector*

This section describes the single-cell injector used for whole-cell loading into the capillary. This device is also useful for sampling directly from tissue cross-sections, as will be described in section 8.7. Because the single-cell injector holds the capillary between the lens and the condenser of an inverted microscope, this device is made of transparent Plexiglas so that the material does not block the light source that illuminates the sample. Figure 8.4 shows the single-cell injector in detail. A three-way solenoid valve (General Valve, Florham Park, NJ) is used in the design of the single-cell injector. When the cell is being loaded by siphoning, this solenoid valve connects the capillary fluidics with a water-filled container 110 cm below the microscope glass slide holding the cell suspension. This height difference generates a partial vacuum that draws the cell into the capillary. An electronic timer precisely controls the period in which the valve will be open for loading the cell. The device shown in Fig. 8.4 is also capable of loading cells by applying a high voltage to the capillary for a short time.

Other devices have been used for single cell sampling. In some instances, these devices are capable of sampling individual cells in intact organs under physiological conditions. For example, there is a collection of techniques referred to as single cell

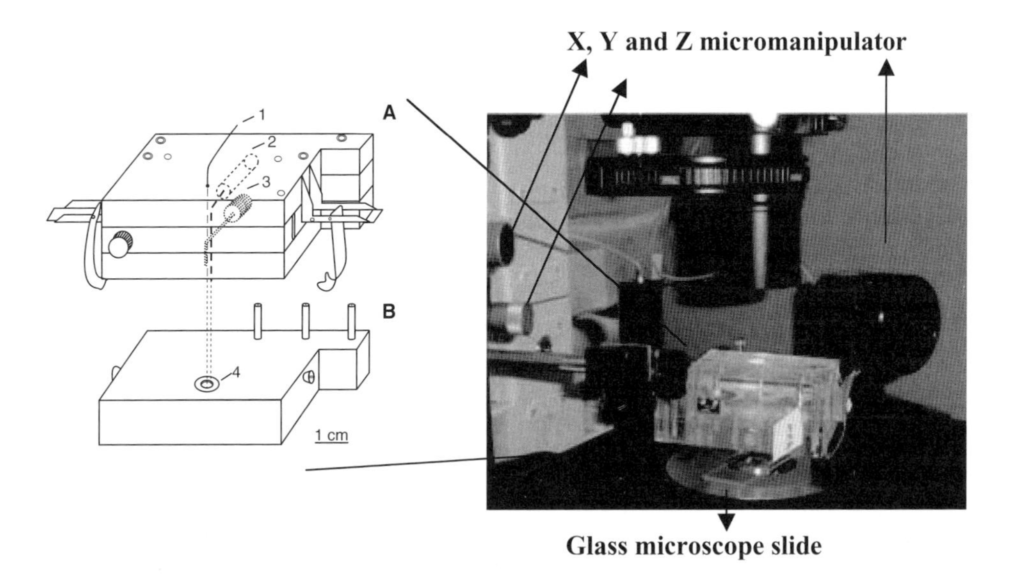


**Fig. 8.4** In the right panel is a picture of a device for loading single cells into a capillary mounted on a microscope pole. The cells are placed on the glass microscope slide and visualized using the microscope. The X,Y and Z micromanipulator is then used to move the capillary on top of the targeted cell. The cell is then introduced into the capillary for further lysis and CE separation by applying negative pressure on the end of the capillary away from the cell. In the left panel, reproduced with permission from ref. [10], is an expanded view of the cell injector. 1, Capillary being inserted into the cell injector; 2, electrode used to apply high voltage for CE separation; 3, gas line for pressure injection or capillary rinsing. The run buffer is poured into an Eppendorf vial and placed at the location marked as 4 in the diagram. After injection, the lower part of the single cell injector (B) will be clamped to the upper part (A), and the CE separation will be initiated by applying high voltage to the electrode. The detection end of the system is the same as in Fig. 8.5.

sampling and analysis (SiCSA) in which a fine oil-filled glass microcapillary mounted on a micromanipulator is used to sample individual plant cells. In this instance, the cell's internal pressure drives the cytoplasm into the glass microcapillary which is then removed from the cell. The collected cytoplasm is then expelled from the microcapillary by increasing the pressure on the oil phase, which responds by delivering the capillary contents to an analytical instrument [41]. Two main limitations of SiCSA are: (i) in SiCSA, cell loading relies on the cell's internal pressure which can be low for small cells, thereby impeding sampling, and (ii) such sampling is not complete since it is mostly cytoplasmic and vacuole material that is being sampled [42]. While SiCSA works best for cells growing on the surface of a plant, confocal microscopy can be used to analyze cells from beneath the epidermis.

### 8.3.3 *Laser-induced fluorescence (LIF) detection*

Detection capability is also an instrumental challenge when analyzing small quantities of a wide range of chemicals in single whole cells. Researchers are trying to meet this challenge by increasing detection sensitivity and lowering the limits of detection (i.e. improving the instrumental signal-to-noise ratio). The detection methods traditionally employed in single cell analysis are absorbance, fluorescence (in particular LIF), and electrochemistry. Both electrochemistry and LIF typically have zeptomole detection limits [2,11,43]. Unfortunately, electrochemistry or LIF detectors can be applied, respectively, only to electroactive or fluorescent molecules. Although fluorescent derivatizing reagents are often used to label nonfluorescent molecules, this labeling step complicates single cell analysis and often results in incomplete reactions and multiple labeling products if there is more than one functional group for labeling per molecule. In this section, we briefly discuss an LIF detector, the sheath-flow cuvette that has been successfully used to detect the contents of single whole cells after CE separation. It was Professor Norman Dovichi who, inspired by flow cytometry, introduced the use of a sheath-flow cuvette as a post-column LIF detection in the 1980s. The reported detection limits have approached single molecule levels [36,44].

Figure 8.5 shows the schematic diagram of a sheath-flow cuvette for LIF detection. The sheath-flow cuvette is usually a 200 μm square chamber that is electrically grounded and houses the end of a fused silica capillary. The sheath-flow buffer enters from the top of the cuvette and exits from the bottom part of the cuvette to a waste reservoir. The flow can be provided either by a high-precision syringe pump or by generating hydrodynamic pressure by holding the sheath fluid in a reservoir that is raised

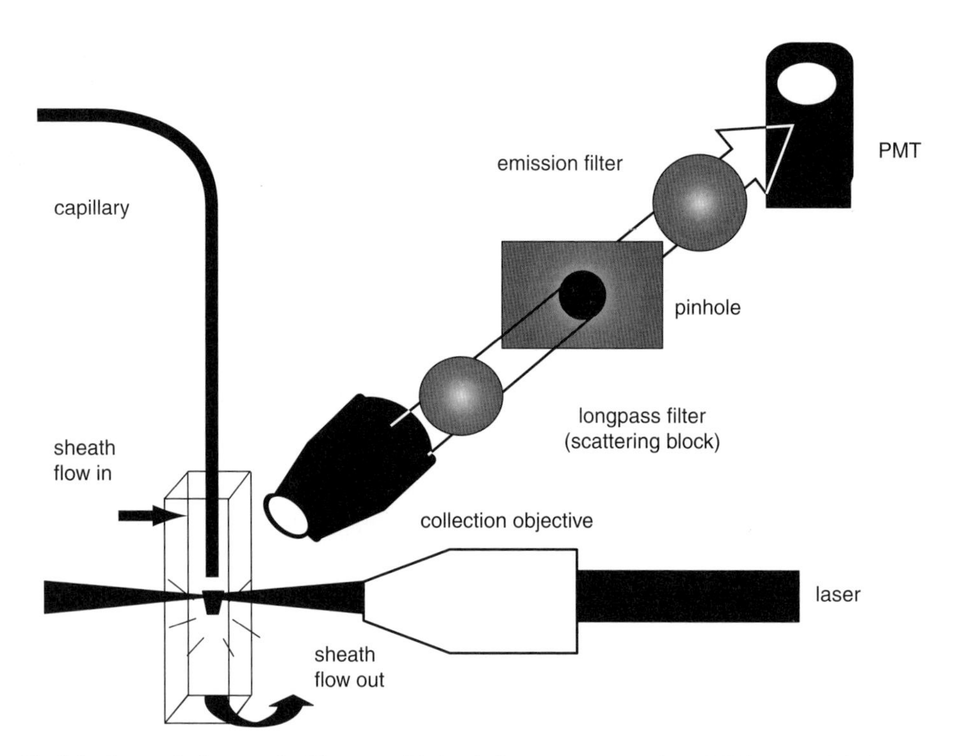


**Fig. 8.5** Schematic diagram of a CE system with sheath-flow cuvette post-column LIF detection. The detection end of the system is shown here. After separation by electrophoresis, the cell contents migrate out of the capillary and cross a laser beam. The fluorescence emission is selected by optical filters that remove scattering and is then detected with a photomultiplier tube.

a few centimeters above the waste reservoir connected to the exit of the sheath-flow cuvette.

After CE separation, the various cellular components migrate out from the capillary and enter the sheath-flow regime. The flowing sheath stream hydrodynamically focuses the sample to a size small enough to cross the laser beam. When the sample passes through the laser beam, excitation will occur and the resulting fluorescence will be collected, as depicted in Fig. 8.5. As opposed to on-column detection, the flat windows of the sheath-flow cuvette help reduce light scattering caused by the highly focused laser beam and the curvature of the capillary surface [35]. Equally important is the lack of an interface between the sheath-flow buffer and CE run buffer which together define a wall-less detector volume where scattering is minimized when these buffers are the same (i.e. they have the same refractive index).

In addition to the optical advantages of using a sheath-flow cuvette as a detector, there are two other benefits. First, a very small detection volume is defined by the hydrodynamically focused capillary flow and the tightly focused laser beam. This small detection volume minimizes the detector dead volume, thereby reducing extra column band broadening. The second advantage is that the cells, which tend to stick to glass, never come in contact with the cuvette windows, preventing contamination and degradation of the optics of the post-column detector.

## 8.4 Single cell gel electrophoresis (comet assay)

A technique related to CE is single cell gel electrophoresis or the comet assay. This qualitative technique is useful for characterizing DNA from single cells based on single strand DNA fragmentation [45,46]. A single cell suspension is embedded in a low melting point agarose gel sandwich on a microscope slide. The cells are lysed by detergents and high salt concentrations at pH 10, and then electrophoresed for a short time at pH $>$13. While lysis removes most of the cell contents, the nuclear DNA remains highly supercoiled. In the presence of an alkaline medium, the DNA unwinds from strand breakage sites. The extensive DNA damage results in smaller fragments that more easily migrate towards an anode. After the separation is complete, the DNA pattern resembles that of a comet tail; the length of the tail is an indication of the extent of DNA damage. Although this method allows for the simultaneous detection of DNA damage in several cells, it cannot be applied to the analysis of other cellular components.

## 8.5 Sampling

### 8.5.1 *Sampling acoustically levitated single cells*

As mentioned in section 8.3, cell sampling is the first step in loading a cell into an analytical instrument. In addition to whole cell sampling from a cell suspension deposited on a microscope slide, sampling can be facilitated by levitating single cells. Previous research has shown that this can be done by combining optical spectroscopy

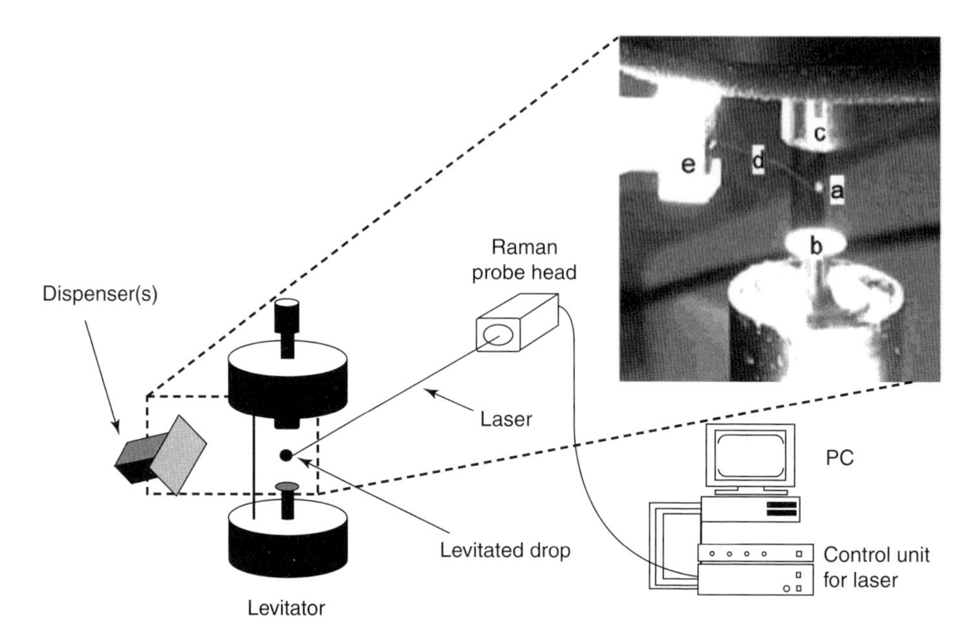


**Fig. 8.6** Schematic representation of an acoustic levitation system for Raman spectroscopy. The insets are: (a) levitated drop, (b) ultrasonic transducer, (c) ultrasonic reflector, (d) dispenser droplet trajectory, and (e) dispenser. This figure has been adapted from the published work of Professor S. Nilsson [48] with permission.

and liquid drop levitation [47–49]. Acoustically levitated drops have already proven to be useful in several analytical areas, such as cell stimulation and inhibition, single-cell analysis, and sample preparation before gas chromatography (GC), or CE/capillary electrochromatography analysis [50–53].

Figure 8.6 shows an instrumental set-up developed in Nilsson's laboratory for Raman spectroscopy using acoustically levitated drops [47–49]. The probe head is kept in a fixed position relative to the ultrasonic levitator. The acoustic levitator generates a standing wave with equally spaced nodes and antinodes by creating multiple reflections between an ultrasonic radiator and a solid reflector, and operates with a frequency of 100 kHz [48]. Sample drops 20–100 nl in volume are positioned in the levitator using an in-house-developed flow-through dispenser or a hamilton syringe. The position where the drop is located is adjusted by moving the levitator.

Acoustically levitated drops containing a single cell can be lysed by injecting the drops with denaturing agents such as SDS, using piezoelectric flow-through dispensers (ejecting 50–100 pl droplets at 1–9000 droplets/s). The contents of the cell can then be introduced into the instrument. This approach to sampling has the advantage of reducing capillary fouling and increasing run-to-run reproducibility.

### 8.5.2 *High throughput single cell sampling and analysis*

In the single cell CE approaches described in sections 8.3.1, 8.3.2 and 8.5.1, manual single cell loading into the capillary presents a bottleneck. To circumvent this bottleneck,

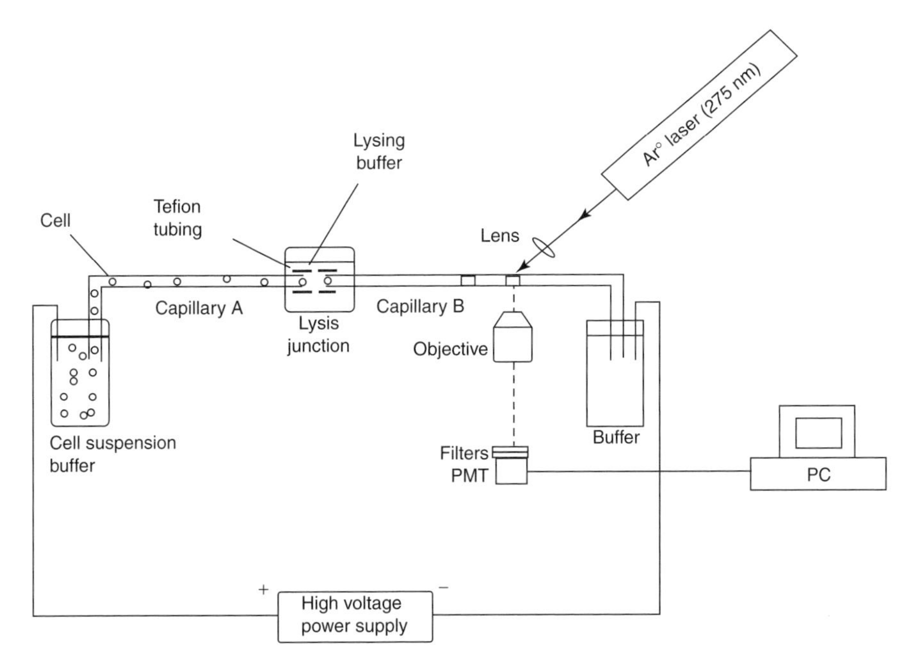


**Fig. 8.7** Diagram showing continuous cell introduction, reproduced with permission from ref. [54]. There are two capillaries that meet at the lysis junction. Electro-osmotic flow is used as a pump to draw the cell suspension continuously from the buffer inlet vial into the junction where the cells mix with the lysing buffer. After individual cells have been lysed, proteins are separated in the separation capillary (second capillary) and detected using on-column LIF detection.

the Lillard laboratory has developed methodology for the continuous automated loading of intact human erythrocytes into a capillary [54]. Figure 8.7 is a schematic representation of this approach. The first of two capillaries is used for continuous cell delivery from the cell suspension with electro-osmotic flow as the driving force to draw the cells into the capillary. The second capillary is used for the electrophoretic separation of the cellular contents. At the junction of the two capillaries, cells are disrupted because they come in contact with a lysing agent and are subject to a mechanical shear buffer that is introduced at this location. This approach uses on-column LIF detection, but a post-column detector such as the one described in section 8.3.3 could also be used. The total analysis time, using continuous cell injection, is approximately 4 min and no microscope is required. This method holds promise as a high throughput technique for analyzing many cells, while extracting information on a cell-to-cell basis and at the same time simplifying the cell sampling process.

### 8.5.3 *Probing the cytoplasm of large cells*

Compared to typical mammalian cells, neurons and oocytes are too large to be directly introduced into the capillary of an analytical instrument. Instead, small samples that are fractions of the total cytoplasm are collected for CE analysis with MS detection

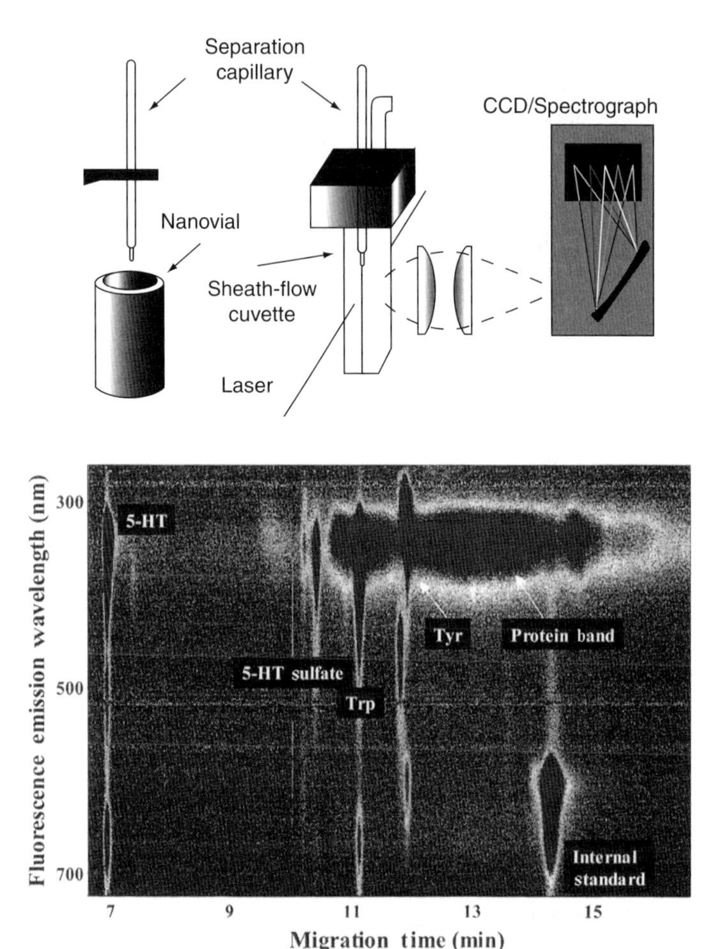


**Fig. 8.8** Schematic diagram of a CE system used for single neuron analysis (top panel). One nanoliter of the homogenized cell is electrokinetically injected into the capillary for separation. Precise positioning with a motion controller system has been used to place the capillary inside the nanovial. After the separation, the cell components migrate towards the sheath-flow cuvette and are excited by a 257 nm UV laser. The emission is collected orthogonally, filtered, and sent to a CCD/spectrograph. Spectra are collected as the analyte travels through the detector volume in the sheath-flow cuvette. The information-rich data from the electropherogram are shown in the lower panel. Courtesy of Professor J. Sweedler.

or fluorescence spectral emission detection. These separation schemes provide a more complete description of the analyte.

In Fig. 8.8, the subsample of an entire neuron cell provides analytical data. Here the cell is homogenized and diluted to 0.3 μl before CE analysis. Using electrokinetic sampling, about 1 nl of the sample is introduced into the capillary, separated by CE, and detected [16,55]. In this instrument, the detector is a sheath-flow cuvette similar to the one in Fig. 8.5, but a charge coupled device (CCD)/spectrograph is used instead of a photomultiplier tube to allow identification of neurotransmitters with overlapping migration times. For the analysis of neurotransmitters such as serotonin, UV laser excitation at 257 nm is also used, eliminating the need for the chemical derivatization of biomolecules. When data from CE are compared with commercially available neurotransmitter standards, it is possible to identify the chemical composition of neuronal cells with high reliability.

In addition to neuronal content, the instrument described in Fig. 8.8 can also be used to obtain dynamic information about neurotransmitter release. Complementing the static chemical inventory of the cell, this information tells us a great deal about the function of a single cell in a network of cells [56]. In this type of experiment, a sampling capillary is often placed close to the cell of interest; a stimulant is applied to the cell and sampling is performed immediately after the subsequent release of the chemicals from the cell. The release of chemicals follows after stimulation as a function of time, either using fast sequential separations or parallel separation methods [19,57].

In another example of time-resolved cell release detection, a cell is placed in a sampling capillary and moved precisely at fixed velocity along the edge of a separation channel. The compounds that are released can be collected very efficiently with good temporal resolution [19]. Recent trends suggest that new sampling, separation, and detection strategies with CE are being developed to obtain better temporal, spatial and chemical information from decreasing amounts of sample [58].

## 8.6 Mass spectrometry for single cells and tissues

### 8.6.1 *Classical hyphenated techniques*

Instrumentation for the MS of single cells can be used (i) to detect cellular components separated by capillary liquid chromatography and CE, or (ii) to analyze single cells directly. Despite its lack of sensitivity, the structural and molecular mass information gained from MS can be a definitive tool in the qualitative identification of compounds. Attempts are therefore being made to improve MS for single cell applications by reducing chemical background noise, for example, by using ultra high-resolution mass spectrometers to separate analyte masses from the background noise. One impressive example of ultra high-resolution MS for CE analysis of whole cell digests is high magnetic field ($>9$ tesla) Fourier transform ion cyclotron resonance mass spectrometry (FTICR-MS) [59,60].

The MS analysis of the chemical contents of a single intact cell has also been carried out using the following approaches: (i) matrix assisted laser desorption/ionization MALDI [61], (ii) direct laser vaporization/ionization [62], and (iii) on-line CE separation, ionization, and detection of cell contents using electrospray ionization (ESI)-MS [59]. MALDI has a faster analysis time because no on-line separation is performed, but additional sample clean-ups and the transfer of the cells to a new buffer system [62] are necessary for successful MS analysis. Furthermore, off-line sampling and lysing is necessary for MALDI [61]. Some of the parameters under current investigation that could improve the use of MS for single cell related techniques include enhancing the inherent sensitivity of the MS instrument and/or that of the ESI processes.

CE on-line separation before ESI-MS has been applied to the analysis of human red blood cells [59,63]. However, this approach was successful only in detecting

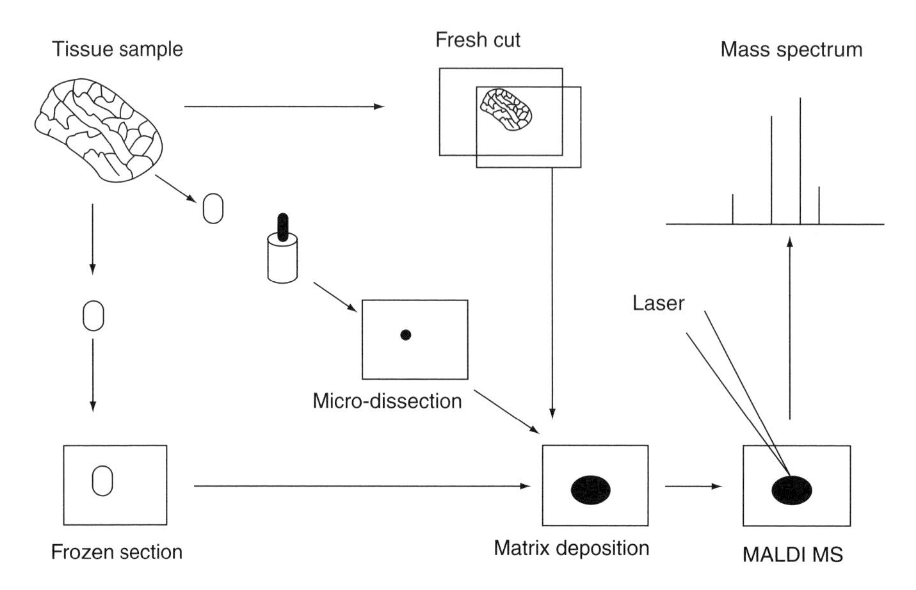


**Fig. 8.9** Schematic representation of three different strategies for sample preparation methods for MALDI analysis. Either a freshly cut tissue sample or frozen sections of a tissue can be placed directly on the MALDI plate for MS analysis. For the third strategy, LCM microdissection is performed to isolate the cells from the heterogeneous population before placeing the sample on the MALDI plate.

hemoglobin because this protein is highly abundant in red blood cells. Other applications may need to wait until the overall signal-to-noise ratio for ESI-MS is increased.

### 8.6.2 *Laser capture microdissection*

When cells are part of a tissue, laser capture microdissection (LCM) facilitates sampling a whole cell that then can be directed to MALDI MS [27,64]. In Fig. 8.9, a flow chart describes several approaches for combining LCM with MALDI MS for the direct sampling of cells from a tissue cross-section. The tissue of interest is sliced into fine cross-sections (e.g. 10 μm thick) using a microtome. As shown in the diagonal path of Fig. 8.9, in LCM a thermoplastic membrane is placed on the tissue and irradiation from a focused infrared (IR) laser pulse melts the membrane. Subsequent removal of the membrane with its adhered cells provides a MALDI target as small as a laser spot size. The microdissected cells are placed directly on a MALDI plate for analysis. Instead of MALDI MS, secondary ion MS can may also be used at this point [65].

In addition to the conventional MALDI instrument called MALDI time-of-flight (TOF) MS, more sophisticated mass spectrometers have begun to appear. Figure 8.10 shows a conventional MALDI instrument and a QqTOF MS/MS. Using the latter instrument, it is feasible to obtain detailed information collected from small samples of tissue cross-sections. Although tissue sampling and analysis either via LCM or direct MALDI analysis are fast and powerful technologies, they cannot fully guarantee that it is a single cell that is being analyzed. Nonetheless, this work represents an important breakthrough in characterizing heterogeneity in tissues and cell populations.

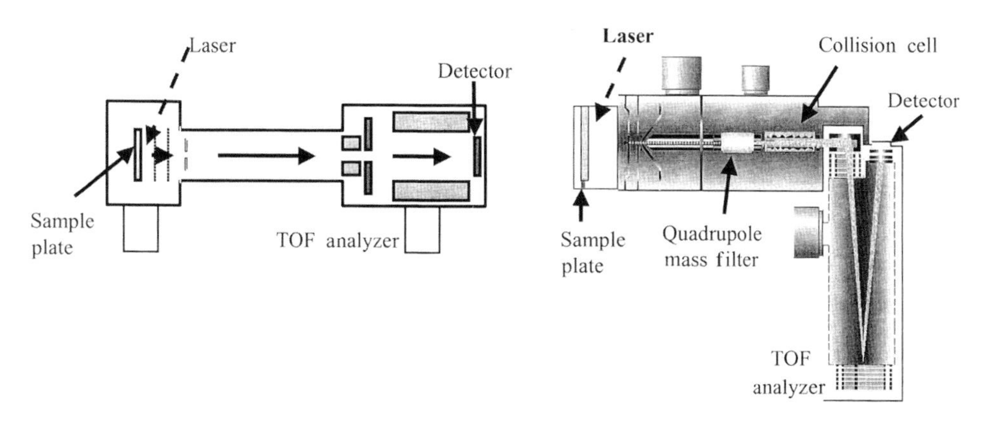


**Fig. 8.10** Block diagrams of MALDI TOF MS (left) and MALDI QqTOF MS/MS systems. In both cases, the sample is microdissected and spotted on MALDI probes to be introduced into the MS. Figure 8.10 is reproduced by courtesy of Professor R.M. Caprioli.

## 8.7 Direct analysis of subcellular compartments by CE

In the Arriaga laboratory, we have been developing instrumentation for the CE analysis of subcellular compartments, either from single cells or tissue cross-sections [66]. Using the same device as for single cell sampling (Fig. 8.4), mitochondria from a 10 μm thick muscle cross-section are loaded directly into the capillary by applying negative pressure. The sampled mitochondria are separated by CE and detected by post-column LIF detection. Figure 8.11 shows an electropherogram indicating narrow events corresponding to individual mitochondria that have become detectable because of their labeling with 10-nonyl acridine orange. Such direct tissue sampling for CE

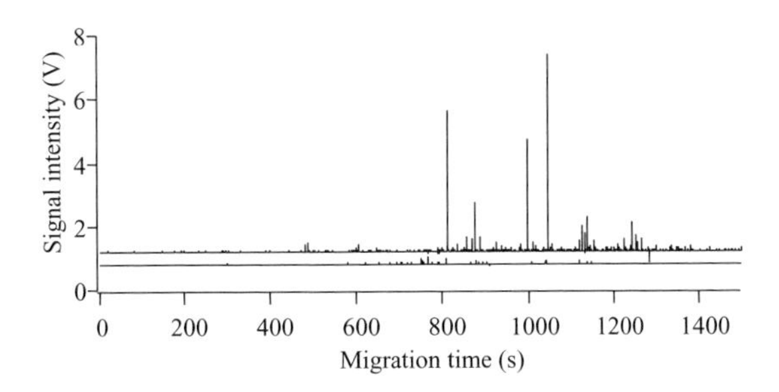


**Fig. 8.11** CE-LIF analysis of mitochondria directly sampled from a tissue cross-section after being fluorescently labeled with 10-nonyl acridine orange. Each sharp event corresponds to one mitochondrion. The lower panel is the control experiment. The X axis is the migration time and the Y axis is the fluorescent intensity.

analysis opens new avenues for describing heterogeneous mitochondrial surface properties (migration time variation) and cardiolipin content (fluorescence intensity) found in muscle tissues. This technology will likely be useful in characterizing age-related and muscle-related diseases, such as the ragged red fibers that are characteristically heterogeneous.

### Acknowledgements

Hossein Ahmadzadeh thanks NIH for support (R01-AG20866). Edgar Arriaga is supported through an NIH K02 Careers Award (K02-AG21453).

### References

1. Sims, C.E., Meredith, G.D., Krasieva, T.B., Berns, M.W., Tromberg, B.J. & Allbritton, N.L. (1998) Laser-micropipet combination for single-cell analysis. *Anal. Chem.*, **70**(21), 4570–4577.
2. Li, H., Sims, C.E., Wu, H.Y. & Allbritton, N.L. (2001) Spatial control of cellular measurements with the laser micropipet. *Anal. Chem.*, **73**(19), 4625–4631.
3. Sims, C.E. & Allbritton, N.L. (2003) Single-cell kinase assays: opening a window onto cell behavior. *Curr. Opin. Biotechnol.*, **14**(1), 23–28.
4. Chiu, D.T., Lillard, S.J., Scheller, R.H., Zare, R.N., Rodriguez-Cruz, S.E., Williams, E.R., Orwar, O., Sandberg, M. & Lundqvist, J.A. (1998) Probing single secretory vesicles with capillary electrophoresis. *Science*, **279**, 1190–1193.
5. Zabzdyr, J.L. & Lillard, S.J. (2001) Measurement of single-cell gene expression using capillary electrophoresis. *Anal. Chem.*, **73**(23), 5771–5775.
6. Han, F. & Lillard, S.J. (2002) Monitoring differential synthesis of RNA in individual cells by capillary electrophoresis. *Anal. Biochem.*, **302**(1), 136–143.
7. Krylov, S.N. & Dovichi, N.J. (2000) Single-cell analysis using capillary electrophoresis: influence of surface support properties on cell injection into the capillary. *Electrophoresis,* **21**(4), 767–773.
8. Krylov, S.N., Arriaga, E., Zhang, Z., Chan, N.W., Palcic, M.M. & Dovichi, N.J. (2000) Single-cell analysis avoids sample processing bias. *J. Chromatogr. B, Biomed. Sci. Appl.*, **741**(1), 31–35.
9. Krylov, S.N., Arriaga, E.A., Chan, N.W., Dovichi, N.J. & Palcic, M.M. (2000) Metabolic cytometry: monitoring oligosaccharide biosynthesis in single cells by capillary electrophoresis. *Anal. Biochem.*, **283**(2), 133–135.
10. Krylov, S.N., Starke, D.A., Arriaga, E.A., Zhang, Z., Chan, N.W., Palcic, M.M. & Dovichi, N.J. (2000) Instrumentation for chemical cytometry. *Anal. Chem.*, **72**(4), 872–877.
11. Hu, S., Lee, R., Zhang, Z., Krylov, S.N., Dovichi, N.J. (2001) Protein analysis of an individual *Caenorhabditis elegans* single-cell embryo by capillary electrophoresis. *J. Chromatogr. B, Biomed. Sci. Appl.*, **752**(2), 307–310.
12. Hu, S., Jiang, J., Cook, L.M., Richards, D.P., Horlick, L., Wong, B. & Dovichi, N.J. (2002) Capillary sodium dodecyl sulfate–DALT electrophoresis with laser-induced fluorescence detection for size-based analysis of proteins in human colon cancer cells. *Electrophoresis,* **23**(18), 3136–3142.
13. Dovichi, N.J. & Pinkel, D. (2003) Analytical biotechnology. Tools to characterize cells and their contents. *Curr. Opin. Biotechnol.,* **14**(1), 3–4.
14. Page, J.S., Rubakhin, S.S. & Sweedler, J.V. (2002) Single-neuron analysis using CE combined with MALDI MS and radionuclide detection. *Anal. Chem.*, **74**(3), 497–503.
15. Stuart, J.N. & Sweedler, J.V. (2003) Single-cell analysis by capillary electrophoresis. *Anal. Bioanal. Chem.*, **375**(1), 28–29.
16. Stuart, J.N., Zhang, X., Jakubowski, J.A., Romanova, E.V. & Sweedler, J.V. (2003) Serotonin catabolism depends upon location of release: characterization of sulfated and gamma-glutamylated serotonin metabolites in *Aplysia californica. J. Neurochem.*, **84**(6), 1358–1366.
17. Kennedy, R.T., Oates, M.D., Cooper, B.R., Nickerson, B. & Jorgenson, J.W. (1989) Microcolumn separations and the analysis of single cells. *Science* **246**, 57–63.

18. Kennedy, R.T. & Jorgenson, J.W. (1989) Quantitative analysis of individual neurons by open tubular liquid chromatography with voltammetric detection. *Anal. Chem.*, **61**(5), 436–441.
19. Cannon, D.M., Jr., Winograd, N. & Ewing, A.G. (2000) Quantitative chemical analysis of single cells. *Annu. Rev. Biophys. Biomol. Struct.*, **29**, 239–263.
20. Chiu, D.T., Hsiao, A., Gaggar, A., Garza-Lopez, R.A., Orwar, O. & Zare, R.N. (1997) Injection of ultrasmall samples and single molecules into tapered capillaries. *Anal. Chem.*, **69**(10), 1801–1807.
21. Chen, G. & Ewing, A.G. (1997) Chemical analysis of single cells and exocytosis. *Crit. Rev. Neurobiol.*, **11**(1), 59–90.
22. Chiu, D.T. & Zare, R.N. (1998) Assaying for peptides in individual *Aplysia* neurons with mass spectrometry. *Proc. Natl Acad. Sci. U.S.A.*, **95**(7), 3338–3340.
23. Bhattacharya, S.H., Gal, A.A. & Murray, K.K. (2003) Laser capture microdissection MALDI for direct analysis of archival tissue. *J. Proteome. Res.*, **2**(1), 95–98.
24. Craven, R.A., Totty, N., Harnden, P., Selby, P.J. & Banks, R.E. (2002) Laser capture microdissection and two-dimensional polyacrylamide gel electrophoresis: evaluation of tissue preparation and sample limitations. *Am. J. Pathol.*, **160**(3), 815–822.
25. Rekhter, M.D. & Chen, J. (2001) Molecular analysis of complex tissues is facilitated by laser capture microdissection: critical role of upstream tissue processing. *Cell. Biochem. Biophys.*, **35**(1), 103–113.
26. Tanji, N., Ross, M.D., Cara, A., Markowitz, G.S., Klotman, P.E. & D'Agati, V.D. (2001) Effect of tissue processing on the ability to recover nucleic acid from specific renal tissue compartments by laser capture microdissection. *Exp. Nephrol.*, **9**(3), 229–234.
27. Xu, B.J., Caprioli, R.M., Sanders, M.E. & Jensen, R.A. (2002) Direct analysis of laser capture microdissected cells by MALDI mass spectrometry. *J. Am. Soc. Mass Spectrom.*, **13**(11), 1292–1297.
28. Tagawa, Y., Yasutake, T., Ikuta, Y., Oka, T. & Terada, R. (2003) Chromosome 8 numerical aberrations in stage II invasive ductal carcinoma: correlation with patient outcome and poor prognosis. *Med. Oncol.*, **20**(2), 127–136.
29. Reich, O., Purstner, P., Klaritsch, P., Haas, J., Lahousen, M., Tamussino, K. & Winter, R. (2003) Prognostic significance of preoperative DNA flow cytometry in surgically-treated cervical cancer. *Eur. J. Gynaecol. Oncol.*, **24**(1), 13–17.
30. Molnar, B., Bocsi, J., Karman, J., Nemeth, A., Pronai, L., Zagoni, T. & Tulassay, Z. (2003) Immediate DNA ploidy analysis of gastrointestinal biopsies taken by endoscopy using a mechanical dissociation device. *Anticancer Res.*, **23**(1B), 655–660.
31. Hu, Y.P., He, L.X., Yu, D., Xia, H.S., Ke, Y.H., Yang, L., Jiang, H., Xiao, Z.H., Fu, X.Y. & Fan, Y.H. (2003) DNA content and its correlation with biological behavior of non-small cell lung cancer. *Zhonghua Zhong Liu Za Zhi,* **25**(1), 55–58 (in Chinese).
32. Crossland-Taylor, P.J. (1953) A device for counting small particles suspended in a fluid through a tube. *Nature* **171**, 37–38.
33. Van Dilla, M.A., Trujillow, T.T., Mullaney. P.F. & Coulter, J.R. (1969) Cell microfluorometry: a method for rapid fluorescence measurement. *Science* **163**, 1213–1214.
34. Haugland, R.P. (2002) *Handbook of Fluorescent Probes and Research Products*, 9th edn. Molecular Probes, Eugene, Oreg.
35. Ahmadzadeh, H. & Dovichi, N.J. (2000) *Instrumentation, Capillary Coating and Labeling Chemistry for Capillary Electrophoresis with Laser Induced Fluorescence Detection.* University of Alberta, Edmonton, Alberta.
36. Chen, D.Y., & Dovichi, N.J. (1994) Y octomole detection limit by laser-induced fluorescence in capillary electrophoresis. *J. Chromatogr. B, Biomed. Appl.* **657**(2), 265–269.
37. Anderson, B.B. & Ewing, A.G. (1999) Chemical profiles and monitoring dynamics at an individual nerve cell in *Planorbis corneus* with electrochemical detection. *J. Pharm. Biomed. Anal.*, **19**(1-2), 15–32.
38. Zimmerman, U. & Neli, G. (1996) *Electroporation of Cells*. CRC Press, Boca Raton, Fla.
39. Yong, H. & Boris, R. (2001) Microfabricated electroporation chip for single cell membrane permeabilization. *Sensors Actuators A,* **89**, 242–249.
40. Zimmermann, U., Friedrich, U., Mussauer, H., Gessner, P., Hamel, K. & Sukhorukov, V. (2000) Electromanipulation of mammalian cells: fundamentals and applications. *IEEE Trans. Plasma Sci.* **28**(1), 72–82.
41. Tomos, A.D. & Sharrock, R.A. (2001) Cell sampling and analysis (SiCSA): metabolites measured at single cell resolution. *J. Exp. Bot.*, **52**, 623–630.
42. Susumu, H., Yasuko, N.-O., Kazutaka, T. & Atsushi, T. (1997) Analysis of the components of the fluid in a single plant cell by high-performance capillary electrophoresis. *Anal. Sci.*, **13**(4), 557–564.
43. Baldwin, R.P. (2000) Recent advances in electrochemical detection in capillary electrophoresis. *Electrophoresis*, **21**(18), 4017–4028.
44. Dovichi, N.J., Martin, J.C., Jett, J.H., Trkula, M. & Keller, R.A. (1984) Laser-induced fluorescence of flowing samples as an approach to single-molecule detection in liquids. *Anal. Chem.*, **56**(3), 348–354.

45. Johnson, L.A. & Ferris, J.A.J. (2002) Analysis of postmortem DNA degredation single cell gel electrophoresis. *Forensic Sci. Int,* **126,** 43–47.
46. Rojas, E., Lopez, M.C., & Valverde, M. (1999) Single cell gel electrophoresis assay: methodology and applications. *J. Chromatogr. B, Biomed. Sci. Appl.*, **722**(1-2), 225–254.
47. Santesson, S., Andersson, M., Degerman, E., Johansson, T., Nilsson, J. & Nilsson, S. (2000) Airborne cell analysis. *Anal. Chem.*, **72**(15), 3412–3418.
48. Santesson, S., Johansson, J., Taylor, L.S., Levander, I., Fox, S., Sepaniak M., *et al.* (2003) Airborne chemistry coupled to Raman spectroscopy. *Anal. Chem.,* **75**(9), 2177–2180.
49. Santesson, S., Cedergren-Zeppezauer, E.S., Johansson, T., Laurell, T., Nilsson, J. & Nilsson, S. (2003) Screening of nucleation conditions using levitated drops for protein crystallization. *Anal. Chem.* **75**(7), 1733–1740.
50. Petersson, M., Nilsson, J., Wallman, L., Laurell, T., Johansson, J. & Nilsson, S. (1998) Sample enrichment in a single levitated droplet for capillary electrophoresis. *J. Chromatogr. B, Biomed. Sci. Appl.,* **714**(1), 39–46.
51. Santesson, S., Degerman, E., Johansson, T., Nilsson, J. & Nilsson, S. (2001) Bioanalytical chemistry in levitated drops. *Am. Lab.* (Shelton, Conn.) **33**(3), 13–14.
52. Nilsson, S., Santesson, S., Degerman, E., Johansson, T., Laurell, T., Nilsson, J. & Johansson, J. (2000) Airborne chemistry for biological micro analysis. In: *Micro Total Analysis Systems 2000, Proceedings of the 4th Micro Total Analysis Systems Symposium,* (ed. A. Van den Berg, W. Olthuis & P. Bergveld) Enschede, The Netherlands, pp. 19–24.
53. Nilsson, J., Laurell, T., Petersson, M., Santesson, S. & Johansson, T. (1999) Flow-through microdispenser for liquid handling in a levitated-droplet based analyzing system. In: *Proceedings of the International Conference on Microreaction Technology,* 18–21 April 1999, pp. 320–26, Frankfurt. Springer-Verlag, Berlin.
54. Chen, S., Lillard, S.J. (2001) Continuous cell introduction for the analysis of individual cells by capillary electrophoresis. *Anal. Chem.*, **73**(1), 111–118.
55. Fuller, R.R., Moroz, L.L., Gillette, R. & Sweedler, J.V. (1998) Single neuron analysis by capillary electrophoresis with fluorescence spectroscopy. *Neuron,* **20**(2), 173–181.
56. Lillard, S.J. & Yeung, E.S. (1997) Temporal and spatial monitoring of exocytosis with native fluorescence imaging microscopy. *J. Neurosci. Methods,* **75**(1), 103–109.
57. Kennedy, R.T., German, I., Thompson, J.E., Witowski, S.R. (1999) Fast analytical-scale separations by capillary electrophoresis and liquid chromatography. *Chem. Rev.*, **99**(10), 3081–3132.
58. Woods, L.A. & Ewing, A.G. (2003) Etched electrochemical detection for electrophoresis in nanometer inner diameter capillaries. *Chemphyschem* **4**(2), 207–211.
59. Hofstadler, S.A., Severs, J.C., Smith, R.D., Swanek, F.D. & Ewing, A.G. (1996) Analysis of single cells with capillary electrophoresis electrospray ionization Fourier transform ion cyclotron resonance mass spectrometry. *Rapid Commun. Mass Spectrom.* **10**(8), 919–922.
60. Hofstadler, S.A., Swanek, F.D., Gale, D.C., Ewing, A.G. & Smith, R.D. (1995) Capillary electrophoresis–electrospray ionization Fourier transform ion cyclotron resonance mass spectrometry for direct analysis of cellular proteins. *Anal. Chem.,* **67**(8), 1477–1480.
61. Li, L., Golding, R.E. & Whittal, R.M. (1996) Analysis of single mammalian cell lysates by mass spectrometry. *J. Am. Chem. Soc.*, **118**(46), 11662–11663.
62. Fung, E.N. & Yeung, E.S. (1998) Direct analysis of single rat peritoneal mast cells with laser vaporization/ionization mass spectrometry. *Anal. Chem.,* **70**(15), 3206–3212.
63. Cao, P. & Moini, M. (1998) Analysis of peptides, proteins, protein digests, and whole human blood by capillary electrophoresis/electrospray ionization-mass spectrometry using an in-capillary electrode sheathless interface. *J. Am. Soc. Mass Spectrom.*, **9**(10), 1081–1088.
64. Chaurand, P. & Caprioli, R.M. (2002) Direct profiling and imaging of peptides and proteins from mammalian cells and tissue sections by mass spectrometry. *Electrophoresis*, **23**(18), 3125–3135.
65. Todd, P.J., Schaaff, T.G., Chaurand, P. & Caprioli, R.M. (2001) Organic ion imaging of biological tissue with secondary ion mass spectrometry and matrix-assisted laser desorption/ionization. *J. Mass Spectrom.*, **36**(4), 355–369.
66. Ahmadzadeh, H., Johnson, R., Thompson, L. & Arriaga, E. (2004) Direct sampling from muscle cross-sections for electrophoretic analysis of individual mitochondria. *Anal. Chem.,* **76**(2), 315–321.

# 9 Studies on animal to instrument hyphenation: development of separation-based sensors for near real-time monitoring of drugs and neurotransmitters

MALONNE DAVIES, BRYAN HUYNH, SARA LOGAN, KATHLEEN HEPPERT and SUSAN LUNTE

## 9.1 Introduction

This chapter considers issues related to the collection and analysis of biologically important compounds from live animals. In the first section, classical methods for sampling and analyzing blood and tissue are addressed. New *in vivo* procedures that are less invasive and permit continuous sampling of blood and tissues are then described. Analytical considerations related to *in vivo* sampling are also discussed.

In techniques using sampling from live animals, the sample is the animal itself. Accordingly, attention is focused on issues relating to the management of animals under conditions of sampling. Of necessity, this includes ethical and regulatory issues relating to animal health and welfare.

Integrated sampling–analysis systems that employ microdialysis as the sampling method are emphasized in the second half of this chapter. Many examples are given of microdialysis coupled with both nonseparation- and separation-based detection systems. Finally, the future of integrated sampling–analysis systems is discussed, especially with regard to the development of miniaturized separation-based sensors for near real-time measurements of drugs and neurotransmitters.

## 9.2 Obtaining *in vivo* biological samples by classical techniques: limitations, challenges and other considerations

The reliance of investigators on classical sampling techniques has been a function of accessibility and practicality. Blood samples, for example, can be obtained with relative ease and in volumes sufficient for the analytical requirements of most instruments. This is predicated on the fact that the animal used for the study is large enough to obtain the necessary number of samples. The sampling frequency in this case is dependent on the total blood volume of the animal. If comparison of analyte concentration in blood and tissues is desired, the animal must be sacrificed and the various organs/tissues harvested and prepared for analysis. If temporal information regarding organs/tissue distribution is desirable, this will greatly increase the number of animals needed for a given study because replicate animals are required for each time-point and for each dosing regime or formulation.

The analyte concentration in blood and tissue samples usually represents the total drug or analyte concentration. In many cases, it is the extracellular concentration (the unbound fraction in blood/plasma) that is physiologically or therapeutically relevant and, therefore, this is the value of interest. To obtain the value for the free concentration of an analyte, the degree of protein binding can be determined *in vitro* and used to calculate the relative free concentration.

In general, classical sampling methods treat the animal as a 'black box' from which discrete samples are collected at fixed times. They also, to some degree, assume the contents of the 'black box' to be homogeneous. Pharmacokinetic models also frequently suppose that analyte concentration in the target tissue is identical to the free concentration in the central compartment (blood/plasma). This assumption is not always valid as site selective sampling can often reveal more information related to the function of a specific organ.

### 9.2.1 *Animal considerations*

In all *in vivo* studies, the design of the experimental sampling protocol must take into account the absolute need to use an animal model, the humane treatment of the animal, and all guidelines instituted by federally regulated animal care committees. Consideration of species choice, animal size and awake vs anesthetized studies are discussed below.

9.2.1.1 *Issues relating to animal size and species choice* The traditional experimental designs for adsorption, metabolism, distribution and elimination (ADME) investigations treat the experimental animal, the sampling technique(s), and the analytical method as separate, even unrelated, modules of the study. Each component can be selected and/or optimized independently of the other components with few constraints. The major constraints considered in the design include the sample volume required by the analytical method, the temporal profile required, and the animal model to be used. The animal model selected must have sufficient blood volume to supply the desired number of samples over the time-course of the experiment while meeting guidelines that limit the total amount of blood that can be collected. These guidelines limit the amount of blood withdrawn in a week to 10% and in a month to 20% of the total volume of the animal [1]. Any amount over 20% should be restricted to terminal experiments under anesthesia. For mice, 10% of the blood volume is approximately 0.234 ml and for rats 2.34 ml (assuming an average mass of 0.030 kg with 78 ml/kg of whole blood in mice and 0.30 kg with 78 ml/kg of whole blood in the rat). The small volumes available virtually preclude serial sampling in mice, meaning that a separate animal must be used for each time-point to define the pharmacokinetic profile. In addition, to obtain statistically relevant data, several mice must be used at each time-point. This greatly increases the number of mice needed for one study and, consequently, the cost of the experiment. Larger animals do not have as many constraints on blood sampling; however, they are generally more difficult and expensive to maintain, and have limitations when online analysis is desired.

The choice of animal for a particular experiment is based on a number of parameters. As stated above, small animals generally have limitations on sampling. However, the benefits of using these animals often outweigh those of using larger animals. In particular, small animals are extremely useful in metabolic or online experiments since these experiments usually require very intricate collection systems for each animal. Even though small animals are the most desirable when considering expense (especially in crossover experiments where expenses accrue rapidly), some studies require either too many animals or too much sample to meet the requirements of the analytical method.

For example, with many pharmacokinetic/pharmacodynamic studies, a large number of blood samples must be taken in a short period of time to obtain the necessary data to complete the calculations required for these experiments. Many samples are also required in crossover studies, but these can be obtained from separate animals over a longer period of time and this allows individual animals to recover between sampling. If it is important to run an experiment under normal physiological conditions and a recovery period is not acceptable, then it is essential to go to a larger animal. However, some laboratories still find sacrificing large numbers of small animals more practical than fulfilling the housing requirements of larger animals. The important part of animal selection is to be sure that the requirements of both the analytical tool and the animal are fulfilled to ensure good results.

A last consideration is with respect to hyphenated techniques where the animal is connected online to the analysis system. In this case, the integration of the animal to the analytical instrumentation must take into account both animal size and movement.

9.2.1.2 *Intact animal studies: anesthetized vs recovered* In addition to the animal model, another important point in designing intact animal studies is whether the animal will need to be anesthetized throughout the experiment. It is well established that virtually any anesthesia will influence physiological parameters, generally slowing heart and respiration rates, and lowering blood pressure and body temperature [2,3]. These, in turn, influence hepatic, renal, gastrointestinal, and neurological functions. If the animal is kept under anesthesia, this will limit the total duration of the experiment to a few hours. Furthermore, it may preclude crossover designs.

In contrast, animals that are not anesthetized or that recover from anesthesia before the experiment offer a potentially longer experimental window and usually allow crossover designs. However, if blood sampling is carried out without anesthesia, for example by repeated tail vein puncture, the potential impact on physiological factors of stress induced by repeated handling must be considered. In the case of surgical procedures, the postoperative recovery time needed is dependent on a number of parameters including the anesthetic used, the surgical procedure(s) involved, and the time required for the surgery. The actual recovery period may be a trade-off between the optimal time needed for full recovery and the viability time of surgically implanted devices such as catheters or probes.

For long-term studies with awake, freely moving animals, where fluid or other connections (e.g. catheters, tubing, electrical leads) are involved, special containment systems are often necessary to prevent tangling and/or chewing of the connections. It may also be important to consider the experimental animals' diurnal rhythm during the study, depending on the type of information that is desired. For example, rats are nocturnal while laboratory technicians are (generally assumed to be) active during the day (or in light). If the drug or condition being investigated is related to the wake/sleep cycle, the timing of the experimental protocol must take this into account.

### 9.2.2 *Classical methods of blood sampling*

Blood samples are generally labor-intensive to collect, process, and prepare for analysis. There are several established techniques for manual sampling of blood from rats. Sites for blood collection include the lateral tail vein, the ventral tail artery, the dorsal metatarsal or lateral saphenous vein, tail tip removal, the retro-orbital plexus, the jugular vein and the heart. Typically, the lateral tail vein is used for small volumes of blood. For frequent sampling, these techniques require either anesthesia or placing the animal in a restraining device. The number of samples that can be obtained from one animal is very limited for some of these techniques.

For repeated sampling over long periods of time, surgical placement of an indwelling catheter in a vein or artery is recommended. Sampling is accomplished by drawing a heparinized locking solution, followed by blood, through the catheter into a syringe. The syringe is then exchanged with another to collect the sample. After the sample is obtained, a new syringe is used to flush the catheter with fresh heparin solution.

As stated previously, animal-use regulations limit the relative volume of blood that can be sampled. In addition, there are physiological constraints. Removal of too much blood at one time can lead to a change in the physiology of the animal, modify the pharmacokinetics of a drug, and have adverse effects on the animal's health. In particular, the stress of blood removal will cause the concentrations of cortisol and other stress hormones to increase. This can affect the concentrations of other hormones and neurotransmitters, including epinephrine and norepinephrine.

In most cases, blood samples must be treated before analysis to remove cells and proteins. Usually, the samples need to be processed immediately following collection to avoid chemical and/or enzymatic degradation of the analyte in whole blood or plasma. A typical sample treatment would include, at a minimum, a centrifugation step to separate plasma from blood cells. The isolated plasma typically has an analytical internal standard added, followed by the addition of a protein precipitation reagent. After centrifugation, the supernatant is removed for analysis. The supernatant may then need to undergo even more sample preparation steps, including such things as solid phase extraction (SPE), preconcentration, or chemical derivatization of the analytes of interest.

### 9.2.3 *Classical methods of tissue sampling*

When information regarding the distribution and disposition of drugs, environmental contaminants, and endogenous compounds is desired, tissues or organs are generally collected from euthanized animals. To avoid enzymatic degradation of the analytes of interest, tissues must be harvested and processed quickly. Because of the complex nature of tissue samples, the preparation steps are much more labor-intensive and complex than those used for blood samples. Tissue samples are usually homogenized, followed by centrifugation and/or protein precipitation and SPE or liquid-phase extraction (LPE). If temporal information regarding the distribution of the drug or substance is desired, then the number of animals sacrificed for the study will increase dramatically along with the analyst's workload. The sample preparation steps mix extracellular and intracellular components so that the total analyte content in the tissue is determined, resulting in a loss of useful compartmental information. A final consideration with tissue distribution studies is the cost of the animals. If multiple animals are to be sacrificed at each time-point, the expense can become excessive, especially if special breeds such as knockout mice are used for the study.

## 9.3 Modern *in vivo* sampling techniques

Automated blood sampling, ultrafiltration and microdialysis are three relatively new sampling methods that exhibit several advantages over classical techniques for *in vivo* monitoring. Each of these is discussed in more detail in sections 9.3.1, 9.3.2. and 9.3.3, respectively. The automated blood sampling system is an instrument that collects blood samples from a rat without the need for human intervention. Ultrafiltration is a second sampling method that uses a vacuum to draw the sample from the extracellular space of the tissues. There have been a limited number of reports of online analysis using this technique. The last technique, microdialysis, can be used to sample blood and a variety of tissues (e.g. brain) and has been shown to be very compatible with online analysis. This section focuses on the basic principles of each of these modern sampling methods. The issues involved with coupling microdialysis to an online analysis system will be discussed together with examples of coupling microdialysis online with the analysis system.

### 9.3.1 *Automated blood sampling*

*In vivo* pharmacokinetic studies in the preclinical phase of drug discovery and development are generally labor-intensive. Manual blood sampling, whether from repeated tail vein punctures or collection via an indwelling cannula, requires well-trained animal technicians. Often, it is not possible to obtain an entire concentration–time profile from a single animal. Furthermore, as discussed in section 9.2.1.1, the amount of blood that can be removed from a single animal at any given time is limited. Small volume samples are extremely difficult to draw and are generally subject to more error during the processing and preparation steps.

Recently, an automated blood sampling system (Culex®, Bioanalytical Systems, Inc., West Lafayette, IN) has become commercially available [4–6]. This device consists of a collection of instruments that work together to collect serial blood samples from an awake, freely moving animal (usually a rat). The system is set up so that animal activity can be monitored together with the collection of both urine and feces. Diagrams of the system are shown in Figs. 9.1 and 9.2. Desired sampling volumes (10–200 μl), intervals (5 min or longer), and sampling times can be set using the software.

To sample with the automated blood sampling system, a volume of heparinized saline is placed in the collection vial. Blood is then drawn via the implanted cannula into a reservoir. The last portion of the drawn blood is then dispensed via sterile tubing into sealed refrigerated (3°C) vials for analysis. The remaining blood is returned to the animal together with a calculated amount of sterile saline to compensate for the volume of blood removed. The system is then flushed with saline before taking the next sample.

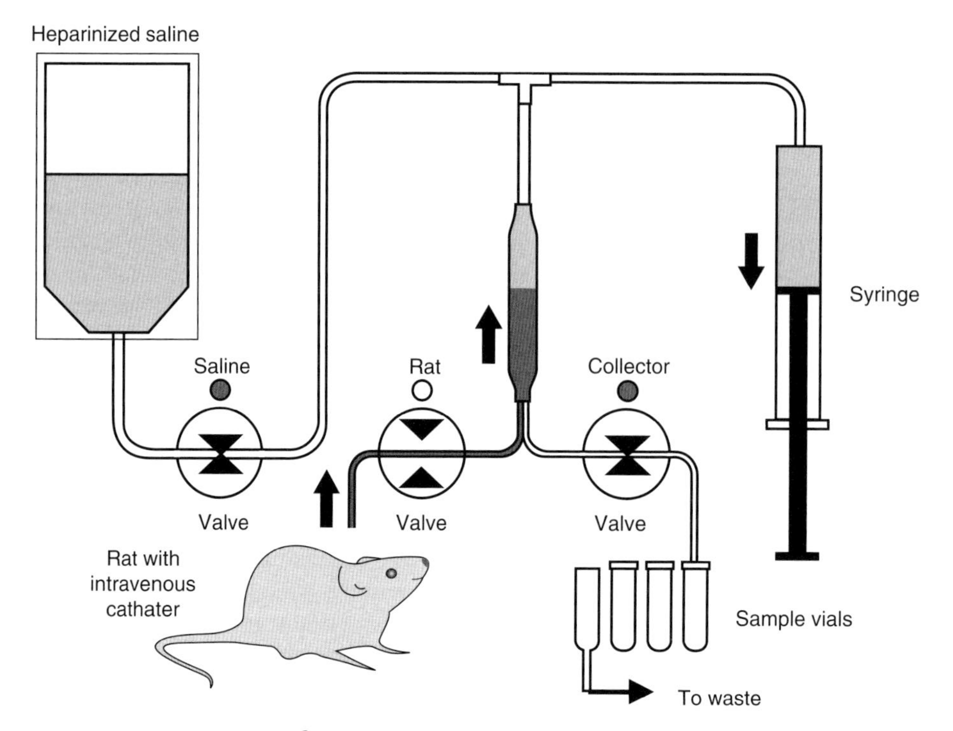


**Fig. 9.1** Schematic of the Culex® blood sampling system. Heparinized saline is used in the blood transfer process to enable serum or plasma collection. (Reproduced with permission from ref. [5].)

Culex experiments are generally performed using a large Plexiglas metabolism bowl seated on a motorized turntable. The turntable is designed to counteract the rat's movements, allowing the rat to move freely in the bowl while protecting fluid lines from twisting and crimping. The turning of the bowl can be monitored and the data collected to record the rat's movements. The bowl is also designed to collect urine and feces for total metabolism studies. This animal containment system can be used for microdialysis or any awake animal experiments that require the rat to be connected via cables, tubing or wires to an analytical device.

A significant aspect of the automated blood sampling system is that an entire concentration–time profile can readily be obtained from a single animal. It is also possible to obtain pharmacokinetic, pharmacodynamic and metabolism data simultaneously. Lastly, the overall quality of data is better because the stress (to both the animal and to the technician) associated with the repeated handling of the animal to collect samples is eliminated. Plasma levels of corticosterone and catecholamines, which are commonly used as indicators of stress in rats, were reported to be lower with the automated sampling system than with manual sampling [7]. In addition, Cronier *et al.* [8] demonstrated that the automated system provided more accurate pharmacokinetic parameters with more consistent data than did manual blood sampling. Because the entire concentration–time profile was obtained from each animal, fewer animals were required for any given study.

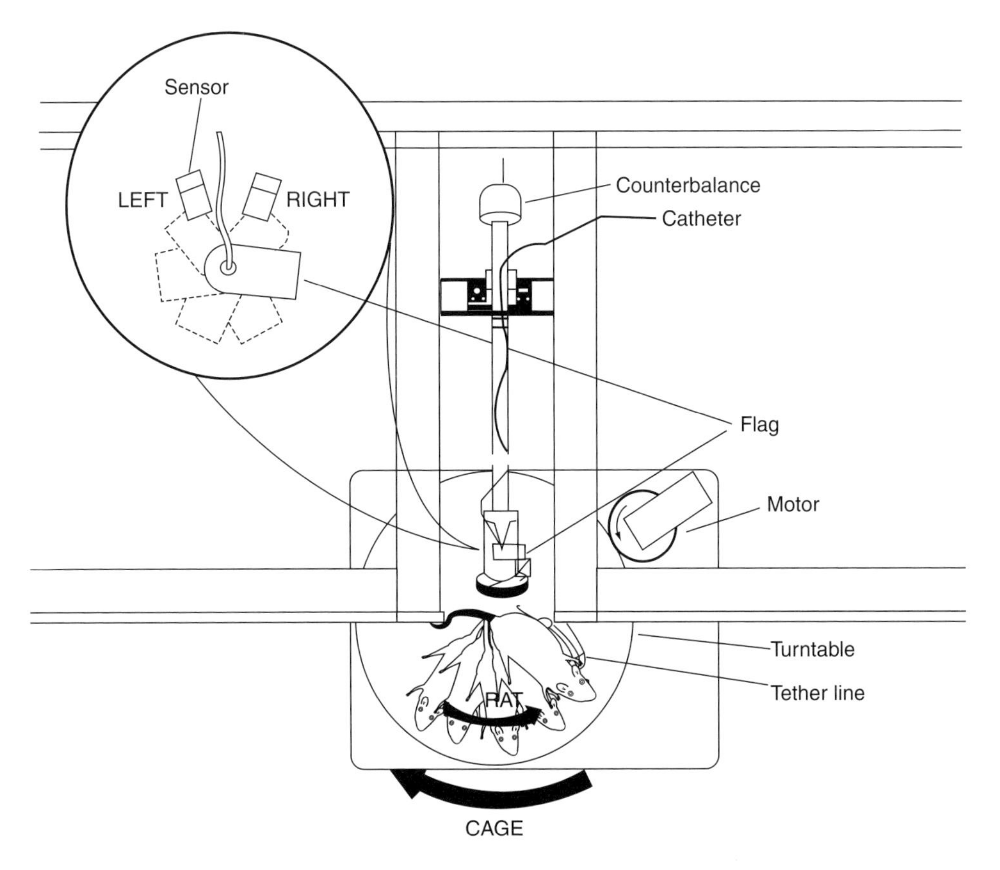


**Fig. 9.2** Diagram of the Raturn system, which allows movement of the animal during sampling. Catheter tubing is connected to the tether line. This tether line is attached to a counterbalance arm that responds to the animal's movements by pivoting or triggering a sensor that rotates the turntable. This system permits the catheter lines to be free of kinks and prevents the animal from chewing on the tubing. (Reproduced with permission from ref. [6].)

The system design is such that it is possible to carry out microdialysis sampling concurrently with blood sampling. Xie *et al.* [9] used the automated sampling system to simultaneously obtain adsorption, distribution, metabolism and elimination information, together with blood–brain barrier penetration and pharmacodynamic data. Automation is used not only to facilitate the blood sampling, but also sample preparation and analysis, improving throughput without sacrificing data quality.

### 9.3.2 *Ultrafiltration (UF)*

UF is a sampling method that uses one or more hollow dialysis loops connected to a small bore impermeable tube [10]. A diagram of an ultrafiltration probe is shown in Fig. 9.3. The loops are placed in the tissue of interest and a negative pressure gradient is used to create a flux of the extracellular fluid (ECF), containing low molecular mass

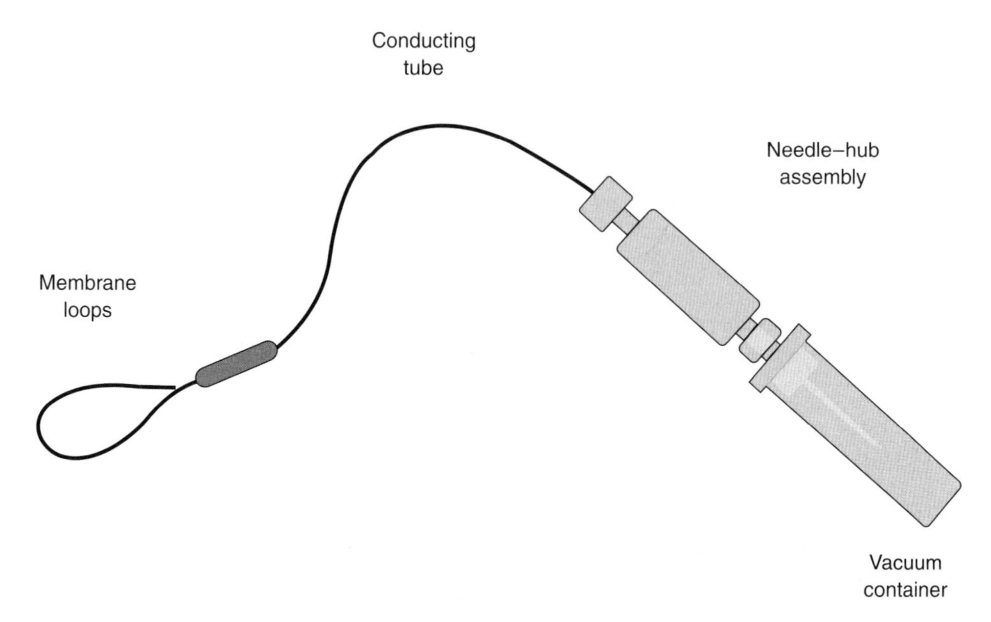


**Fig. 9.3** Ultrafiltration probe with needle hub assembly and vacutainer used to supply negative pressure. A peristaltic pump can also supply the needed pressure gradient. The number and length of membrane loops are varied to optimize sampling from different sizes of animals or from different tissues. (Reproduced with permission from Bioanalytical Systems, West Lafayette, IN.)

compounds, across the porous hydrophobic membrane. This forced transport results in analyte recoveries of greater than 90% for small molecules.

UF probes have been used in animals of various sizes, ranging from mice to horses. Since the UF probe involves the removal of ECF from the site of implantation, it must be implanted in tissue with high fluid turnover to minimize fluid loss. Therefore, this technique is most often used with larger animals since they can better tolerate fluid loss. In addition, the collection vial can be attached to a vest or harness so that the animal does not need to be tethered. In UF, the rate of fluid flux across the probe can change over time. Therefore, ultrafiltration is best suited for studies that do not require high temporal resolution.

UF sampling has been applied to pharmacokinetics of drugs and to monitor endogenous compounds. Subcutaneous space is the most common target site; however, saliva, blood and bone have also been sampled using UF [11–17]. The space available for the probe and the high fluid turnover make these ideal sites for this technique. Most of the time, analysis of UF samples is performed off-line; however, there have been reports of coupling UF with a biosensor for the continuous monitoring of glucose and lactate [18,19].

### 9.3.3 *Microdialysis sampling*

Microdialysis sampling is accomplished by implanting a probe consisting of a hollow fiber dialysis membrane into the organ or biological matrix of interest [20,21]. The

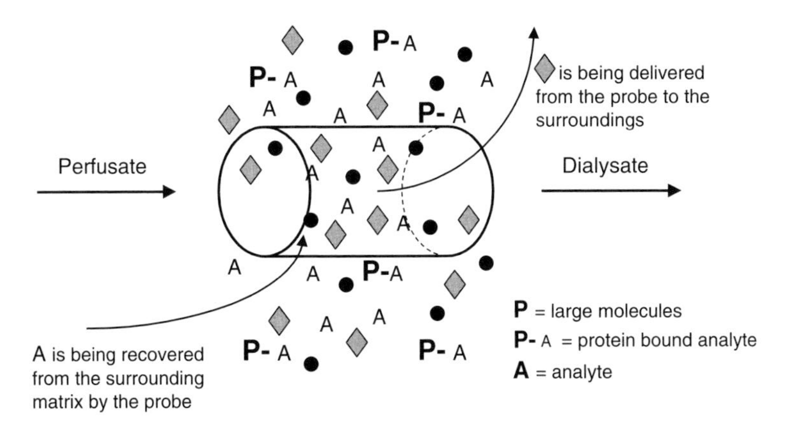


**Fig. 9.4** The microdialysis process [83].

dialysis membrane is affixed to pieces of narrow-bore tubing that serve as inlet and outlet conduits for the probe. An aqueous solution that closely matches the ionic composition of the ECF, termed the perfusate, is pumped slowly through the probe. If the perfusate is correctly matched to the ECF, there should be no net exchange of water or ions across the membrane. Low molecular mass compounds, such as drugs and neurotransmitters, will diffuse into or out of the probe lumen in response to concentration gradients, and are pumped to the fraction collector or analysis system.

A diagram of the process occurring in the microdialysis probe is shown in Fig. 9.4. For clarity, the diagram shows only one analyte (A), although several compounds are typically sampled at the same time in a microdialysis experiment. Large molecules such as proteins (P) and small molecules bound to proteins (P-A) are excluded by the membrane. Those molecules entering the lumen of the membrane are swept along by the perfusate and exit the probe. The solution leaving the probe, termed the dialysate, is collected for analysis. A good analogy to a microdialysis probe is a blood vessel, in that it can both deliver and remove compounds from the local area. Delivery of the parent compound via the probe permits the study of local metabolism without systemic involvement. Figure 9.5 shows a basic diagram of a microdialysis system for sampling from an awake, freely moving animal [22].

Microdialysis is an established *in vivo* tool in the neurosciences, and has been used *in vitro* and in peripheral tissues for a variety of applications of pharmacological and metabolic interest. Samples obtained by microdialysis represent a local profile of low molecular mass hydrophilic substances in the area surrounding the probe. For compounds of pharmaceutical interest, the dialysate reflects the free fraction of the compound of interest, which is the therapeutically active portion of the dose. Microdialysis sampling does not change the net fluid balance in the surrounding matrix or tissue, so samples can be collected continuously for hours or days from a single animal. Most importantly, each animal can serve as its own control, reducing the number of experimental animals needed.

Typically, manual sampling *in vitro* from various incubation mixtures or *in vivo* from blood or other fluids involves the manipulation of small volume samples and

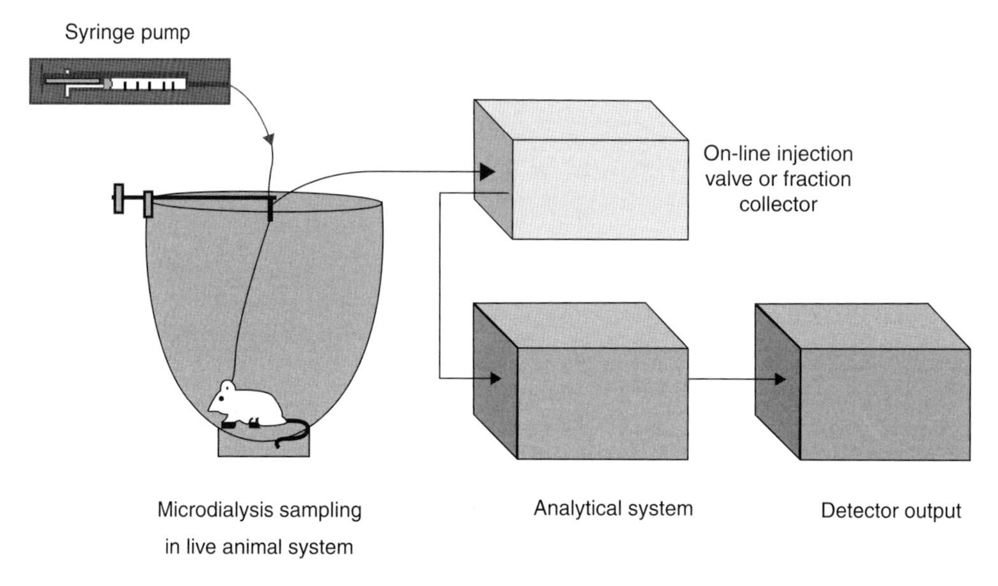


**Fig. 9.5** General diagram of microdialysis system for sampling from an awake freely moving animal. The analysis may be performed on a variety of analytical systems, and coupling between the analytical and microdialysis systems may be on- or off-line. (Reproduced with permission from ref. [22]. Copyright 1999 Wiley-Liss, Inc., a subsidiary of John Wiley & Sons, Inc.)

multiple step clean-up procedures. Elimination products and tissue samples also require extensive sample clean-up before analysis. Because microdialysis sampling provides protein-free samples, labor intensive sample clean-up is not needed. Enzymes are also excluded from the dialysate sample; thus, there is no further enzymatic degradation of the sample.

Microdialysis probes are generally very small, usually about 300 μm o.d. with an active window 2–10 mm in length. This small size causes minimal perturbation to the tissue and makes it possible to use this technique in awake freely moving animals, while still maintaining the integrity of tissues, organs, and systems. Several different probe geometries have been devised to facilitate *in vivo* sampling from various sites. Those commonly used are shown in Fig. 9.6. The designs are of two general types: parallel and serial. Parallel perfusion probes are constructed of rigid or flexible cannula material, and may have either a side-by-side or concentric arrangement. Probes with rigid cannulae are especially suited for intracerebral sampling, while those with flexible cannulae work well for venous sampling. Serial perfusion probes include linear and loop styles. They are typically used for sampling soft peripheral tissue such as skin, muscle, tumor, and liver. The shunt probe suspends a linear probe within a larger tube, allowing sampling from flowing fluids (such as bile or blood) without permanent diversion of the fluid [23,24].

While the small size of the microdialysis probe causes minimal perturbation to the tissue, the surgery to implant the probe is invasive. Experimental animals are anesthetized during probe implantation, and the anatomical location of the target tissue dictates the duration of anesthesia and the severity of the surgical invasion. For example, probe

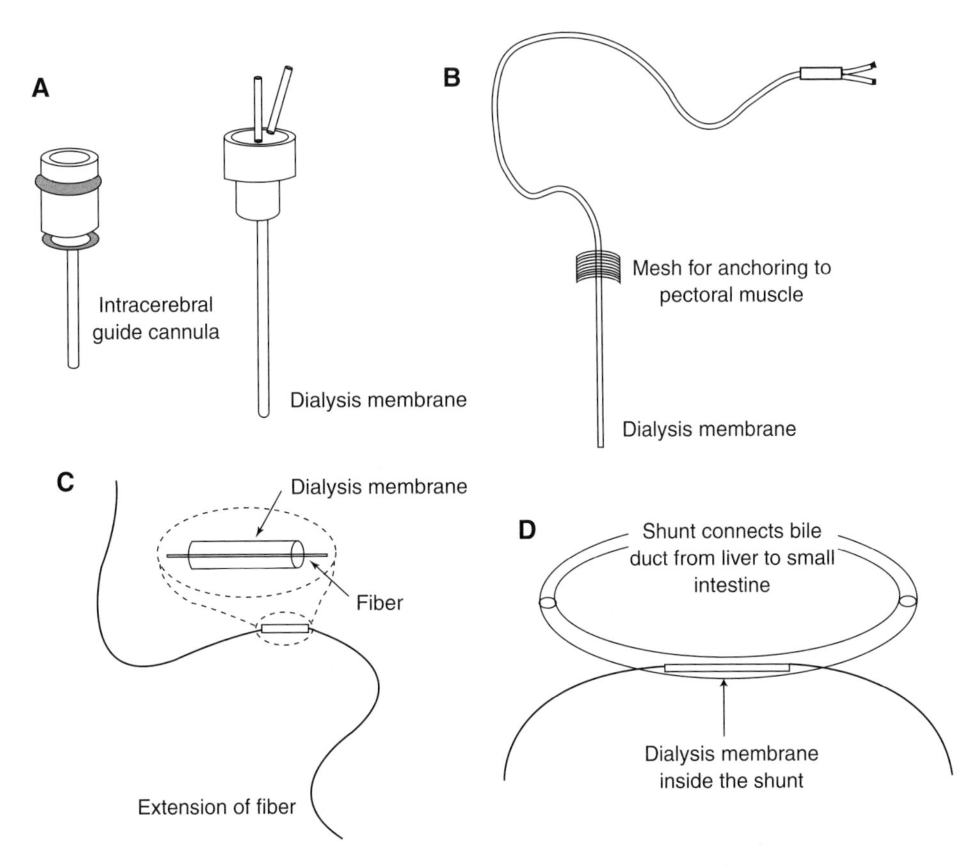


**Fig. 9.6** Typical microdialysis probe designs (probes not drawn to scale). (A) Concentric cannula probe and intracerebral guide for sampling from brain tissue. (B) Flexible concentric cannula probe for intravenous use. (C) Linear probe for peripheral tissues such as muscle, liver, or dermis. (D) Flow-through or shunt probe for sampling from rat bile duct. (Reproduced with permission from Bioanalytical Systems, West Lafayette, IN.)

implantation into the dermis or muscle is much less disturbing to the animal than implantation in the liver or bile duct.

9.3.3.1 *Analyte recovery* Microdialysis sampling is not typically performed under equilibrium conditions, so the concentration of the analyte in the dialysate is some fraction of its actual concentration in the surrounding sample matrix. The relationship between the concentration of analyte in the dialysate and that of the sample matrix may be thought of as the extraction efficiency (EE) of the probe. Among the parameters that influence EE are temperature, perfusate flow rate, chemical and physical properties of the dialysis membrane, probe geometry, membrane surface area, and properties of the analyte. The diffusion rate of the analyte within the matrix also affects EE. *In vivo* uptake into cells, metabolic rate, extent of tissue vascularization, and blood flow will influence diffusion through the tissue [25–27]. Fortunately, under normal conditions of microdialysis sampling, these parameters remain constant; thus, a steady state is rapidly achieved, although equilibrium is not established. Therefore, the EE of the probe for a given set of parameters is constant, and the direction of net flux of the analyte across the membrane is determined by the concentration gradient of the analyte. In practice, the EE of the probe may be determined by recovery or delivery experiments.

A major concern in the wider adoption of microdialysis sampling for pharmaceutical studies is the calibration of the microdialysis probes. Microdialysis is generally performed under nonequilibrium conditions where the recovery of a compound depends on many parameters. Recoveries ranging from less than 1% to more than 99% can be achieved. It is not always clear whether a high or low recovery is preferable. High recoveries are favored from the analytical perspective because higher concentrations are obtained. A high recovery also results in less depletion of the concentration around the probe because the situation is closer to equilibrium, i.e. the concentrations inside and outside the probe are the same. However, high recovery is typically achieved by using either very slow perfusion rates or long dialysis membranes. Using a very slow perfusion rate leads to a loss of temporal resolution because of the relatively large sample volume requirement of most analytical methods. Higher recoveries can be obtained with large sampling windows; however, a loss of spatial resolution is the drawback. Low recoveries, on the other hand, result in less concentrated samples for analysis and a greater decrease in the analyte concentration around the microdialysis probe. However, these limitations may be offset by the fact that higher perfusion rates and shorter dialysis membranes can be used.

While an *in vitro* determination of recovery provides a crude estimate of the behavior of a microdialysis probe with a given compound, extrapolating such a value to an *in vivo* experiment is impossible [28–31]. This is because the nature of the tissue to be sampled and its interactions with the analyte affect the recovery. For many experiments, accurate calibration of the microdialysis probe is not necessary. For example, if the desired information is the relative change in concentration brought about by some experimental manipulation, only the concentration independence and stability of the recovery need be known. However, if a relative distribution of a compound to various sites is the information desired, then the behavior of each probe used must be normalized versus the other probes. Only when an absolute concentration is needed must each probe be calibrated for an accurate *in vivo* recovery. As this is often the most difficult step in a microdialysis experiment, careful experimental design should be practiced. If the desired information can be obtained from relative rather than absolute concentrations, the calibration issue is greatly simplified.

9.3.3.2 *Calibration of microdialysis probes* For systems in which the sample is a flowing fluid (for example, the blood and bile), mass transfer in the sample is dominated by diffusion through the probe membrane [32]. Therefore, for hydrodynamic systems, an *in vitro* calibration can provide accurate recovery values, assuming that the perfusion flow rate and sample temperature are well controlled and the characteristics of the membrane do not change during the experiment. Because the membranes are hydrophilic and used in an aqueous environment, physical changes in the membrane are not likely. However, one must always be concerned that physiological responses to probe implantation may result in changes to the membrane environment. For example, there have been reports of the formation of clots around microdialysis probes implanted intravenously [16]. Careful experimental design will usually prevent such problems, but one must always keep this possibility in mind.

A more difficult case is the use of microdialysis sampling in tissues, particularly those that are poorly perfused. In most cases, it is the transport of the analyte through the tissue and not the probe that determines the recovery of the microdialysis probes [25,31]. Under these conditions, a probe calibration performed *in vitro* may not be valid *in vivo*. To correct for this, a number of approaches to the *in vivo* calibration of microdialysis probes have been described. The two most common methods are: (i) addition of an internal standard to the perfusate, which is commonly known as 'retrodialysis' [33,34], and (ii) estimating the equilibrium condition by adding varying concentrations of analyte to the perfusate; this is known as the method of zero net flux [35,36]. Alternatively, very slow perfusion rates can be employed. Under these conditions, analyte recovery is close to 100%, obviating the need for probe calibration [37].

Before undertaking *in vivo* calibration, the experimentalist should consider what information is desired from the microdialysis experiment. The need for *in vivo* calibration will generally fall into one of three categories. In the first, two states (such as basal versus excited or formulation A versus formulation B) are compared. In these cases, changes in analyte concentration usually provide sufficient information. Here, one does not need an exact EE value but, rather, a constant EE over the time-course of the experiment. In this case, the behavior of the probe can be tracked using retrodialysis. In the second category, one needs to know only the order of magnitude of the analyte concentration in ECF: in other words, is the analyte present at 10 ng/ml or 100 ng/ml? *In vitro* calibration may be sufficient because it usually differs by not more than two or three times the *in vivo* value. If the EE is 60% *in vitro* and 30% *in vivo*, the concentration of analyte in the *in vivo* dialysate corrected for the *in vitro* EE will be within the order of magnitude of the analyte concentration in ECF. In the third category, where an accurate *in vivo* concentration is required, the *in vivo* EE must be determined for the analyte in the target tissue. The approaches that can be used to obtain this information include very low flow rates, zero net flux, and *in vivo* delivery of the analyte [38–40].

## 9.4 Analytical methodologies for online microdialysis sampling

A major advantage of microdialysis sampling is that it provides protein-free aqueous samples that can be analyzed by a wide variety of analytical techniques. Non-separation-based methods can be used to monitor a single analyte, while separation-based methods allow the detection of multiple analytes in each sample. This section will provide a brief overview of the considerations that must be heeded when coupling microdialysis sampling to various analytical techniques.

### 9.4.1 *Nonseparation-based sensors*

Continuous measurements of bioactive compounds can be obtained using nonseparation-based detection systems such as biosensors [41–43]. In this case,

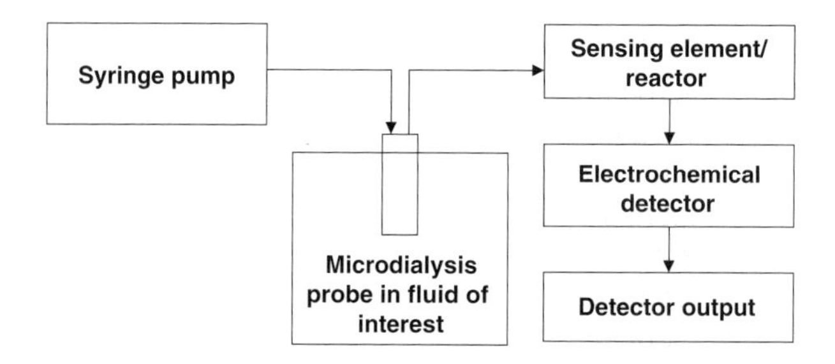


**Fig. 9.7** Typical online set-up for biosensors coupled with microdialysis sampling. The system consists of a syringe pump, a microdialysis probe, a sensor or an enzyme reactor, followed by an electrochemical detector and output readout.

microdialysis provides an alternative to implantable sensors. When online biosensors are used with microdialysis sampling, the apparatus typically consists of a syringe pump, a microdialysis probe, a sensor or an enzyme reactor, followed by an electrochemical detector (Fig. 9.7). The enzyme provides reaction specificity and generates a redox-active species that can be detected electrochemically.

Although the response time of an online biosensor system coupled to microdialysis sampling is slower than that of an implantable modified microelectrode, the former is more quantitative, provides a higher conversion of analytes, has a longer lifetime, and is easier to calibrate. One other disadvantage of implantable biosensors that is commonly encountered is fouling by proteins over time. Microdialysis sampling overcomes this problem because of the size-exclusion nature of the dialysis membrane. In addition, the development of an *in vivo* implantable biosensor generally means that the sensor itself must be very small. When a microdialysis probe is used for sampling, the analysis is performed outside the body, and the biosensor can be much larger.

An example of a biosensor being coupled to microdialysis sampling is demonstrated by Osborne *et al.* [44] who coupled a microdialysis sampling system to a thin-layer radial flow cell that utilized ring and split-disk plastic film carbon electrodes (PFCEs) for the detection of glucose and lactate. The PFCEs were coated with a layer of redox polymer horseradish peroxidase and oxidase enzyme, followed by a cellulose acetate coating, to enable simultaneous electrochemical detection of glucose and lactate at 0 mV (vs Ag/AgCl). This system was used to monitor glucose and lactate simultaneously in an awake restrained rat during local stimulation with veratridine.

Biosensors are not always used for continuous online monitoring. In some cases, a valve is used to inject a discrete sample plug into a flow injection system. The biosensor is then used as a detector for the flow injection system [45–47]. Advantages of this approach include the ability to perform online sample dilution or derivatization, and the capability to periodically recalibrate the biosensor with standard solutions.

Other nonseparation-based methods have been reported, including immunoassays [48–50] and mass spectrometry (MS) [51–53]. Online interfacing of microdialysis and MS has been reported for the analysis of tris(2-chloroethyl) phosphate in rat plasma [54], together with the utilization of microdialysis for online sample clean-up of nucleic acid [55], protein and peptide samples [56].

### 9.4.2 *Separation-based sensors*

The combination of microdialysis to liquid chromatography (LC) or capillary electrophoresis (CE) results in a separation-based sensor that can be used to monitor several substances simultaneously. The advantage of this approach over conventional sensors is that the sampling, separation, and detection steps can all be optimized for the particular analytes of interest. LC and CE are the most common separation methods used in conjunction with microdialysis. Lastly, the detector that provides the best sensitivity and selectivity for the particular analytes of interest can be chosen. Examples include electrochemical detection (ECD) for electroactive neurotransmitters and laser-induced fluorescence (LIF) with derivatization for amino acids.

9.4.2.1 *Liquid chromatography (LC)* Microdialysates consist primarily of relatively small hydrophilic analytes in highly ionic aqueous samples. In general, these characteristics have made LC the analytical method of choice to couple to microdialysis sampling [57,58]. Reversed-phase or ion-exchange are the modes of LC that are most compatible with direct injection of aqueous microdialysis samples. The mode of chromatography chosen for analysis is dependent on the physiochemical properties of the analyte. The type of column used (length, particle size, and internal diameter) is determined by the sampling interval desired and the required sensitivity. In a typical LC assay, 5–10 μl of sample are needed. This means that the temporal resolution is 5–10 min if a perfusion flow rate of 1 μl/min is employed and the analysis time is less than 10 min. If lower microdialysate flow rates are used to increase recovery, temporal resolution is further decreased.

Because microdialysis samples are protein-free, they are amenable to direct injection into a chromatographic system and, therefore, direct coupling of the two systems is possible. Microdialysis sampling has been coupled online to microbore LC for pharmacokinetic and neurochemical investigations in awake freely moving animals. The first online systems employed a six-port injection valve and have been used extensively for monitoring a variety of compounds including neurotransmitters and drug substances [59–63]. However, this approach has the disadvantage that some of the dialysate is shunted to waste while the separation is performed. This results in a loss of some chemical and temporal information.

To take advantage of the continuous chemical information provided by microdialysis sampling, all of the dialysate should be analyzed. This is particularly important for pharmacokinetic studies, where several of the calculated parameters depend on the area under the curve (AUC). If this collection requirement is met, then the area under the curve is simply the sum of the individual dialysate samples, because each of these is an integral over the collection time interval. However, if some of the dialysate is not collected, then the trapezoidal rule must be used to estimate the AUC. In these cases, the online system employing a six-port valve is not appropriate because only a single sample loop is used.

Another consideration in online analysis is sample carryover resulting from incomplete flushing of the injection valve. If conventional LC columns with a flow rate of 1 ml/min are employed, a 5 μl sample loop is flushed in approximately 3 s. However, if a microbore (or capillary) LC column with a flow rate of 50 μl/min is used, the time required to flush the loop with ten volumes of mobile phase becomes 1 min. Even if the separation time is not the limiting factor, 1 min of chemical information is lost during

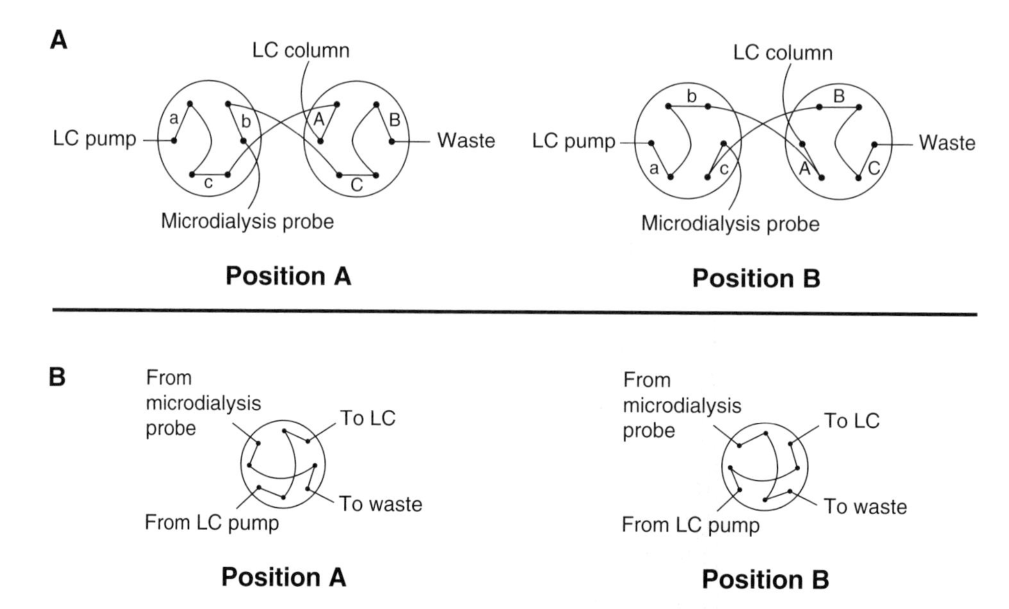


**Fig. 9.8** Schematic diagram of: (A) dual six-port online injection valve system; (B) eight-port online injection valve system. (Reproduced with permission from ref. [64]. Copyright 1995 Elsevier.)

the process. To overcome these problems, online systems capable of analyzing all the dialysates continuously have been developed; these use two six-port or one eight-port valve [64].

Figure 9.8A shows the configuration using two six-port valves for the injection system. When the valves are in position A, the injection volume consists of groove c and loop 1; groove b and loop 2 are filling with the next dialysate sample. In position B, the roles are reversed and the injection volume is defined by groove b and loop 2. One drawback of this arrangement is that the loops in both valves must be exactly the same size. The system can be simplified through the use of an eight-port valve containing two loops. This configuration is shown in Fig. 9.8B. The response time for both systems was very good, with the eight-port valve system providing better precision than the dual six-port valve system.

Although it is usually possible to obtain the chemical information more rapidly with an online system, monitoring actually occurs in *near real-time* due to the time delay between the actual sampling at the dialysis membrane and analysis by the analytical system. This delay is dependent on the length and diameter of the tubing used between the animal and the injection valve. For studies with awake freely moving animals, approximately 70 cm of tubing is used together with a liquid swivel with a volume of 1.4 μl. At 1 μl/min this leads to a time delay of approximately 15 min, precluding the analysis.

The online system described above was capable of rapidly following changes in the dialysate concentration and provided near real-time analysis of the plasma concentration of analytes. The system was evaluated by monitoring the pharmacokinetics of acetaminophen in awake freely moving rats. Figure 9.9A shows a chromatogram of a plasma dialysate obtained 30 min after dosing. In this case, acetaminophen and its metabolites were separated by reversed-phase LC in approximately 6 min.

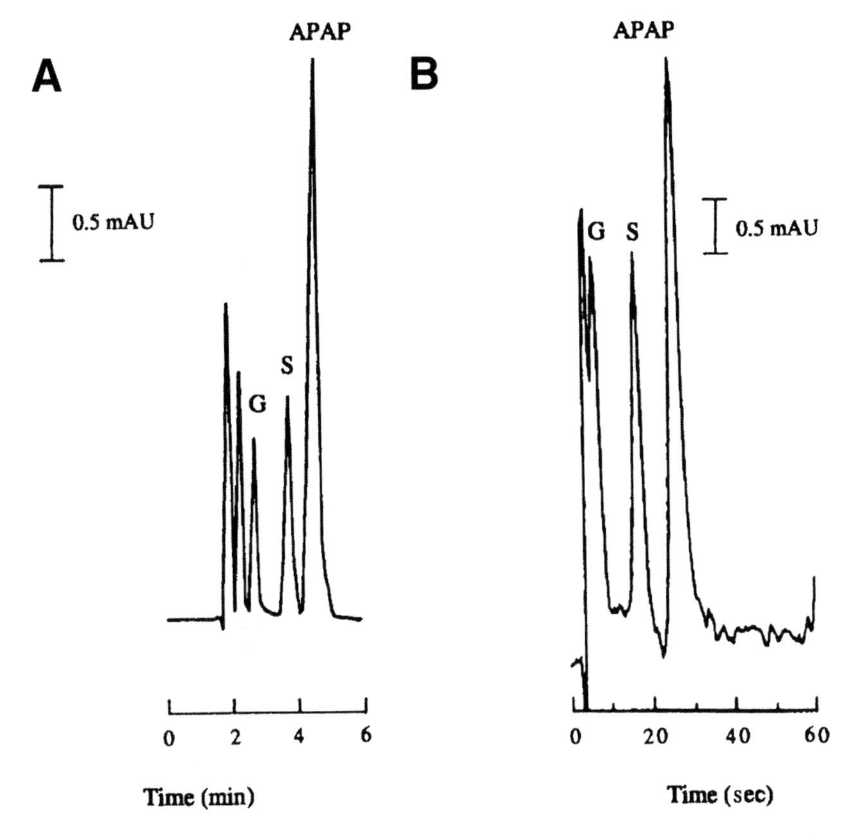


**Fig. 9.9** Chromatograms of blood dialysate samples after doses of APAP: (A) conventional LC; (B) fast microbore LC. Peak identities are as follows: APAP, acetaminophen; G, acetaminophen-4-*O*-glucuronide; S, acetaminophen-4-*O*-sulfate. (Reproduced with permission from refs. [64,65]. Copyright 1995 Elsevier.)

The temporal resolution of the microdialysis-LC experiments is defined by both the injection volume and the analysis time. If better temporal resolution is desired, it is necessary to inject smaller volumes and speed up the analysis time. This can be accomplished by using short small diameter columns. An example of this approach is the use of fast microbore LC coupled with microdialysis, where intravenous microdialysis sampling was accomplished using a flexible microdialysis probe; a 14 mm × 1 mm i.d. octadecylsilica (ODS) microbore column was employed for the separation [65]. Under these conditions, acetaminophen and its two main metabolites, the glucuronide and sulfate conjugates, could be resolved in 30 s. This fast separation allowed online microdialysis sampling to be conducted with a 1 min sampling interval. Figure 9.9B shows the chromatogram obtained for an intravenous dialysate using the microbore LC approach. The pharmacokinetic curve obtained for acetaminophen can be seen in Fig. 9.10. The AUC can be calculated directly from the chromatograms.

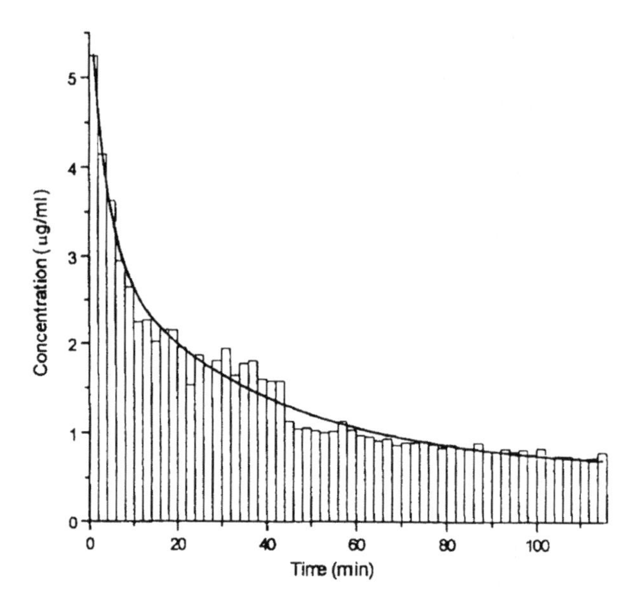


**Fig. 9.10** The pharmacokinetic curve obtained for acetaminophen (APAP) generated from online chromatographic separations of APAP and its two main metabolites, the glucuronide and sulfate conjugates. (Reproduced with permission from ref. [65]. Copyright 1995 Elsevier.)

The microbore LC setup was further employed to monitor the pharmacokinetics of caffeine, as shown in Fig. 9.11. In this case, caffeine, theobromine, and paraxanthine could be resolved in less than 1 min. Separations of caffeine and its metabolites in plasma were conducted sequentially on this time-scale (Fig. 9.11A) and used to construct the pharmacokinetic curve for caffeine (Fig. 9.11B).

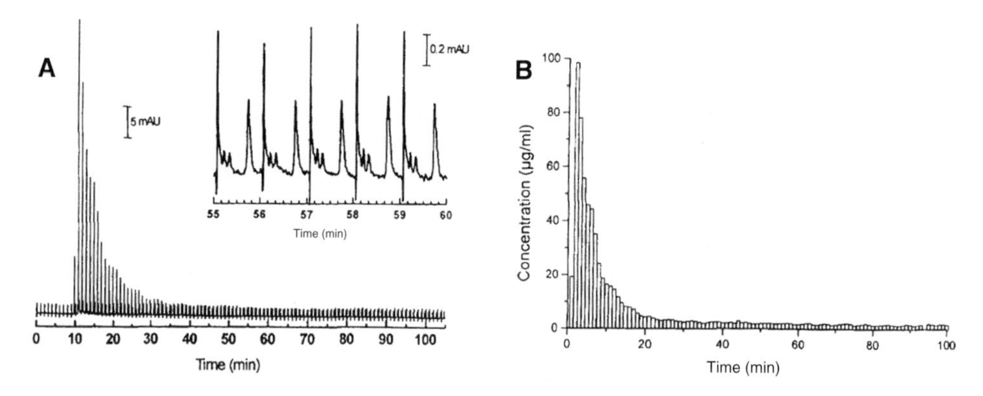


**Fig. 9.11** (A) Detector output from the online microdialysis sampling, fast microbore LC system monitoring plasma concentration of caffeine and its main metabolites. Caffeine was administered intravenously (30 mg/kg) at 10 min. The inset shows an expanded scale of output from 55 to 60 min. (B) Pharmacokinetic curve for caffeine generated from detector output. (Reproduced with permission from ref. [65]. Copyright 1995 Elsevier.)

In one study, microdialysis coupled to LC with electrochemical detection (LC-EC) was used to monitor the concentrations of norepinephrine in the myocardial interstitium of conscious freely moving rats [66]. In these studies, linear microdialysis probes were implanted in the tissue in the periphery of the left descending coronary artery. Norepinephrine (NE) has been reported to be a useful parameter for estimation of cardiac function. The basal steady-state concentration of NE in myocardial dialysate of awake freely moving rats was found to be approximately 0.17 ng/ml. The system was used to investigate the effects of the NE reuptake inhibitor desipramine (DMI) on the extracellular concentrations of norepinephrine.

An online microdialysis microbore HPLC method was also used to investigate the pharmacokinetics of the antitumor agent tirapazamine and its metabolites, 3-amino-1,2,4-benzotriazine-1-*N*-oxide (SR4317) and 3-amino-1,2,4-benzotriazine (SR4330), in the muscle and blood of rats previously administered an intraperitoneal dose of tirapazamine [67]. A linear probe and flexible probe were used for tissue and blood sampling, respectively. The use of an online microdialysis HPLC system permitted the determination of tirapazamine, SR4317 and SR4330 in blood and muscle tissue of rats with high temporal resolution.

More recently, the Kennedy group has coupled capillary LC-MS to microdialysis sampling. In this case, 5 μm diameter reversed-phase columns were used in conjunction with electrospray ionization (ESI)-MS for peptide monitoring and discovery in microdialysis samples. The system was used to monitor neuropeptides from the globus pallidus of anesthetized rats [68]. Met-enkephalin and leu-enkephalin were detected in the dialysates and could be monitored with a temporal resolution of less than 30 min. In addition, several unknown peptides present in the dialysate were identified.

9.4.2.2 *Capillary electrophoresis (CE)* CE is another separation method that is well suited for the analysis of microdialysis samples. The low sample volume requirement of CE is compatible with the small volumes generated by microdialysis sampling. Typical injection volumes in CE are 1–10 nl; however, 1–5 μl samples are generally required due to difficulties associated with the physical manipulation of submicroliter volumes, in particular, problems with surface tension, evaporation, and reduction in separation efficiency.

One disadvantage of CE is its incompatibility with high ionic strength samples. In CE, the optimal procedure to achieve stacking, or compression of the injection zone, is to prepare the sample in an injection buffer that is tenfold more dilute than the background electrolyte. The high ionic strength of the microdialysis sample causes antistacking and, therefore, lowers detection sensitivity. In contrast, LC is amenable to high ionic strength samples.

There have been several reports of the online coupling of microdialysis and capillary electrophoresis with UV and LIF detection [69–72]. The first reported coupling of microdialysis with CE and LIF detection involved the separation of an investigational antineoplastic, SR 4233, from its main metabolite, SR 4317 [73]. Figure 9.12 shows a schematic of the online microdialysis/CE system employed in this study; using a high-speed micellar electrokinetic chromatography (MEKC) separation, resolution of SR 4233 from SR 4317 was achieved in less than 60 s, which allowed the online system to achieve a 90 s temporal resolution for determining the pharmacokinetics of SR 4233

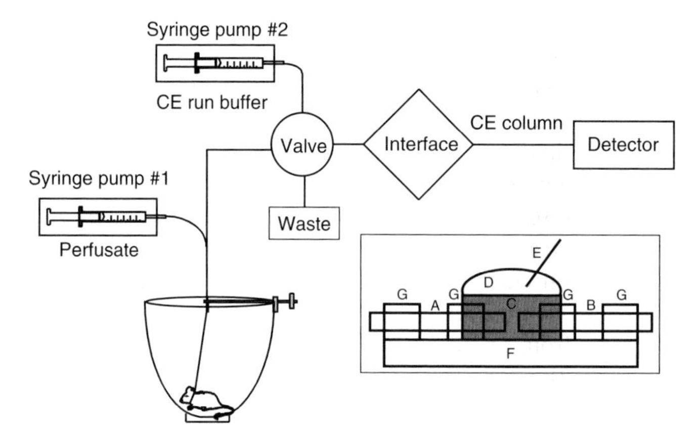


**Fig. 9.12** Representation of the online microdialysis/capillary electrophoresis system used for the analysis of SR4233. Inset: the interface schematic for injection onto the CE system. Components: A, transfer capillary; B, CE capillary; C, CE run buffer reservoir; D, CE run buffer; E, CE ground electrode; F, microscope slide; G, guide tubing. (Reproduced with permission from ref. [73]. Copyright 1994 American Chemical Society.)

*in vivo*. During the online study, a 4 mg/kg intravenous injection of SR 4233 was administered to the rat. A pharmacokinetic curve (Fig. 9.13) was then constructed, and a 15.3 $\pm$ 1.0 min half-life of elimination and 1.1 $\pm$ 0.2 min half-life of distribution were determined for SR 4233.

Online microdialysis-CE has been employed for the fast monitoring of amino acids in brain. The original report by Zhou *et al.* [71] employed online derivatization and

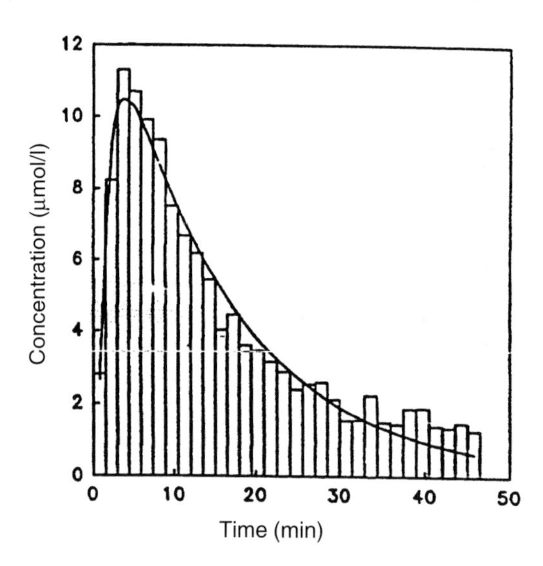


**Fig. 9.13** Typical pharmacokinetic curve from a 4 mg/kg intraperitoneal dose of SR 4233 obtained using the online microdialysis/CE system. The bars represent the microdialysis data, while the curve is the exponential fit to the data. (Reproduced with permission from ref. [73]. Copyright 1994 American Chemical Society.)

LIF detection. The system was capable of near-real-time analysis of aspartate and glutamate with a temporal resolution of less than 2 min in awake freely moving animals. The instrument consists of a microdialysis sampling system, an online reactor, an injection interface, and a CE-LIF system. Primary amine analytes were derivatized with naphthalenedialdehyde/cyanide (NDA/CN) following microdialysis sampling using an online reactor to produce fluorescent cyanobenz(f)isoindole (CBI) derivatives. The reaction takes approximately 1 min. The derivatized sample then travels to a microinjection valve, which alternately sends CE running buffer and reacted microdialysis sample to the CE column via an injection interface. The interface allows a controllable volume of 10–20 nl to be injected onto the CE separation capillary. Separation of aspartate and glutamate from the other amino acids present in the microdialysis sample was achieved within 70 s. The performance of this system was demonstrated in an anesthetized rat by monitoring ECF levels of aspartate and glutamate released in brain after stimulation with high concentrations of $K^+$.

More recently, Kennedy and coworkers have used a flow-gated interface to couple microdialysis to CE [72,74]. The primary difference between the flow-gated interface and the previously described system is that the injection is made by turning a voltage on and off instead of using a valve to send a plug of analyte to a continuously operating CE system. Because there is less dead volume in the gating interface than in the microinjection valve, it is possible to perform microdialysis at very slow flow rates while still maintaining good temporal resolution. Injection of analyte onto the CE capillary with this interface is accomplished by stopping the gating flow, turning off the electrophoresis voltage, and allowing the dialysate to build up near the CE inlet. After a specified delay period, a low voltage is applied for a few seconds to inject the sample. The voltage is then turned off, the cross flow resumed by opening the valve, and the separation voltage is applied to the capillary. A schematic of a flow-gated interface is shown in Fig. 9.14.

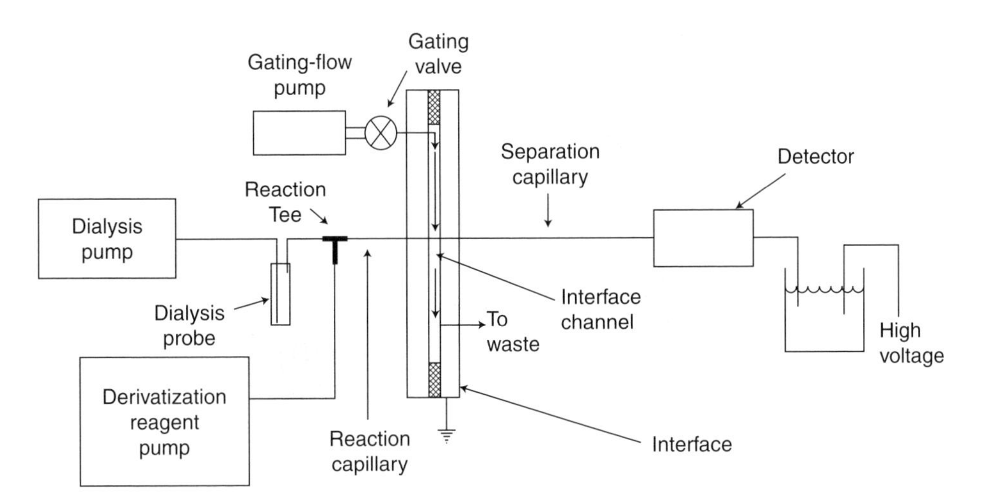


**Fig. 9.14** Microdialysis/CZE-LIF system with online derivatization. Injection was achieved using a flow-gated scheme. (Reproduced with permission from ref. [72]. Copyright 1996 American Chemical Society.)

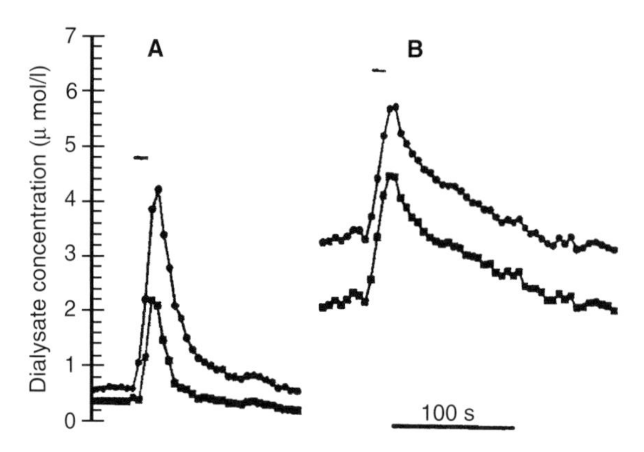


**Fig. 9.15** Monitoring of *in vivo* glutamate and aspartate when electrical stimulation has been applied. Profiles were generated from glutamate (●) and aspartate (■) peaks obtained from individual electropherograms. The bar indicates electrical stimulation at: (A) basal conditions; (B) 20 min after addition of the glutamate uptake inhibitor, L-*trans*-pyrrolidine-2,4-dicarboxylic acid, to the perfusate. (Reproduced with permission from ref. [75]. Copyright 1997 American Chemical Society.)

To further reduce the dead volume in the system, experiments are generally performed on anesthetized rats to minimize the amount of tubing between the experimental animal and the analytical system. The system was initially employed for measuring ascorbate in brain at perfusate flow rates of 79 nl/min. Later, high temporal resolution monitoring of aspartate and glutamate *in vivo* was accomplished using CE-LIF [75]. Sampled dialysate at a flow rate of 1.2 µl/min was derivatized with *o*-phthalaldehyde and β-mercaptoethanol (OPA/βME) using a mixing T and allowed to react before flow-gated injection into the CE column. The temporal resolution was limited by the separation time. With a sampling rate of 5–7 s, the temporal response was found to be in the order of 10–14 s for this system. Finally, the *in vivo* monitoring of glutamate and aspartate in the rat striatum was carried out under basal conditions and during electrical stimulation of the prefrontal cortex (Fig. 9.15).

A similar approach was employed for the rapid determination of D,L-aspartate and glutamate in rat brain tissue samples [76]. The method involved use of β-cyclodextrin in the separation buffer for capillary electrophoresis and an optically gated injection scheme. The amino acids were derivatized with *o*-phthalaldehyde and β-mercaptoethanol before the separation. Figure 9.16 shows a diagram of the system. The reaction time was optimized to be of the order of 35 s. This reaction capillary was directly connected to the separation capillary and optical gating/LIF apparatus. The microdialysis probe was initially placed into sample vials of known concentrations to evaluate and optimize performance; this was followed by placement of the probe into the tissue homogenates of various localized areas in the rat brain.

CE with ECD has also been coupled to microdialysis (MD-CEEC). This detection system provides higher sensitivity than ultraviolet (UV) detection and is comparable to LIF. Furthermore, the number of potential analytes that can be studied at high sensitivity is greatly increased because more compounds are electroactive than are naturally

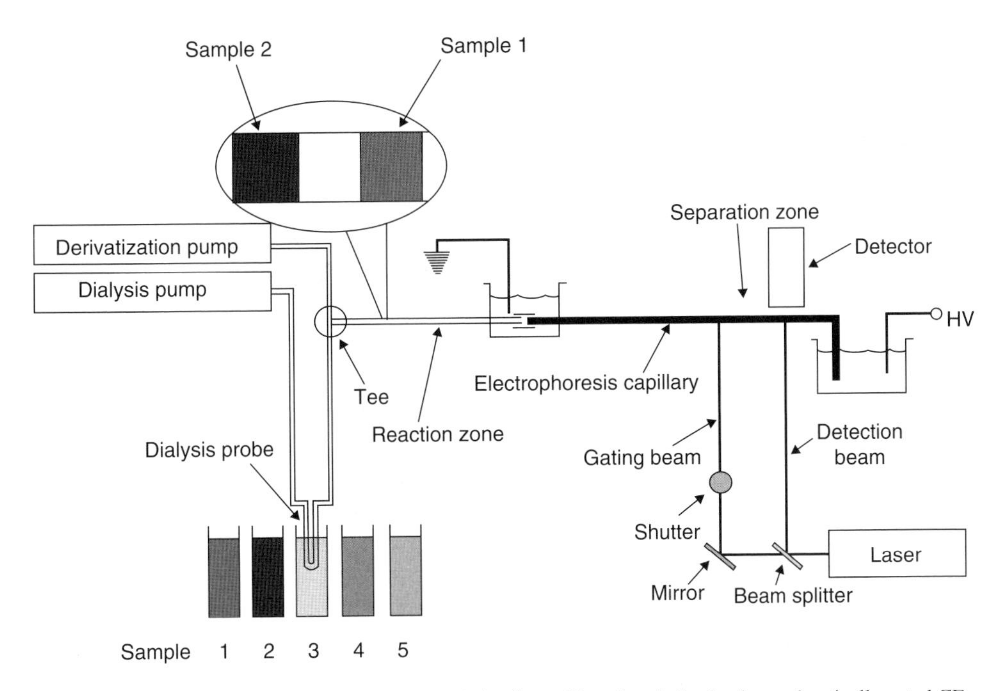


**Fig. 9.16** Representation of microdialysis coupled online with online derivatization and optically gated CE-LIF. (Reproduced with permission from ref. [76]. Copyright 1999 American Chemical Society.)

fluorescent. Lastly, this system has the greatest potential to be miniaturized. However, the design of the MD-CEEC system is challenging because both the animal and the detector must be at ground. This is accomplished by placing a second ground before the injection valve. A schematic of the system is shown in Fig. 9.17 and a diagram of the interface used is presented in Fig. 9.18. A linear microdialysis probe was used for these studies. Using this set-up, nicotine and two of its electroactive metabolites were separated by the CE system [77]. This system was then used to monitor the transdermal delivery of nicotine over a 24 h period (Fig. 9.19).

9.4.2.3 *Future directions in separation-based sensors* The future of the separation-based sensor is to miniaturize the analytical system so it is the same scale as the sampling system. By reducing the scale of the separation system, the run times are shortened, allowing faster and more efficient analyses to be conducted. This, in turn, leads to investigations with improved temporal resolution. Both electrophoresis and chromatography systems have been miniaturized on chips in a planar format. Some preliminary work toward this end has been described by several groups [78–81].

Harrison's group [78] has developed and evaluated intricate design manifolds to allow the introduction of large volume samples into microfluidic devices. Such a design is necessary if microdialysis sampling is to be connected to microchip CE. The design integrated a large sample introduction channel (SIC) with low flow resistance that

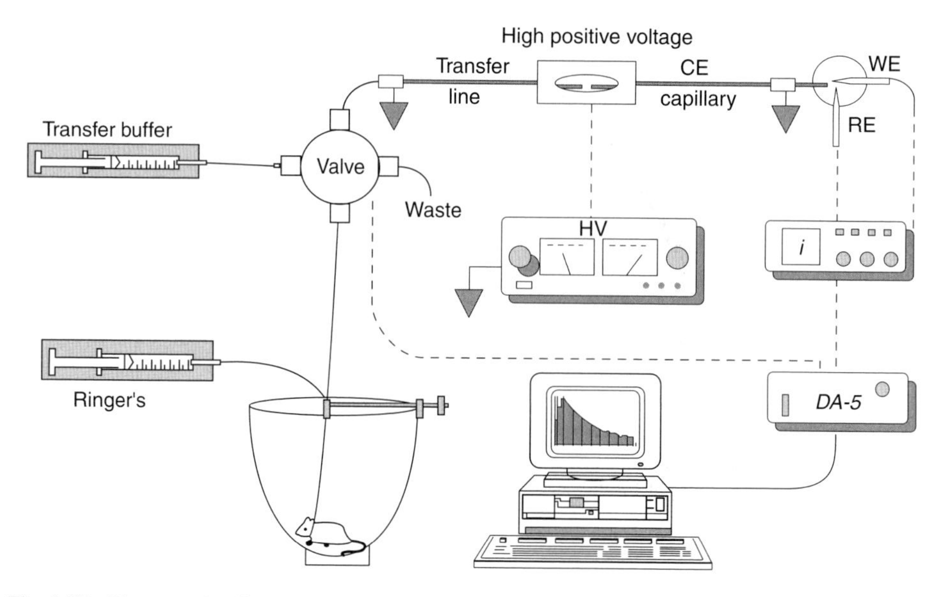


**Fig. 9.17** Diagram of online microdialysis-CEEC system. (Reproduced with permission from ref. [77]. Copyright 1999 Elsevier.)

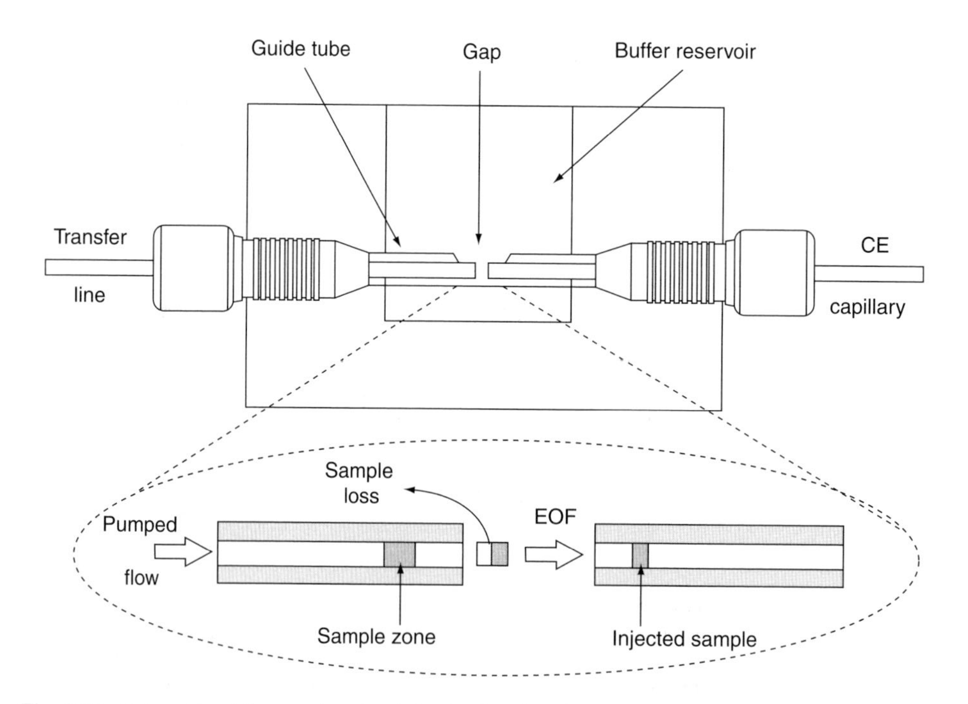


**Fig. 9.18** Schematic of the gap–junction injection interface for the online microdialysis-CEEC system. (Reproduced with permission from ref. [77]. Copyright 1999 Elsevier.)

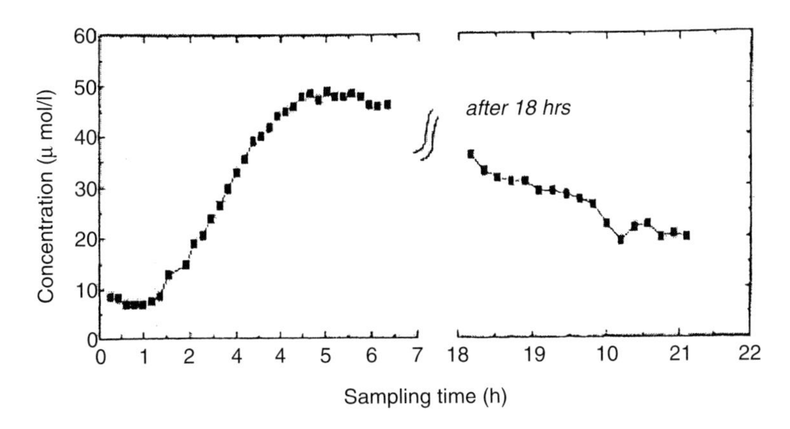


**Fig. 9.19** Time-course of nicotine in microdialysis following patch administration. (Reproduced with permission from ref. [77]. Copyright 1999 Elsevier.)

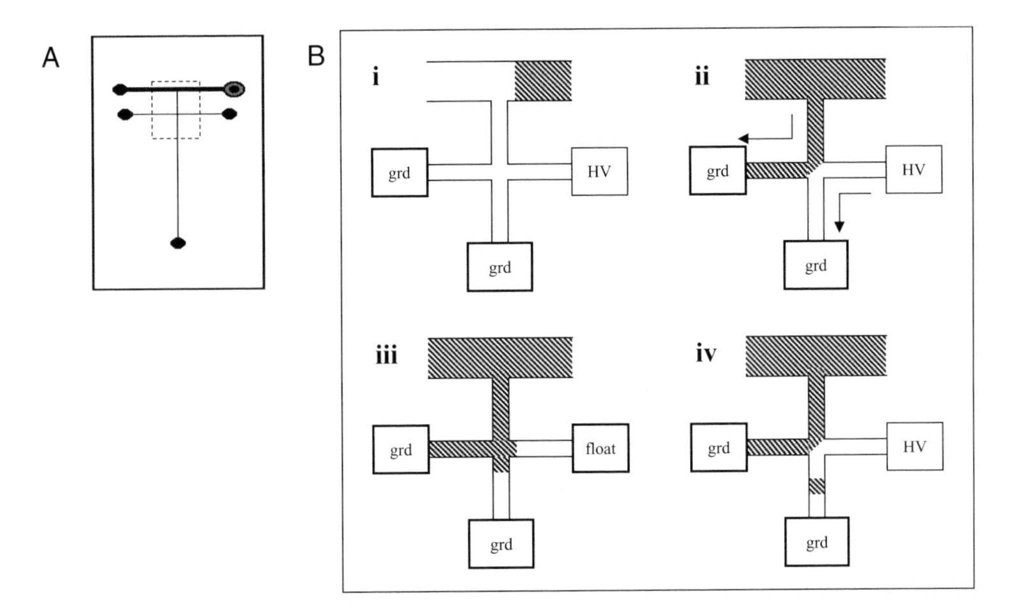


**Fig. 9.20** (A) Layout of microchip capillary electrophoresis device for flow-through sampling. (B) Injection scheme for device: (i) sample begins to fill the sampling channel hydrodynamically; (ii) sample enters the microchip T and is electrokinetically directed towards SW reservoir (threshold voltage); (iii) to inject, microchip T is filled by floating the B reservoir electrode for 1–2 s; (iv) voltage is turned back on and sample plug is introduced down the separation channel. (Adapted and used with permission from ref. [80]. Copyright 2001 American Chemical Society.)

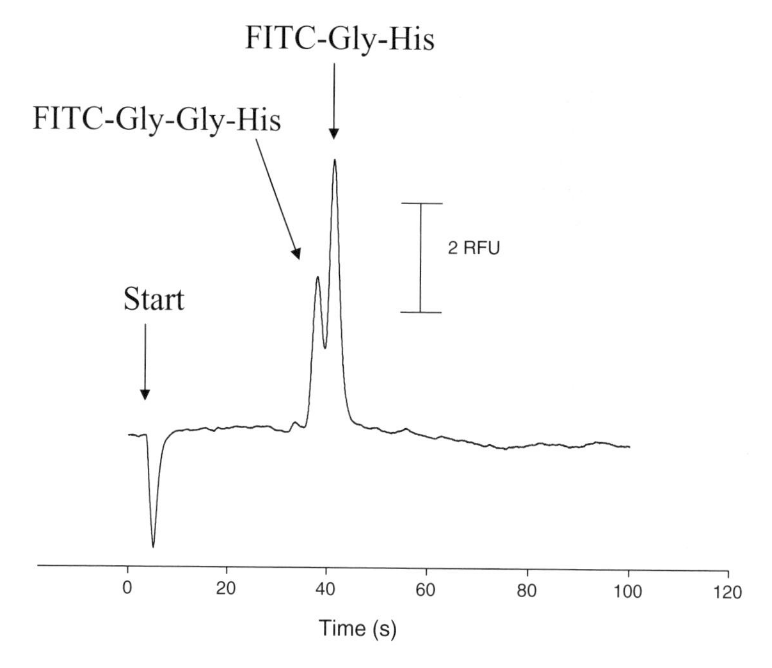


**Fig. 9.21** *In vitro* separation of a simple fluorescently labeled mixture of two peptides, Gly-His and Gly-Gly-His. A microchip capillary electrophoresis device based on the design shown in Fig. 9.20 was coupled to a 2mm microdialysis brain probe and used for the separation, with laser-induced fluorescence as the detection mode. The concentrations of analytes were 15 μmol/l and 10 μmol/l for FITC-Gly-His and FITC-Gly-Gly-His, respectively [82].

would allow real-world samples to enter and flow through the chip device without perturbing the remainder of the contents of the microfluidic device. The subsequent electrokinetic microchip channels that were connected to the SIC would be much smaller in dimension and pose as highly resistant to fluid flow so that hydrodynamically pumped sample (1 ml/min flow rate) would preferentially flow into the SIC. Two device design variations were evaluated experimentally using dyes and dye-derivatized analytes, and theoretically using a mathematical treatment based on Kirchhoff's laws of electrical resistance.

Chen's group [79] has described a simple microchip CE device design that permitted flow-through sampling to occur without perturbing separation components (Fig. 9.20). By using an electrokinetic gating scheme, the hydrodynamically pumped sample flow is directly introduced into the chip and is pushed away from the separation channel. The device injection scheme was described, and the gating voltage was evaluated for flow rates of 5 μl/min. The device was also evaluated for continuous consecutive injections of constant and varying concentrations of rhodamine B. Finally, a separation of rhodamine B and Cy3 was conducted. This device was further modified and optimized to couple with sampling from a microdialysis brain probe to perform a simple *in vitro* separation of a fluorescently labeled peptide mixture (Fig. 9.21) [82].

## 9.5 Conclusions

Conventional blood and tissue sampling methods generally perturb the physiological system and require large sample volumes or numbers of animals. Several strategies are presented that minimize the sample volumes required (and, therefore, the quantity of animals needed) and allow continuous monitoring of the concentration of compounds in biological fluids and tissues. The exact sampling method chosen is dependent on the type of information desired. UF and automated blood sampling remove fluids from the animal, while microdialysis sampling produces no net fluid loss, permitting continuous sampling of the ECF over long periods of time.

The use of these new sampling methods generates a new problem, the ability to analyze the small sample volumes obtained from microdialysis sampling and UF. Typically, volumes sampled using these techniques are less than those required for traditional analytical methodologies such as LC. One solution to this problem has been to couple the online sampling technique directly with separation methods that have low sample volume requirements, including microbore or capillary LC and CE. The result is a separation-based sensor that can be used for near-real-time monitoring of drugs and neurotransmitters. In the future, these separation-based sensors will be miniaturized for portable online monitoring of multiple analytes simultaneously.

## Acknowledgements

The authors gratefully acknowledge the editorial assistance of Nancy Harmony in the preparation of the manuscript. The financial support of NIH and NSF is also acknowledged.

## References

1. University of Nebraska Medical Center at Omaha (UNMC/UNO). *IACUC Guidelines for Care and Use of Live Vertebrate Animals, Appendix J: Guidelines for Bleeding, Antibody Production and Hybridomas.* www.unmc.edu/Education/Animal/guide/appenJ1.html.
2. Bowling Green State University Institutional Animal Care and Use Committee (2002). *Guidelines for the use of Anesthesia, Analgesia, and Tranquilizers.* www.bgsu.edu/offices/spar/orc/iacuc/ReferenceMaterials.html.
3. Waynforth, H.B. & Flecknell, P.A. (1992). *Experimental and Surgical Technique in the Rat*, 2nd edn. Harcourt Brace Jovanovich, San Diego.
4. Gunaratna, C., Cregor, M. & Kissinger, C. (2000). Pharmacokinetic and pharmacodynamic studies in rats using a new method of automated blood sampling. *Curr. Sep.*, **18**(4), 153–157.
5. Peters, S., Hampsch, J., Cregor, M., Starrett, C., Gunaratna, G. & Kissinger, C. (2000). Culex ABS Part I: Introduction to automated blood sampling. *Curr. Sep.*, **18**(4), 139–145.
6. Bohs, B., Cregor, M., Gunaratna, C. & Kissinger, C. (2000). Culex ABS Part II: Managing freely-moving animals and monitoring their activity. *Curr. Sep.*, **18**(4), 147–151.
7. Kissinger, C., Hilt, R., Gunaratna, C., Solomon, B., Zhu, Y., Zhou, Q., Cregor, M., Turner, J., Kissinger, P. T., Prow, M. & Skeeter, M. Effects of blood sampling technique on pharmacokinetics. www.bioanalytical.com/info/poster/pdf/CBK-01.pdf.
8. Cronier, D., Denault, M., Chavent, J., Laurent, M., Gaixet, M., Achart-Arribert, M. & Bolze, S. (2003). Application of an automated blood sampling method to pharmacokinetic studies in rodents in drug discovery. *Drug Met. Rev.*, **35**(S1), 143.

9. Xie, F., Bruntlett, C. S., Zhu, Y., Kissinger, C. & Kissinger, P.T. (2003). Good preclinical bioanalytical chemistry requires proper sampling from laboratory animals: automation of blood and microdialysis sampling improves the productivity of LC/MSMS. *Anal. Sci.*, **19**, 479–485.
10. Janle, E. M. & Kissinger, P. T. (1996). Microdialysis and ultrafiltration, *Adv. Food Nutr. Res.*, 183–196.
11. Linhares, M. C. & Kissinger, P. T. (1993). Pharmacokinetic monitoring in subcutaneous tissue using *in vivo* capillary ultrafiltration probes. *Pharm. Res.*, **10**(4), 598–602.
12. Linhares, M. C. & Kissinger, P. T. (1993). Determination of endogenous ions in intercellular fluid using capillary ultrafiltration and microdialysis probes. *J. Pharm. Biomed. Anal.*, **11**(11–12), 1121–1127.
13. Spehar, A. M., Tiedje, L. J., Sojka, J. E., Janle, E. M. & Kissinger, P. T. (1998). Recovery of endogenous ions from subcutaneous and intramuscular spaces in horses using ultrafiltrate probes. *Curr. Sep.*, **17**, 47–51.
14. Linhares, M. C. & Kissinger, P. T. (1992). Capillary ultrafiltration: *in vivo* sampling probes for small molecules. *Anal. Chem.*, **64**(22), 2831–2835.
15. Janle, E. M., Ostroy, S. & Kissinger, P. T. (1992). Monitoring the progress of streptozocin diabetes in the mouse with the ultrafiltrate pobes. *Curr. Sep.*, **11**(1-2), 17–19.
16. Ash, S. R., Poulos, J. T., Rainier, J. B., Zopp, W. E., Janle, E. & Kissinger, P. T. (1992). Subcutaneous capillary filtrate collector for measurement of blood glucose. *Asaio J.*, **38**(3), M416–M420.
17. Ash, S. R., Rainier, J. B., Zopp, W. E., Truitt, R. B., Janle, E. M., Kissinger, P. T. & Poulos, J. T. (1993). A subcutaneous capillary filtrate collector for measurement of blood chemistries. *Asaio J.*, **39**(3), M699–M705.
18. Tiessen, R. G., Kaptein, W. A., Venema, K. & Korf, J. (1999). Slow ultrafiltration for continuous *in vivo* sampling: application for glucose and lactate in man. *Anal. Chim. Acta.*, **379**, 327–335.
19. Tiessen, R. G., Tio, R. A., Hoekstra, A., Venema, K. & Korf, J. (2001). An ultrafiltration catheter for monitoring of venous lactate and glucose around myocardial ischemia. *Biosens. Bioelectron.*, **16**(3), 159–167.
20. Ungerstedt, U. (1984). Measurement of neurotransmitter release by intracranial dialysis, In: *Measurement of Neurotransmitter Release In Vivo* (ed. C.A. Marsden), pp. 81–105. John Wiley & Sons, New York.
21. Robinson, T. E. & J. B. Justice, Jr. (eds.) (1991). *Microdialysis in the Neurosciences*. Techniques in the Behavioral and Neural Sciences, Vol. 7 (ed. J. P. Huston), Elsevier, New York.
22. Hansen, D. K., Davies, M. I., Lunte, S. M. & Lunte, C. E. (1999). Pharmacokinetic and metabolism studies using microdialysis sampling. *J. Pharm. Sci.*, **88**(1), 14–27.
23. Scott, D.O. & Lunte, C. E. (1993). *In vivo* microdialysis sampling in the bile, blood, and liver of rats to study the disposition of phenol. *Pharm. Res.*, **10**(3), 335–342.
24. Dempsey, E., Diamond, D., Smyth, M. R., Malone, M. A., Rabenstein, K., McShane, A., McKenna, M., Keaveny, T. V. & Freaney, R. (1997). *In vitro* optimisation of a microdialysis system with potential for on-line monitoring of lactate and glucose in biological samples. *Analyst*, **122**(2), 185–189.
25. Morrison, P. F., Bungay, P. M., Hsiao, J. K., Ball, B. A., Mefford, I. N. & Dedrick, R. L. (1991). Quantitative microdialysis: analysis of transients and application to pharmacokinetics in brain. *J. Neurochem.*, **57**(1), 103–119.
26. Smith, A. D. & Justice, J. B. (1994). The effect of inhibition of synthesis, release, metabolism and uptake on the microdialysis extraction fraction of dopamine. *J. Neurosci. Methods*, **54**(1), 75–82.
27. Stenken, J. A., Lunte, C. E., Southard, M. Z. & Stahle, L. (1997). Factors that influence microdialysis recovery. Comparison of experimental and theoretical microdialysis recoveries in rat liver. *J. Pharm. Sci.*, **86**(8), 958–966.
28. Stahle, L., Segersvard, S. & Ungerstedt, U. (1991). A comparison between three methods for estimation of extracellular concentrations of exogenous and endogenous compounds by microdialysis. *J. Pharmacol. Methods*, **25**(1), 41–52.
29. Lindefors, N., Amberg, G. & Ungerstedt, U. (1989). Intracerebral microdialysis: II. Mathematical studies of diffusion kinetics. *J. Pharmacol. Methods*, **22**(3), 157–183.
30. Benveniste, H., Hansen, A. J. & Ottosen, N. S. (1989). Intracerebral microdialysis: II. Mathematical studies of diffusion kinetics. *J. Neurochem.*, **52**, 1741–1750.
31. Bungay, P. M., Morrison, P. F. & Dedrick, R. L. (1990). Steady-state theory for quantitative microdialysis of solutes and water *in vivo* and *in vitro*. *Life Sci.*, **46**(2), 105–119.
32. Stenken, J. A., Topp, E. M., Southard, M. Z. & Lunte, C. E. (1993). Examination of microdialysis sampling in a well-characterized hydrodynamic system. *Anal. Chem.*, **65**(17), 2324–2328.
33. Larsson, C. I. (1991). The use of an 'internal standard' for control of the recovery in microdialysis. *Life Sci.*, **49**(13), L73–L78.
34. Yokel, R. A., Allen, D. D., Burgio, D. E. & McNamara, P. J. (1992). Antipyrine as a dialyzable reference to correct differences in efficiency among and within sampling devices during *in vivo* microdialysis. *J. Pharmacol. Toxicol. Methods*, **27**(3), 135–142.
35. Lonnroth, P., Jansson, P. A. & Smith, U. (1987). A microdialysis method allowing characterization of intercellular water space in humans. *Am. J. Physiol.*, **253**(2-1), E228–E231.

36. Lindefors, N., Amberg, G. & Ungerstedt, U. (1989). Intracerebral microdialysis: I. Experimental studies of diffusion kinetics. *J. Pharmacol. Methods*, **22**(3), 141–153.
37. Menacherry, S., Hubert, W. & Justice, J. B., Jr. (1992). *In vivo* calibration of microdialysis probes for exogenous compounds. *Anal. Chem.*, **64**(6), 577–583.
38. Palsmeier, R. K. & Lunte, C. E. (1994). Microdialysis sampling in tumor and muscle: study of the disposition of 3-amino-1,2,4-benzotriazine-1,4-di-*N*-oxide (SR 4233). *Life Sci.*, **55**(10), 815–825.
39. Ault, J. M., Riley, C. M., Meltzer, N. M. & Lunte, C. E. (1994). Dermal microdialysis sampling *in vivo*. *Pharm. Res.*, **11**(11), 1631–1639.
40. Stahle, L., Segersvard, S. & Ungerstedt, U. (1991). Drug distribution studies with microdialysis. II. Caffeine and theophylline in blood, brain and other tissues in rats. *Life Sci.*, **49**(24), 1843–1852.
41. Stamford, J. A. & Justice, J. B. (1996). Probing brain chemistry. *Anal. Chem.*, **68**(11), 359A–363A.
42. Steinkuhl, R., Sundermeier, C., Hinkers, H., Dumschat, C., Cammann, K. & Knoll, M. (1996). Microdialysis system for continuous glucose monitoring. *Sens. Actuators B*, **33**, 19–24.
43. Garris, P. A., Kilpatrick, M., Bunin, M. A., Michael, D., Walker, Q. D. & Wightman, R. M. (1999). Dissociation of dopamine release in the nucleus accumbens from intracranial self-stimulation. *Lett. Nature*, **198**, 67–69.
44. Osborne, P. G., Niwa, O. & Yamamoto, K. (1998). Plastic film carbon electrodes: enzymatic modification for on-line, continuous, and simultaneous measurement of lactate and glucose using microdialysis sampling. *Anal. Chem.*, **70**(9), 1701–1706.
45. Palmisano, F., Centonze, D., Quinto, M. & Zambonin, P. (1996). A microdialysis fiber based sampler for flow injection analysis: determination of L-lactate in biofluids by an electrochemically synthesized bilayer membrane based biosensor. *Biosens Bioelectron*, **11**, 419–425.
46. Csoregi, E., Laurell, T., Katakis, I., Heller, A. & Gorton, L. (1995). On-line glucose monitoring by using microdialysis sampling and amperometeric detection based on 'wired' glucose oxidate in carbon paste. *Mikrochim. Acta.*, **121**, 31–40.
47. Amine, A., Digua, K., Xie, B. & Danielsson, B. (1995). A microdialysis probe coupled with a miniaturized thermal glucose sensor for *in vivo* monitoring. *Anal. Lett.*, **28**, 2275–2286.
48. Muller, M., v Osten, B., Schmid, R., Piegler, E., Gerngross, I., Buchegger, H. & Eichler, H. G. (1995). Theophylline kinetics in peripheral tissues *in vivo* in humans. *Naunyn Schmiedebergs Arch Pharmacol.*, **352**(4), 438–441.
49. Desrayaud, S., Boschi, G., Rips, R. & Scherrmann, J. M. (1996). Dose-dependent delivery of colchicine to the rat hippocampus by microdialysis. *Neurosci. Lett.*, **205**(1), 9–12.
50. Wagstaff, J., Gibb, J. & Hanson, G. (1996). Dopamine D2-receptors regulate neurotensin release from nucleus accumbens and striatum as measured by *in vivo* microdialysis. *Brain Res.*, **721**, 196–203.
51. Andren, P. E., Emmett, M. R., DaGue, B. B., Steulet, A. F., Waldmeier, P. & Caprioli, R. M. (1998). Blood–brain barrier penetration of 3-aminopropyl-*n*-butylphosphinic acid (CGP 36742) in rat brain by microdialysis/mass spectrometry. *J. Mass. Spectrom.*, **33**(3), 281–287.
52. Emmett, M., Andren, P. & Caprioli, R. (1995). Specific molecular mass detection of endogenously released neuropeptides using *in vivo* microdialysis/mass spectroscopy. *J. Neurosci. Methods*, **62**, 141–147.
53. Boismenu, D., Mamer, O., Ste Marie, L., Vachon, L. & Montgomery, J. (1996). *In vivo* hydroxylation of the neurotoxin, 1-methyl-4-phenylpyridinium, and the effect of monoamine oxidase inhibitors: electrospray-MS analysis of intrastriatal microdialysates. *J. Mass Spectrom.*, **31**(10), 1101–1108.
54. Deterding, L. J., Dix, K., Burka, L. T. & Tomer, K. B. (1992). On-line coupling of *in vivo* microdialysis with tandem mass spectrometry. *Anal. Chem.*, **64**(21), 2636–2641.
55. Liu, C., Wu, Q., Harms, A. C. & Smith, R. D. (1996). On-line microdialysis sample cleanup for electrospray ionization mass spectrometry of nucleic acid samples. *Anal. Chem.*, **68**(18), 3295–3299.
56. Wu, Q., Liu, C. & Smith, R. (1996). On-line microdialysis desalting for electrospray ionization mass spectrometry of proteins and peptides. *Rapid Commun. Mass Spectrom.*, **10**, 835–840.
57. Riley, C. M., Ault, J. M. & Lunte, C. E. (1994). On-line microdialysis sampling. In: *Pharmaceutical and Biomedical Applications of Liquid Chromatography* (eds. C. M. Riley, W. J. Lough & I. W. Wainer). pp. 193–239. Elsevier, Oxford.
58. Davies, M. I. & Lunte, C. E. (1997) Microdialysis sampling coupled on-line to microseparation techniques. *Chem. Soc. Rev.*, **26**, 215–222.
59. Sarre, S., Deleu, D., Van Belle, K., Ebinger, G. & Michotte, Y. (1993). *In-vivo* microdialysis sampling in pharmacokinetic studies. *Trends. Anal. Chem.*, **12**(2), 67–73.
60. Wages, S. A., Church, W. H. & Justice, J. B., Jr. (1986). Sampling considerations for on-line microbore liquid chromatography of brain dialysate. *Anal. Chem.*, **58**(8), 1649–1656.
61. Church, W. H. & Justice, J. B., Jr. (1987). Rapid sampling and determination of extracellular dopamine *in vivo*. *Anal. Chem.*, **59**(5), 712–716.
62. Westerink, B. H., De Vries, J. B. & Duran, R. (1990). Use of microdialysis for monitoring tyrosine hydroxylase activity in the brain of conscious rats. *J. Neurochem.*, **54**(2), 381–387.

63. Sabol, K. E. & Freed, C. R. (1988). Brain acetaminophen measurement by *in vivo* dialysis, *in vivo* electrochemistry and tissue assay: a study of the dialysis technique in the rat. *J. Neurosci. Methods*, **24**(2), 163–168.
64. Steele, K. M. & Lunte, C. E. (1995). Microdialysis sampling coupled to on-line microbore liquid chromatography for pharmacokinetic studies. *J. Pharm. Biomed. Anal.*, **13**(2), 149–154.
65. Chen, A. & Lunte, C. E. (1995). Microdialysis sampling coupled on-line to fast microbore liquid chromatography. *J. Chromatogr. A.*, **691**(1-2), 29–35.
66. Gilinsky, M. A., Faibushevish, A. A. & Lunte, C. E. (2001). Determination of myocardial norepinephrine in freely moving rats using *in vivo* microdialysis sampling and liquid chromatography with dual-electrode amperometric detection. *J. Pharm. Biomed. Anal.*, **24**(5-6), 929–935.
67. McLaughlin, K. J., Faibushevich, A. A. & Lunte, C. E. (2000). Microdialysis sampling with on-line microbore HPLC for the determination of tirapazamine and its reduced metabolites in rats. *Analyst.*, **125**(1), 105–110.
68. Haskins, W. E., Wang, Z., Watson, C. J., Rostand, R. R., Witowski, S. R., Powell, D. H. & Kennedy, R. T. (2001). Capillary $LC\text{-}MS^2$ at the attomole level for monitoring and discovering endogenous peptides in microdialysis samples collected *in vivo*. *Anal. Chem.*, **73**(21), 5005–5014.
69. Lada, M. W. & Kennedy, R. T. (1995). Quantitative *in vivo* measurements using microdialysis on-line with capillary zone electrophoresis. *J. Neurosci. Methods*, **63**(1-2), 147–152.
70. Lada, M. W., Schaller, G., Carriger, M. H., Vickroy, T. W. & Kennedy, R. T. (1995). On-line interface bertween microdialysis and capillary zone electrophoresis. *Anal. Chim. Acta.*, **307**, 217–225.
71. Zhou, S. Y., Zuo, H., Stobaugh, J. F., Lunte, C. E. & Lunte, S. M. (1995). Continuous *in vivo* monitoring of amino acid neurotransmitters by microdialysis sampling with on-line derivatization and capillary electrophoresis separation. *Anal. Chem.*, **67**(3), 594–599.
72. Lada, M. W. & Kennedy, R. T. (1996). Quantitative *in vivo* monitoring of primary amines in rat caudate nucleus using microdialysis coupled by a flow-gated interface to capillary electrophoresis with laser-induced fluorescence detection. *Anal. Chem.*, **68**(17), 2790–2797.
73. Hogan, B. L., Lunte, S. M., Stobaugh, J. F. & Lunte, C. E. (1994). On-line coupling of *in vivo* microdialysis sampling with capillary electrophoresis. *Anal. Chem.*, **66**(5), 596–602.
74. Bowser, M. T. & Kennedy, R. T. (2001). *In vivo* monitoring of amine neurotransmitters using microdialysis with on-line capillary electrophoresis. *Electrophoresis*, **22**, 3668–3676.
75. Lada, M. W., Vickroy, T. W. & Kennedy, R. T. (1997). High temporal resolution monitoring of glutamate and aspartate *in vivo* using microdialysis on-line with capillary electrophoresis with laser-induced fluorescence detection. *Anal. Chem.*, **69**(22), 4560–4565.
76. Thompson, J. E., Vickroy, T. W. & Kennedy, R. T. (1999). Rapid determination of aspartate enantiomers in tissue samples by microdialysis coupled on-line with capillary electrophoresis. *Anal. Chem.*, **71**(13), 2379–2384.
77. Zhou, J., Heckert, D. M., Zou, H., Lunte, C. E. & Lunte, S. M. (1999). On-line coupling of *in vivo* microdialysis with capillary electrophoresis/electrochemistry. *Anal. Chim. Acta.*, **379**, 307–317.
78. Attiya, S., Jaemere, A. B., Tang, T., Fitzpatrick, G., Seiler, K., Chiem, N. & Harrison, D. J. (2001). Design of an interface to allow microfluidic electrophoresis chips to drink from the fire hose of the external environment. *Electrophoresis*, **22**, 318–327.
79. Lin, Y. H., Lee, G. B., Li, C. W., Huang, G. R. & Chen, S. H. (2001). Flow-through sampling for electrophoresis-based microfluidic chips using hydrodynamic pumping. *J. Chromatogr. A.*, **937**, 115–125.
80. Chen, S. H., Lin, Y. H., Wang, L. Y., Lin, C. C. & Lee, G. B. (2002). Flow-through sampling for electrophoresis-based microchips and their applications for protein analysis. *Anal. Chem.*, **74**, 5146–5753.
81. Fang, Q., Xu, G. & Fang, Z. (2002). A high-throughput continuous sample introduction interface for microfluidic chip-based capillary electrophoresis. *Anal. Chem.*, **74**(6), 1223–1231.
82. Huynh, B. H., Fogarty, B. A., Martin, R. S. & Lunte, S. M. (2003). Development of a microchip capillary electrophoresis device for discrete introduction of continuous samples at ultra-low flow rates. In: *SmallTalk Conference on Microfluidics, Microarrays and BioMEMS*, San Jose, abstract 016. Association for Laboratory Automation.labautomation.org/Conference PDFs/ST03Book.pdf.
83. Davies, M. I. (1995). *Microdialysis sampling for in vivo hepatic metabolism studies*, PhD thesis, University of Kansas, Lawrence, KS.

# Index